REEDS MARINE ENGINEERING AND TECHNOLOGY

APPLIED THERMODYNAMICS
FOR MARINE ENGINEERS

REEDS MARINE ENGINEERING AND TECHNOLOGY SERIES

3

REEDS MARINE ENGINEERING AND TECHNOLOGY

APPLIED THERMODYNAMICS
FOR MARINE ENGINEERS

Revised by Paul A. Russell

William Embleton

Leslie Jackson

R E E D S

LONDON • OXFORD • NEW YORK • NEW DELHI • SYDNEY

REEDS
Bloomsbury Publishing Plc
50 Bedford Square, London, WC1B 3DP, UK
29 Earlsfort Terrace, Dublin 2, Ireland

BLOOMSBURY, REEDS and the Reeds logo are trademarks of
Bloomsbury Publishing Plc

First published in Great Britain 1963
Second edition 1971
Third edition 1982
Fourth edition 1991
Fifth edition 2016
This edition 2022

A catalogue record for this book is available from the British Library

Library of Congress Cataloguing-in-Publication data has been applied for

ISBN: PB: 978-1-4729-9340-3; eBook: 978-1-4729-9342-7; ePub: 978-1-4729-9341-0

2 4 6 8 10 9 7 5 3 1

Typeset in Myriad Pro 10/14 by Newgen Knowledge Works Pvt. Ltd, Chennai, India
Printed and bound in Great Britain by CPI Group (UK) Ltd, Croydon CR0 4YY

To find out more about our authors and books visit www.bloomsbury.com
and sign up for our newsletters

CONTENTS

PREFACE

This book covers the syllabuses in Applied Heat for all classes of Marine Engineers' Certificates of Competency worked out by the International Maritime Organization's (IMO) Standards of Training, Certification and Watchkeeping for Seafarers (STCW). The examinations administered by individual flag administrations, leading to Certificates of Competency for marine engineering officers, will have similar requirements for learning the information contained in this book.

Basic principles are dealt with commencing at a fairly elementary stage, and each chapter has fully worked examples interwoven into the text. Test examples are set at the end of each chapter for the student to complete as revision, and finally there are some typical examination questions included. The prefix Ⓒ is no longer used to indicate those parts of the text, and some test examples, that are considered to be at the level suitable for the Chief Engineering Officer's qualification. The reason is that most administrations now have the one 'Management' level qualification that combines the older second engineering officer and chief engineering officer levels.

The author has gone beyond the common practice of just supplying bare answers to the test examples and examination questions by providing fully worked step-by-step solutions leading to the final answers.

This latest revision of the book is a major update of previous editions, which takes the material for study through the 2020s and includes the requirements of the STCW 2010 Manila Amendments.

It is vitally important for marine engineers to know about the scientific principles that underpin the operation of marine machinery. The reason for this is they need to be able to diagnose current, developing and potential faults with any machinery before they become serious enough to cause any malfunction which could lead to delay of the vessel or cause injury to staff.

This 'diagnostic' ability means that the engineer needs to draw on knowledge about how a system works and the science that provides the driving principles for the machinery. Understanding the different descriptors of the properties of materials such as mass, energy and work is vital for an engineer to be able to work out from the instruments if the machinery is running correctly.

UNITS AND COMMON TERMS: SYSTÈME INTERNATIONAL d'UNITÉS

SI Units

The SI system has been in use in France since the 1840s and was officially adopted by them at the General Conference on Weights and Measures in 1960. However, the system has now slowly been adopted by more and more nations around the world.

SI units consist of *seven* base units and a set of subunits derived from the base units. For example, the unit of force is newton (N) which is mass × acceleration. The mass in kilograms, the acceleration in metres and the time in seconds are direct SI units. When they come together to measure the force, then the newton is a derived SI unit.

Base SI Units

Quantity and symbol	SI unit name	Symbol used
Length, l	meter	m
Mass, m	kilogram	kg
Time, t	second	s
Electric current	ampere	A
Thermodynamic temperature	kelvin	K
Amount of substance	mole	mol
Luminous intensity	candela	cd

The Derived SI Units together with Names and Symbols

The derived units are given this term as they are combinations of the base units. Some of these combinations have been assigned special names and/or symbols.

Physical property	Name of unit	Symbol of unit	SI base units	Other units used
angle of a two-dimensional circle	radian	rad		
electrical capacitance	farad	F	$m^{-2}\,kg^{-1}\,s^4\,A^2$	C/V
electrical charge	coulomb	C	A s	
electrical conductance	siemens	S	$m^{-2}\,kg^{-1}\,s^3\,A^2$	A/V
electrical inductance	henry	H	$m^2\,kg\,s^{-2}\,A^{-2}$	
electrical potential	volt	V	$m^2\,kg\,s^{-3}\,A^{-1}$	W/A
electrical resistance	ohm	w	$m^2\,kg\,s^{-3}\,A^{-2}$	V/A
force	newton	N	$kg\,ms^{-2}$	
frequency	hertz	Hz	s^{-1}	
illuminance	lux	lx	$m^{-2}\,cd\,sr$	lm/m²
luminous flux	lumen	lm	cd sr	

Physical property	Name of unit	Symbol of unit	SI base units	Other units used
magnetic flux	weber	Wb	m^2 kg s^{-2} A^{-1}	V s
magnetic flux density	tesla	T	kg s^{-2} A^{-1}	Wb/m^2
power	watt	W	kg m^2 s^{-3}	J/s
pressure	pascal	Pa	$kg/(m\ s^2) = (N/m^2)$	
work, energy, heat	joule	J	m^2 kg s^{-2}	N m
acceleration	meter per second squared	m/s^2	m s^{-2}	
area	square meter	m^2	m^2	
coefficient of heat transfer (often used symbol h or U)	watt per square meter kelvin	$W/(m^2\ K)$	kg s^{-3} K^{-1}	
concentration (of amount of substance)	mole per cubic meter	mol/m^3	mol m^{-3}	
current density (often used symbol r)	ampere per square meter	A/m^2	A m^{-2}	
density (mass density)	kilogram per cubic meter	kg/m^3	kg m^{-3}	
electrical charge density	coulomb per cubic meter	C/m^3	m^{-3} s A	
electric field strength	volt per meter	V/m	m kg s^{-3} A^{-1}	
electric flux density	coulomb per square meter	C/m^2	m^{-2} s A	
energy density	joule per cubic meter	J/m^3	m^{-1} kg s^{-2}	
force	newton	N or J/M	m kg s^{-2}	
heat capacity	joule per kelvin	J/K	m^2 kg s^{-2} K^{-1}	
heat flow rate (often used symbol Q or q)	watt	W or J/s	m^2 kg s^{-3}	
heat flux density or irradiance	watt per square meter	W/m^2	kg s^{-3}	
luminance	candela per square meter	cd/m^2	cd m^{-2}	
magnetic field strength	ampere per meter	A/m	A m^{-1}	
modulus of elasticity (or Young's modulus)	giga pascal	GPa	10^{-9} m^{-1} kg s^{-2}	
molar energy	joule per mole	J/mole	m^{-2} kg s^{-2} mol^{-1}	

Physical property	Name of unit	Symbol of unit	SI base units	Other units used
molar entropy (or molar heat capacity)	joule per mole kelvin		J/(mole K)	m^{-2} kg s^{-2} K^{-1} mol^{-1}
moment of force (or torque)	newton meter		N m	m^2 kg s^{-2}
moment of inertia	kilogram meter squared		kg m^2	kg m^2
momentum	kilogram meter per second		kg m/s	kg m s^{-1}
permeability	henry per meter		H/m	m kg s^{-2} A^{-2}
permitivity	farad per meter		F/m	m^{-3} kg^{-1} s^4 A^2
power	kilowatt		kW	10^{-3} m^2 kg s^{-3}
pressure (often used symbol P or p)	kilo pascal		kPa	10^{-3} m^{-1} kg s^{-2}
specific energy	joule per kilogram		J/kg	m^2 s^{-2}
specific heat capacity (or specific entropy, often used symbol c_p, c_v or s)	joule per kilogram kelvin		J/(kg K)	m^2 s^{-2} K^{-1}
specific volume	cubic meter per kilogram		m^3/kg	m^3 kg^{-1}
stress	mega pascal		MPa	10^{-6} m^{-1} kg s^{-2}
surface tension	newton per meter		N/m	kg s^{-2}
thermal conductivity (often used symbol k)	watt per meter kelvin		W/(m K)	m kg s^{-3} K^{-1}
torque	newton meter		N m	m^2 kg s^{-2}
velocity (or speed)	meters per second		m/s	m s^{-1}
viscosity, absolute or dynamic (often used symbol m)	pascal second		Pa s	m^{-1} kg s^{-1}
viscosity, kinematic (often used symbol n)	square meter per second		m^2/s	m^2 s^{-1}
volume	cubic meter		m^3	m^3
wave number	1 per meter		1/m	m^{-1}
work (or energy heat, often used symbol W)	joule		J or N m	m^2 kg s^{-2}

Decimal fractions of SI units are shown as follows:

Prefix of ten	Prefix symbol	Power	Name
pico	p	10^{-12}	Billionth
nano	n	10^{-9}	Thousand millionth
micro	µ	10^{-6}	Millionth
milli	m	10^{-3}	Thousandth
centi	c	10^{-2}	Hundredth
deci	d	10^{-1}	Tenth

Decimal multiples of SI units are shown as follows:

Prefix of ten	Prefix symbol	Power	Name
deca	da	10^{1}	Ten
hecto	h	10^{2}	Hundred
kilo	k	10^{3}	Thousand
mega	M	10^{6}	Million
giga	G	10^{9}	Milliard
tera	T	10^{12}	Billion

The maritime industry is a global industry, and therefore it is important that all marine engineers work to the SI standards. They will need to communicate with engineers from other nations and from other branches of engineering. Working and thinking to a common standard will help to reduce mistakes and/or costly redesign or repair. Starting with mathematics, this module will now explore the knowledge required by the marine engineer and give some practical examples about why this level of knowledge is required.

Students wishing to prepare for an internationally recognised licence of 'Marine engineer' will then be able to complete academic qualifications in the subjects discussed and approach a flag state of their choice for more detail about the licensing requirements in that country.

A further advantage of using the SI system is the standardised way of presenting large numbers. The table titled 'Decimal multiples of SI units' shows the labels given and the method of representing large numbers within mathematical calculations.

LAWS OF THERMODYNAMICS. When observing the natural world there are certain things that happen consistently. When no alternative can be found it can be safely said that the observations become 'a natural law'

Thermodynamics is one such area of study. The laws of thermodynamics are covered in more detail later in this book (page 68) but briefly the laws of thermodynamics show that energy is not destroyed but is converted from one form to another. The natural order is for heat to flow from a 'hot' body to a 'colder' one and generally the natural world becomes more disorganised as its 'entropy' increases.

MASS. This is the name given to the quantity of matter possessed by a body or a substance. The mass of an object is proportional to the volume and the density of that specific object. It is a constant quantity, that is, the mass of a body can be changed only by adding more matter to it or by taking matter away from it.

The common expression, used in engineering and calculations, for mass is m and the 'standard' unit used is the kilogram [kg]. For very large or small quantities, multiples or submultiples of the kilogram are used. Large masses are common in engineering, and these are measured in megagrams [Mg]. One megagram is equal to 10^3 kilograms and called a tonne [t]. (See the 'Decimal fraction' table on page 5.)

Mass can be proportionally accelerated or retarded by an applied force. To maintain a coherent system of units and to enable comparisons of physical properties, a unit of force is chosen which will give unit acceleration to a unit of mass. This unit of force is called the *newton* [N]. Therefore, 1 newton of force acting on 1 kilogram of mass will give it an acceleration of 1 metre per second per second, and this gives rise to the standard equation:

Accelerating force [N] = Mass [kg] × Acceleration [m/s²]

or in symbols: $F = ma$

FORCE OF GRAVITY. All bodies are attracted towards each other; the size of that force of attraction will depend upon the masses of the bodies and their distances apart. Newton's law of gravitation states that this force of attraction is proportional to the product of the masses of the bodies and inversely proportional to the square of the distance apart.

An important example of this characteristic of the physical world is the mass of the earth which, as we can observe, attracts all bodies towards it. This force of attraction, drawing a body towards the centre of the earth, is called the force of gravity and its effect on an object is called the *weight* of the object.

If a body is allowed to fall freely towards the earth, it will fall with an acceleration of approximately 9.81 m/s²; this is termed 'earth's gravitational acceleration' and is represented by the symbol g. We have often heard the term 'g' force. This is based on earth's gravitational acceleration, with 1 'g' being equal to the force of gravity and 2 'g' being twice the force of gravity. Centrifugal force (created in a fuel or lubricating oil purifier, for example) can be described in this way.

Since 1 newton is the force which will give 1 kilogram of mass an acceleration of 1 m/s², then the force in newtons to give m kg of mass an acceleration of 9.81 m/s² is $m \times 9.81$. Therefore, at the earth's surface, the gravitational force on a mass of m kg is mg newtons, or in other words:

$$\text{Weight [N]} = \text{Mass [kg]} \times g \text{ [m/s}^2\text{]}$$

The distance between the centre of gravity of the mass and the centre of gravity of the earth slightly reduces the attractive force between them. Thus, the weight of a mass measured by a spring balance (not a pair of scales which is merely a means of comparing the weight of one mass with another) will vary slightly at different parts of the earth's surface due to the earth not being quite a perfect sphere.

If a body is placed in earth's orbit (where the object is continually rotating around the earth), it will experience a condition known as 'weightlessness'. However, the object has not suddenly changed its form or anything about its physical properties. In fact, at the distance from the earth to orbit, the effect of gravity has only reduced from 9.81 m/s² to approximately 8.7 m/s². This means that an object weighing 1000 N on the earth's surface should still weigh about 890 N in orbit, the balance of its weight (890 N) being opposed by the centrifugal force set up due to the speed of rotation of the object around the earth.

The lesson here is that if, for example, the effect of gravity is removed – as is the case when the weight of an object, such as a piston, is taken by a crane – the rest of the piston's physical properties, including the mass of the piston, remains the same and therefore any sideways movement will still carry the same mass of the object behind the movement. Albert Einstein found that gravity is a relatively weak force that stretches across the universe, but it interacts with all matter in that universe.

INERTIA. This refers to a property possessed by matter which means that it resists a change of motion. The magnitude of inertia possessed by a body depends upon the properties of its mass, and the 'Polar moment of inertia' is used to describe the capacity of a beam to resist bending. Inertia must be specified with respect to a chosen axis. If the mass is at rest, it requires a force to give it motion, and the greater the mass, the greater the force required to give it that same motion. If the mass is already moving, it requires a force to change its velocity or to change its direction, again the force required being proportional to the mass. Therefore, the student can reason that a force is not necessary to keep an object in motion; however under conventional circumstances, an opposing force (usually friction) will bring that body to rest.

WORK. This is done when a force applied on a body/object causes it to move. Work is then measured by the product of the force and the distance through which that force acts. Work is also independent of time, and therefore for the effect to be considered 'work' there must be a force and there must be displacement of an object caused by the force or a component of the force.

The unit of work is the *joule* [J] which is defined as 'the work done when the point of application of a force of 1 newton moves through a distance of 1 metre in the direction in which the force is applied'. Hence, 1 joule is equal to 1 newton-metre, which in symbols is expressed as J = Nm.

$$\text{Work done [J]} = \text{Force [N]} \times \text{Distance moved [m]}$$

The joule is a small unit and even moderate quantities of work will need to be expressed in kilojoules [1 kJ = 10^3 J] and larger quantities of work will be expressed in megajoules [1 MJ = 10^6 J].

POWER. This is where time comes in as power is the rate at which the work is completed, that is, the quantity of work done in a given time, or it is a measure of the work/time ratio. The SI unit of power is the *watt* [W] which is equal to the rate of 1 joule of work being done every second. In symbols, W = J/s = Nm/s.

$$\text{Power [W]} = \frac{\text{Work done [J]}}{\text{Time [s]}}$$

Again, the watt is a relatively small unit and will only be seen when describing low-powered output such as that produced by people. It will be much more normal, in engineering, to express power in kilowatts [1 kW = 10^3 W] and megawatts [1 MW = 10^6 W] as these are usually more convenient units. For example, the maximum power output from the main engine of a VLCC (very large crude carrier) tanker might be described at 45 MW and the overall power plant of a large cruise ship might be as much as 95 MW. (Power also equals force × velocity.)

ENERGY. This is a measure of the capacity for carrying out work, and it is calculated by the size or amount of work completed. Confusingly this means that energy is also expressed in the same units as work, that is, joules, kilojoules and megajoules.

Another useful measure of energy is the *kilowatt-hour* [kWh]. This, as the name implies, represents the energy used or the work done when 1 kilowatt of power is exerted continually for 3600 seconds or 1 hour.

$$\text{Energy} = \text{Power} \times \text{Time}$$
$$1 \text{ kWh} = 1000 \text{ watts} \times 3600 \text{ seconds}$$
$$= 1000 \left[\text{J/s} \right] \times 3600 \left[\text{s} \right]$$
$$= 3.6 \times 10^6 \text{ J}$$
$$= 3.6 \text{ MJ}$$

EFFICIENCY. This is the ratio of the work output from a machine relative to the work put into it, and, as this is measured in a set time frame, it is also the ratio of the output power to the input power. All machines use some energy to drive their internal components and overcome friction, and therefore the power output from a machine is always less than the input. Hence the efficiency is always less than unity (1).

The symbol for efficiency is the Greek letter 'eta' which in lower case is shown as η and may be expressed as a fraction or as a percentage.

$$\eta = \frac{\text{Output power}}{\text{Input power}}$$

Example 1.1. A mass of 1600 kg is lifted by a winch through a height of 25 m in 30 seconds. Calculate (i) the work done, and, if the efficiency of the winch is 60%, find (ii) the input power in kW and (iii) the energy consumed in kWh.

$$\text{Force} \left[\text{N} \right] \text{ to lift mass against gravity}$$
$$= \text{Weight of mass} = mg$$
$$= 1600 \times 9.81 \text{ newtons}$$
$$\text{Work done} \left[\text{J} = \text{Nm} \right] = \text{Force} \left[\text{N} \right] \times \text{Distance} \left[\text{m} \right]$$
$$= 1600 \times 9.81 \times 25$$
$$= 392400 \text{ J} = 3924 \text{ kJ} \quad \text{Ans. (i)}$$

$$\text{Output power} \left[\text{kW} = \text{kJ/s} \right] = \frac{\text{Work done [kJ]}}{\text{Time [s]}}$$
$$= \frac{392.4}{30} = 13.08 \text{ kW}$$
$$\text{Efficiency} = \frac{\text{Output power}}{\text{Input power}}$$
$$\text{Input power} = \frac{13.08}{0.6} = 21.8 \text{ kW} \quad \text{Ans. (ii)}$$

$$\text{Energy consumed} = \frac{392.4}{0.6} = 654 \text{ kJ}$$
$$1 \text{ kWh} = 3.6 \text{ MJ}$$
$$654 \text{ kJ} = \frac{0.654}{3.6} = 0.1817 \text{ kWh} \quad \text{Ans. (iii)}$$

Alternatively,

$$\text{Energy } [\text{kWh}] = \text{Power } [\text{kW}] \times \text{Time } [\text{h}]$$
$$= 21.8 \times \frac{30}{3600}$$
$$= 0.1817 \text{ kWh}$$

PRESSURE. This is a measure of the force that is being applied over a unit area. Force, as we have seen, is expressed in newtons [N] and area is measured in square metres [m²]. Therefore the fundamental unit of pressure is the newton per square metre [N/m²], with the symbol representing pressure usually being p and the derived SI unit being the pascal.

Pressure within liquids and gases is caused by the excited state of the molecules contained within the substance. The excitation of the molecules causes them to move, hitting each other as well as coming up against the solid wall of their enclosure. It can be appreciated by students that adding energy in the form of heat will raise the kinetic energy of the molecules and increase the speed of collisions with the sides of the container, thus increasing the force exerted and the pressure registered.

The rising pressure of liquids and gases can reach high values which are expressed in multiples of the basic unit of force. For example, the steam pressure in low pressure boilers is often in the region of 7×10^5 N/m² and in high pressure boilers it could be 6×10^6 N/m². The former can be conveniently written as 700 kN/m² and the latter as 6000 kN/m² or 6 MN/m².

Another very convenient unit of pressure commonly used is 'bar'. This has the advantage of being simple and therefore easier to think about and remember, since 1 bar is approximately equal to 1 atmosphere of pressure (1 atm = 1.013 bar). One bar is 10^5 N/m², which is 100 kN/m², and hence the working pressure of the boilers given in the paragraph above would be stated as 7 bar and 60 bar, respectively. These are common working pressures for marine auxiliary steam boilers (7 bar) and main superheated steam boilers (60 bar) used where the main propulsion plant of the ship is a steam turbine.

Pressures in internal combustion (IC) engines vary from a little above or below 1 bar during the air charging (or scavenging) period to as much as 200 bar $= 200 \times 10^5$ N/m² $= 20.000$ kN/m² during the peak period of combustion.

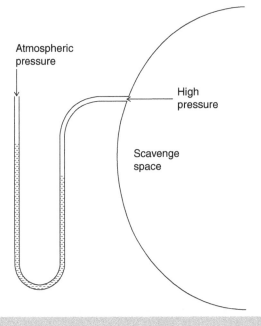

Figure 1.1 *Manometer*

Low pressures are usually measured in millimetres of mercury [mmHg], and very small pressures and pressures approaching a vacuum in millimetres of water [mm water]. The original instrument used to measure these pressures was the *manometer* which is still used for some applications (Figure 1.1). This is a glass U-tube partially filled with mercury or water; one end is connected to the source of pressure and the other end is open to the atmosphere. The difference between the levels of liquid in the two legs indicates the difference in pressure between the source being measured and the atmosphere. This measurement is taken in millimetres.

Consider a manometer containing mercury, if we take the density of mercury to be 13.6×10^3 kg/m^3 and the force of gravity on a mass of 1 kilogram as 9.80665 newtons, which is just a more accurate figure for the standard value of gravitational acceleration than the 9.81 usually acceptable for general engineering calculations, then the weight of 1 cubic metre of mercury is $13.6 \times 10^3 \times 9.80665$ newtons = 133.3 kN. This means that a column of mercury 1 metre high exerts a pressure of 133.3 kN on 1 square metre. Therefore a column of mercury 1 millimetre high is equivalent to a pressure of 133.3 N/m^2.

Similarly, each millimetre of water pressure is equal to 9.80665 N/m^2, which is usually taken as 9.81 N/m^2. This is due to the density of water being equal to 1 kg/m^3. Small pressures may also be expressed in millibars [mbar]:

$$1 \text{ mbar} = 1 \text{ bar} \times 10^{-3} = 10^5 \times 10^{-3} \text{ N/m}^2 = 100 \text{ N/m}^2$$

The mercury-filled *barometer* works on the principle of the atmospheric pressure supporting a column of mercury. The vertical column of mercury left standing up the tube, with the vacuum above, is supported by the outside atmospheric pressure and is therefore a measure of the pressure of the atmosphere. As the atmospheric pressure rises and falls, the level of the supported column of mercury rises and falls accordingly.

For example, if the column of mercury supported by the atmospheric pressure is 760 mm, then the atmospheric pressure will be:

$$760 \times 133.3 = 1.013 \times 10^5 \text{ N/m}^2 = 1.013 \text{ bar or } 101.3 \text{ kN/m}^2$$

This illustration is interesting as it helps students to visualize basic concepts and understand the operating principle behind the most straightforward of instruments. It must be accepted, however, that the world of engineering is moving ever closer to electrical/electronic control of system operation in real time.

With this in mind it can be seen that the liquid in glass measuring instruments is now being replaced by pressure sensors. Piezoelectric sensors, for example, work on having a change in electrical resistance which is proportional to the change in their crystalline structure caused by the pressure change.

The change is detected by an electronic circuit, compensated for temperature difference and the final signal output as a measurement of the pressure being exerted on the sensor.

GAUGE PRESSURE AND ABSOLUTE PRESSURE. Pressure is the measured variable of a force acting upon an open or closed system in all directions. It is helpful to think about measuring pressure in one of three different ways based upon where the reference point is set.

With absolute pressure the starting point is at zero pressure (0 pascals) or a (practical) vacuum. Therefore, measuring the pressure of a system may need to include the effect of atmospheric pressure to give an 'absolute' value of the true pressure.

Gauge pressure, as its name suggests, is a reading given by a gauge measuring the pressure inside a closed system and therefore will not include atmospheric pressure.

Differential pressure (ΔP) is simply the measured pressure difference between two different systems or the two different parts of one system.

Most pressure-recording instruments will simply measure the pressure within a system. The pressure so recorded is actually a *gauge pressure* and the word 'gauge' or 'g' should be used to indicate this aspect. Thus, if a pressure gauge reads 2000 kN/m², the pressure should be stated as 2000 kN/m² gauge, meaning that this is the pressure over and above the atmospheric pressure.

A true pressure, one that is measured above a perfect vacuum, is called the *absolute pressure*, and this is the value which is usually used in calculations. The absolute pressure is therefore obtained by adding the atmospheric pressure to the gauge pressure, the gauge pressure being read from the pressure gauge and the atmospheric pressure obtained from the barometric reading.

Therefore, we have

$$P_{atm} = \text{Atmospheric pressure}$$

$$P_g = \text{Gauge pressure}$$

$$P_s = \text{Pressure overall in the system}$$

As an example, if the pressure of a fluid is 550 kN/m² gauge, and the barometer stands at 758 mmHg, then:

$$\text{Atmospheric pressure} = 758 \times 133.3$$
$$= 1.01 \times 10^5 \text{ N/m}^2$$
$$= 101 \text{ kN/m}^2$$
$$\text{Absolute pressure} = \text{Gauge pressure} + \text{Atmospheric pressure}$$
$$= 550 + 101$$
$$= 651 \text{ kN/m}^2$$

The final calculated pressure could be written as 651 kN/m² absolute. However, it is usual to omit the word 'absolute' and take it for granted that if the word 'gauge' does not follow the value of the pressure, then it means that it is an absolute pressure. This will be the practice throughout this book.

Mechanical pressure gauges are not always perfectly accurate; it is also difficult to read a gauge to an accuracy of 1 or 2 kN/m² and for this reason it is common, when exact accuracy is not essential, to assume the atmospheric pressure to be 100 kN/m². In the above example, if the barometer reading was not known, the absolute pressure would be taken as: 550 + 100 = 650 kN/m² with very little difference in the final result of a calculation.

If the manometer was to be used as a vacuum gauge, as may be the case for a steam condenser, the level of the mercury in the leg connected to the condenser will be higher than the level in the leg open to atmosphere. The difference in level indicates the pressure *below* atmospheric and is written 'mmHg vacuum'. For example, if the gauge reads 600 mmHg on vacuum and the barometer reads 758 mmHg, then:

$$\text{Pressure below atmospheric} = 600 \times 133.3 = 8 \times 10^4 \text{ N/m}^2$$

$$\text{Atmospheric pressure} = 758 \times 133.3 = 1.01 \times 10^5 \text{ N/m}^2 = 101 \text{ kN/m}^2$$

Therefore absolute pressure in the condenser is 80 kN/m² *below* 101 kN/m² which is 21 kN/m² or, more simply calculated:

$$\begin{aligned}
\text{Abs. press.} &= (\text{Barometer mmHg} - \text{Vacuum gauge mmHg}) \times 133.3 \\
&= (758 - 600) \times 133.3 \\
&= 158 \times 133.3 \\
&= 2.1 \times 10^4 \text{ kN/m}^2 = 21 \text{ kN/m}^2
\end{aligned}$$

Please note that 1 kN/m² = 1 KPa and kN/m² has been used throughout this book to remind students of the detail that is involved within the unit. However, students may find that their own institution uses KPa in place of kN/m².

VOLUME. This has the basic derived SI unit of the cubic metre [m³]. However another term commonly used is the litre [l]. This is equal in volume to 1 cubic decimetre and is usually used to measure fluid (although the capacities of smaller IC engines are categorised in litres). The litre is:

$$1 \text{ m}^3 = 10^3 \text{ dm}^3; \text{ therefore } 10^3 \text{ litres} = 1 \text{ m}^3$$

The millilitre [ml] is 1×10^{-3} litre and therefore equal in volume to 1 cubic centimetre. Whereas the basic unit of density is kilogram per cubic metre [kg/m³], densities of liquids are sometimes expressed in grams per millilitre [g/ml] and densities of solids in grams per cubic centimetre [g/cm³].

SPECIFIC VOLUME. This is the volume occupied by unit mass and the basic unit is cubic metre per kilogram [m³/kg]; thus, the specific volume is the reciprocal of density. In certain cases, specific volume may be expressed in cubic metres per tonne [m³/t] and litres per kilogram [l/kg].

TEMPERATURE and HEAT. Knowing exactly what 'temperature' represents in science or engineering seems quite straightforward; however there are some pitfalls to consider. The most important concept is that heat and temperature are not the same, although of course they are interlinked. Heat is a measure of the transfer of energy from one object/substance/system to another, and temperature is a measure of an object's 'internal energy' or the intensity of heat that it possesses at the given moment that it is measured using a thermometer.

On board a ship it is still not unusual to measure temperature using the mercury (or other liquid) in glass thermometer. This consists of a glass tube of very fine bore with a bulb at its lower end; the bulb and tube are exhausted of air, partially filled with mercury and hermetically sealed at the top end. When the thermometer is placed in a substance whose temperature is to be measured, the mercury takes up the same temperature and expands (if heated) or contracts (if cooled) and the level, which rises or falls as a consequence, indicates on the thermometer's scale the degree of heat intensity the substance possesses.

The advantages of mercury are that it does not wet the bore of the glass and therefore none of the liquid is lost from the main pool, due to it sticking to the glass, as the temperature falls. It can be used over a wide useful range of temperature as its freezing point is low, about −38°C, and boiling point high, about 358°C.

Alcohol is sometimes used but it is required to be coloured so that the liquid can be seen clearly. Although its freezing point is in the region of −115°C and so it can be useful for measuring very low temperatures, its boiling point is also low, about 78°C, and therefore it has limited use.

The system of measurement: Initially the 'metric' system took the point at which pure water froze into ice and marked this zero. It then considered the point at which pure water boils into steam at atmospheric pressure and assigned 100 divisions between the two points; hence the 'centi' (100) and 'grade' (scale). The centigrade label was used until 1948 when the term was changed to Celsius.

The Celsius scale is derived from the two fixed points of absolute zero and the 'triple point' of water. The unit representing a temperature reading on the Celsius scale is °C and the symbol for temperature is θ. The incremental divisions on the Celsius scale are the same as in the centigrade scale and align with the Kelvin thermodynamic measuring system (see page 1).

The Fahrenheit scale was used on instruments in older ships' engine rooms. To convert a *temperature interval* from Fahrenheit to Celsius:

$$\text{Interval on Celsius scale} = \frac{5}{9} \times \text{Interval on Fahrenheit}$$

For example, if a body is heated through 153°F, this is:

$$153 \times \frac{5}{9} = 85°C$$

To convert a *temperature* reading from Fahrenheit to Celsius:

$$\text{Reading on the Celsius scale} = (F - 32) \times \frac{5}{9}$$

For example, a temperature of 77°F is equivalent to:

$$(77 - 32) \times \frac{5}{9} = 25°C$$

It is not expected that students will encounter the need to know this conversion. However, there are some places that retain the Fahrenheit measurement system, and therefore this illustration has been retained in this version of the book.

ABSOLUTE ZERO TEMPERATURE. As long as the external pressure remains constant all gases expand at practically the same rate when heated through the same range of temperature. They also contract at the same rate when cooled.

The rate of expansion or contraction of a perfect gas is approximately 1/273 of its volume taken at 0°C when heated or cooled at constant pressure through 1°C.

If the linear contraction graph is projected backwards, then *absolute zero temperature* can be calculated. Measured accurately this value on the Celsius scale is –273.15°C.

The record temperature reduction known is one that has been created here on earth. In 2003 at the Massachusetts Institute of Technology a cloud of sodium atoms was cooled to 0.45 nanokelvin. This is lower than the average found anywhere in the rest of the known universe, which is 2.7 K. The temperature of the supercooled magnets in the Large Hadron Collider, which is part of the particle physics laboratory in Cern, Switzerland, is –271.3°C or 1.86 K.

As the temperature of absolute zero is approached, matter starts to take on different characteristics. For example, at 1.175 K aluminium becomes superconducting and loses its resistance to electrical current flow. Students will be able to recognize that this has interesting possibilities for engineering and, in fact, superconducting electric propulsion motors are already a reality. The problem is the low temperature that must be achieved before the metal becomes superconducting. If high temperature superconducting metals could be developed then super-efficient high powered motors could follow.

In practice it is not possible to cool a gas down to absolute zero due to 'Heisenberg's uncertainty principle' which says that if we precisely know a particle's speed, that is, zero, then the less we know about its position (is it in a fixed position or not). However, as a gas approaches absolute zero its properties change and some gases have the ability to become superfluids. This is where all of their friction is lost.

Students will now be able to see from the above that temperatures can be expressed as *absolute* quantities, that is, stating the degrees of temperature above the level of absolute zero, by adding 273.15 to the ordinary Celsius thermometer reading.

In science and engineering absolute temperature is referred to as *thermodynamic temperature*. The symbol for this is *T* and the unit is the kelvin, which is represented by K, and the interval on the kelvin scale is the same as on the Celsius scale:

Thermodynamic temperature = Celsius temperature + 273

In symbols,

$$T\,[K] = \theta\,[°C] + 273$$

(Note that the degree symbol is not used with Kelvin.)

VOLUME FLOW. This is the volume of a fluid flowing past a given point in unit time. The basic unit is cubic metres per second [m³/s], and other convenient units are cubic metres per hour [m³/h], cubic metres per minute [m³/min] and litres per hour [l/h].

For example, if a fluid is flowing at a velocity of *v* metres per second full bore through a pipe of internal diameter *d* metres, the quantity flowing in cubic metres per second is:

$$\text{Volume flow }[m^3/s] = \text{Area }[m^2] \times \text{Velocity }[m/s]$$
$$= 0.7854d^2 \times \dot{v}$$

$\dot{v}$ is used to indicate volume flow rate.

MASS FLOW. This is the mass of fluid flowing past a given point in unit time, the basic unit being kilograms per second [kg/s]. Since density is the mass per unit volume, then:

$$\text{Mass flow [kg/s]} = \text{Volume flow }[m^3/s] \times \text{Density }[kg/m^3]$$

Mass flow may also be expressed in other convenient units such as tonnes per hour [t/h] and kilograms per hour [kg/h]; m is used to indicate mass flow rate. Mass flow is taking on a greater importance in the marine field due to the measurement of loading bunker fuel. Measuring volume or volume flow does not take into account density and therefore the mixture being loaded could contain impurities such as water or solids, which would not show up when it is taken on board. However, if the mass flow of the fuel is measured, then the on board staff will have a greater confidence that the correct grade and quality is being loaded.

SWEPT VOLUME OR STROKE VOLUME. This is the volume moved through by a piston in the cylinder of a reciprocating engine, pump or air/gas compressor etc., and is the product of the piston area [m²] and the stroke of the piston [m]. If the arrangement of the reciprocating pistons in their cylinders within a machine is vertical, then the 'stroke of the piston' is the term used to describe its movement from its lowest point of travel to its highest. One movement in one direction is called one stroke (Figure 1.2).

The space left between the piston at its top dead centre (TDC), which is the uppermost position in the cylinder or the top of its stroke, and the cylinder head is termed the *clearance volume*. This may be expressed as the actual volume of the clearance space, or expressed as a fraction or ratio of the stroke volume, for example, 13:1 or 20:1.

SYSTEM. A *system* is the term given to matter that is under investigation and is within a defining *boundary*, and the region outside the boundary is termed the *surroundings*. The boundary may be an imaginary enclosure, or it may be a physical one, such as the cylinder wall, cylinder head and piston of an IC engine. The boundary then encloses the mixture of gases within. Energy can be transferred across the boundary from one system to another.

If there is no transfer of matter across the boundary, that is, if no substance can enter the system or leave it during the investigation, it is called a *closed system*, with energy

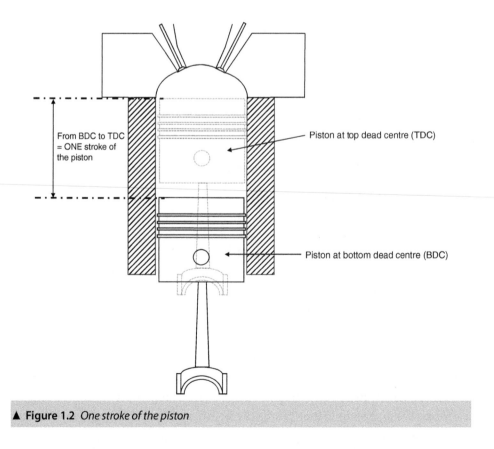

From BDC to TDC = ONE stroke of the piston

Piston at top dead centre (TDC)

Piston at bottom dead centre (BDC)

▲ **Figure 1.2** *One stroke of the piston*

only being transferred across the boundary and whatever changes taking place to the substance happening within the boundary. This system is also termed a 'non-flow' process because no matter flows into or out of it. A simple example would be the inflation of a hot air balloon. Energy is added and the air inside expands but it is trapped by the balloon's envelope.

If there is a flow of matter through the boundary, it is called an *open system*. If the mass flow entering the system is equal to the mass flow leaving so that at any time the quantity of matter within the system is constant, the series of changes to the matter is referred to as a 'steady-flow' process.

An example of a closed system within which a non-flow process takes place is the cylinder of an air compressor and an IC engine. A steam turbine is an example of an open system in which a steady-flow process takes place, as is a nozzle and a uniformly cycling reciprocating unit such as a piston pump.

A 'cycle' is a re-occurring set or a series of connected processes which a system undergoes with the final state of the system at the end of the cycle being exactly the same as it was at the beginning of the cycle.

ENTHALPY is a measure of the amount of energy in a thermodynamic system – see page 70.

ENTROPY is not an easy concept to grasp. Accepted wisdom currently often describes entropy as a measure of 'randomness' and this is increasingly being applied to different aspects of human activity. Initially derived from the second Law of Thermodynamics where the natural order of matter is to move from being 'in-order' to becoming more 'random' with the passage of time.

However, what does this really mean, in a practical sense, for marine engineers? One train of thought is to think about a gas in a container. If energy is added to the gas, the molecules will become more ordered, and the entropy of the gas will fall. From this point the natural action will be for the gas to move toward a less ordered state – in this case by giving up some of its energy to the surrounding area (see Chapter 11 for more details).

Engineering officers, licensed to be in charge of a merchant ship's power plant, will be required to hold a certificate of competency issued by a recognised member of the maritime assembly of the United Nations, called the International Maritime Organization (IMO).

Traditionally, these certificates of competency have focused on the propulsion systems of steam (turbine) and/or motor (diesel engine). However, modern ships need to be as efficient as possible and power systems that use heat as the energy source, will focus on using as much of the heat as possible. Therefore, the topic of 'waste heat recovery' has been elevated to a more central role in the design of modern machinery and the areas of 'steam or motor' have sometimes become blended together.

Propulsion systems that had a diesel engine or a steam boiler/turbine system, directly driving the propeller, convert considerably less than 40% of the heat energy into useful work. This means that there is a significant amount of heat wasted, and by combining a diesel engine with a steam plant useful increases in efficiency can be realised.

These developments mean that it is now important for all marine engineers to understand the basic ideas of thermodynamics that relate to the 'steam cycle'.

Test Examples 1

1. A pump discharges 50 tonnes of water per hour to a height of 8 m, the overall efficiency of the pumping system being 69%. Calculate:

   (i) the output power

   (ii) the input power

   (iii) the energy consumed by the pump in 2 hours, expressed in kWh

   (iv) and in MJ.

2. (i) Express a pressure of 20 mm water in N/m^2 and mbars.

   (ii) Express a pressure of 750 mmHg in kN/m^2 and bars.

3. A condenser vacuum gauge reads 715 mmHg when the barometer stands at 757 mmHg. State the absolute pressure in the condenser in kN/m^2 and bars.

4. Although °F are not SI units it is still useful for students to know the conversions. Convert the following temperature readings from °F to °C:

   140°F   5°F   −31°F   −40°F

5. Oil flows full bore at a velocity of 2 m/s through a nest of 16 tubes in a single pass cooler. The internal diameter of the tubes is 30 mm and the density of the oil is 0.85 g/ml. Find the volume flow in l/s and the mass flow in kg/min.

2

HEAT

Heat is a form of energy associated with the movement of the molecules which constitute the heated body. The molecules become agitated and move more rapidly as more energy is added to a substance. Heat is transferred from one substance to another due to the temperature difference between the two substances; it is interchangeable with other forms of energy and can be made available for doing work and producing mechanical and/or electrical power.

One property that engineers should take note of is that it takes time for heat to transfer from one substance to another. This feature is often responsible for practical applications/machines not being able to reach the maximum efficiency that is theoretically possible.

The basic unit of all energy, including heat, is the joule [J]. As already stated on page 8 this basic unit is small and therefore, units of heat are not only expressed in joules but also in multiples of the joule, the most common multiple being the kilojoule [kJ] and the megajoule [MJ]. The symbol representing the quantity of heat is Q.

Specific Heat

Specific Heat (or specific heat capacity) relates to the ability of different substances to take up and retain heat. The specific heat of a substance is the quantity of heat required to raise the temperature of unit mass of the substance by 1 degree. The units of specific heat are therefore heat units per unit mass per unit temperature. The symbol for specific heat is c.

Problems in engineering, involving specific heat, will use the kilojoule as the most appropriate size of heat unit for a unit mass of 1 kilogram, and hence the specific heat is usually expressed in kilojoules per kilogram per kelvin; in symbols this is kJ/kg K. Note also that 1 Celsius degree of temperature interval on the thermometer scale is the same as 1 kelvin degree of temperature interval on the absolute scale; hence the above units could be written as kJ/kg°C, and this is becoming more of a common presentation.

It follows from the above definition of specific heat that the quantity of heat energy transferred to a substance, to raise its temperature, is the product of the mass of the substance, its specific heat and its rise in temperature.

In symbols this means that:

$$Q \text{ [kJ]} = m \text{ [kg]} \times c \text{ [kJ/kg K]} \times (T_2 - T_1) \text{ [K]}$$

Different substances have different specific heat values. Also, the specific heat of any one particular substance is not always a constant value over a large range of temperatures; the variation in specific heat, however, is usually small, and an average value over the temperature range under consideration, may be used for most practical purposes. For example, the specific heat of water decreases from 4.21 kJ/kg K at 0°C to 4.178 kJ/kg K at 35°C and increases thereafter with rise in temperature, being 4.219 kJ/kg K at 100°C. Between 0°C and 100°C therefore a mean value is usually taken as 4.2.

Example 2.1. Calculate the quantity of heat to be transferred to 2.25 kg of brass to raise its temperature from 20°C to 240°C, taking the specific heat of the brass to be 0.394 kJ/kg K.

Temperature increase = 240 − 20 = 220°C = 220 K

$$\begin{aligned} Q \text{ [kJ]} &= m \text{ [kg]} \times c \text{ [kJ/kg K]} \times \left(T_2 - T_1 \text{[K]}\right) \\ &= 2.25 \times 0.394 \times 220 \\ &= 195 \text{ kJ Ans.} \end{aligned}$$

The characteristics of gases vary considerably at different temperatures and pressures, and heat may be transferred under an infinite number of different conditions; consequently the specific heat can have an infinite number of different values. Two important conditions are transferring heat to or from a gas which is at constant pressure, and transferring heat while its volume is constant. The specific heat of a gas at constant pressure is represented by c_p and at constant volume by c_v. This is dealt with in detail later.

$$\text{Note: } (T_2 - T_1) \text{ [K]} = (\theta_2 - \theta_1) \text{ [°C]}$$

MECHANICAL EQUIVALENT OF HEAT. This refers to the relationship between mechanical energy and heat energy. It was determined by James Prescott Joule, one of the first scientists to demonstrate that heat was a form of energy, using apparatus which generated heat by the expenditure of mechanical work. When work is done in overcoming friction, the mechanical energy expended is converted into heat energy.

The force required to overcome sliding friction between two bodies is the product of the coefficient of friction (μ) and the normal force between the surfaces of the bodies.

Example 2.2. A shaft runs at a rotational speed of 50 rev/s in oil-cooled bearings of 178 mm diameter. The force between the surfaces of the shaft journals and bearings is 2.67 kN and the coefficient of friction is 0.04. Find (i) the friction force at the surface of the journals, (ii) the mechanical energy expended in friction per revolution, (iii) the power loss due to friction, (iv) the temperature rise of the oil if the volume flow through the bearings is 18 l/min, the specific heat of the oil being 2 kJ/kg K and its density 0.9 g/ml.

$$\text{Friction force} = \mu \times \text{Normal force between surfaces}$$
$$= 0.04 \times 2.67 \times 10^3$$
$$= 106.8 \text{ N} \quad \text{Ans.(i)}$$

Work done to overcome friction per revolution $[J = Nm]$
$$= \text{Friction force } [N] \times \text{Circumference of journal } [m]$$
$$= 106.8 \times \pi \times 0.178$$
$$= 59.7 \text{ J} \quad \text{Ans.(ii)}$$

Power expended $[W = J/s]$
$$= \text{Energy per revolution} \times \text{rev/s}$$
$$= 59.7 \times 50$$
$$= 2985 \text{ W} = 2.985 \text{ kW} \quad \text{Ans.(iii)}$$

Density of oil $= 0.9$ g/ml $= 0.9$ kg/litre
Mass flow of oil $[kg/s] = \text{Volume flow } [1/s] \times \text{Density } [kgA]$
$$= \frac{18}{60} \times 0.9 = 0.27 \text{ kg/s}$$
$$Q \ [kJ/s] = m[kg/s] \times c[kJ/kg \ K] \times (T_2 - T_1) \ [K]$$
$$2.985 = 0.7 \times 2 \times \text{Temp. rise}$$
$$\text{Temp. rise} = 5.527 \text{ k or } 5.527°C \quad \text{Ans.(iv)}$$

WATER EQUIVALENT. The water equivalent of a mass of a substance is the method of measuring the energy contained in the matter that makes up the containment system (a metal container for example) for the process that is under consideration. It is the mass of water that would require the same heat transfer as the mass of the containment substance (steel or aluminium for example) to cause the same change of temperature.

For example, taking an aluminium vessel of mass 2 kg, and the specific heat of aluminium to be 0.912 kJ/kg K:

$$Q = \text{Mass} \times \text{Specific heat} \times \text{Temperature change}$$

$$\text{For water, } Q_W = m_W \times c_W \times \left(T_2 - T_1\right)_W$$

$$\text{For aluminium, } Q_A = m_A \times c_A \times \left(T_2 - T_1\right)_A$$

Since Q is to be the same quantity of heat,

$$Q_W = Q_A$$

$$m_W \times c_W \times \left(T_2 - T_1\right)_W = m_A \times c_A \times \left(T_2 - T_1\right)_A$$

and the temperature change is to be the same:

$$m_W \times c_W = m_A \times c_A$$

$$\therefore m_W = m_A \times \frac{c_A}{c_W}$$

Taking the specific heat of water to be 4.2 kJ/kg K, the water equivalent of this mass of aluminium is:

$$m_W = 2 \times \frac{0.912}{4.2}$$

$$= 0.4343 \text{ kg}$$

That is to say, 0.4343 kg of water would require the same amount of heat transferred to it as the 2 kg of aluminium to raise it through the same range of temperature. It is important to know the water equivalent of laboratory calorimeters. When water is contained in a vessel, the temperature of the vessel is the same as that of the water inside, and when the temperature of the water is changed, the temperature of the vessel changes with it. The vessel can therefore be considered as an extra mass of water equal to the water equivalent of the vessel. Without this additional part of the experiment the input energy would be too high and the calculation would be inaccurate.

Example 2.3. The mass of a copper calorimeter is 0.28 kg and it contains 0.4 kg of water at 15°C. Taking the specific heat of copper to be 0.39 kJ/kg K, calculate the heat required to raise the temperature to 20°C.

$$\text{Water equivalent of calorimeter} = 0.28 \times \frac{0.39}{4.2} = 0.026 \text{ kg}$$

Heat received by water and calorimeter:

$$Q \, [\text{kJ}] = m \, [\text{kg}] \times A \, [\text{kJ/kg K}] \times \left(T_2 - T_1\right) \, [\text{K}]$$

$$= \left(0.4 + 0.026\right) \times 4.2 \times \left(20 - 15\right)$$

$$= 0.426 \times 4.2 \times 5$$

$$= 8.946 \text{ kJ} \quad \text{Ans.}$$

When two substances at different temperatures are mixed together, heat will transfer from the hotter substance to the colder until both become the same temperature. Unless otherwise stated, it is assumed that no heat is transferred to or from an outside source during the mixing process, and therefore the quantity of heat absorbed by the colder substance is all at the expense of the loss of heat by the hotter substance.

Example 2.4. In an experiment to find the specific heat of lead, 0.5 kg of lead shot at a temperature of 51°C is poured into an insulated calorimeter containing 0.25 kg of water at 13.5°C and the resultant temperature of the mixture is 15.5°C. If the water equivalent of the calorimeter is 0.02 kg, find the specific heat of the lead.

Heat received by water and calorimeter when their temperature is raised from 13.5°C to 15.5°C:

$$Q = m \times c \times (T_2 - T_1)$$
$$= (0.25 + 0.02) \times 4.2 \times (15.5 - 13.5)$$
$$= 0.27 \times 4.2 \times 2 \text{ kJ}$$

Heat lost by lead in cooling from 51°C to 15.5°C:

$$Q = m \times c \times (T_3 - T_2)$$
$$= 0.5 \times c \times (51 - 15.5)$$
$$= 0.5 \times c \times 35.5 \text{ kJ}$$

Heat transferred from the lead is equal to the heat received by the water and calorimeter:

$$0.5 \times c \times 35.5 = 0.27 \times 4.2 \times 2$$
$$c = \frac{0.27 \times 4.2 \times 2}{0.5 \times 35.5}$$
$$= 0.1278 \text{ kJ/kg K} \quad \text{Ans.}$$

LATENT HEAT. This is the heat which supplies the energy necessary to overcome the binding forces of attraction between the molecules of a substance as it changes its physical state from a solid into a liquid, or from a liquid into a vapour. The change takes place without any change of temperature. This means that there will be an input of heat energy for no change in observed temperature.

The process of changing the physical state from a solid into a liquid is called *melting* or *fusion*, and the quantity of heat required to change unit mass of the substance from solid to liquid at the same temperature is the *latent heat of fusion*.

For example, the latent heat of fusion for ice is 335 kJ/kg at 0°C. This means that 1 kilogram of ice at 0°C would require 335 kilojoules of heat transferred to it to completely melt it into 1 kilogram of water at 0°C. Also, 1 kilogram of water at 0°C would require to lose 335 kilojoules of heat to completely freeze into ice at 0°C.

The process of changing the physical state of a substance from a liquid into a vapour is called *boiling* or *evaporation*, and the quantity of heat to bring about this change at constant temperature to unit mass is the *latent heat of evaporation.*

The latent heat of evaporation of water at atmospheric pressure is 2256.7 kJ/kg. This means that 1 kilogram of water at 100°C would require 2256.7 kilojoules of heat to completely boil it into 1 kilogram of steam at 100°C. Also, 1 kilogram of steam at 100°C would require to lose 2256.7 kilojoules of heat to completely condense it into 1 kilogram of water at 100°C.

The temperature at which a liquid boils and the latent heat of evaporation depend strictly upon the pressure – the higher the pressure, the higher the boiling point and the smaller the amount of latent heat required to evaporate it. For example, at atmospheric pressure, the temperature at which water boils is 100°C and the latent heat of evaporation is 2256.7 kJ/kg; at a pressure of 15 bar (1500 kN/m²) the boiling point is 198.3°C and the latent heat 1947 kJ/kg; at 30 bar (3000 kN/m²) the boiling point is 233.8°C and the latent heat 1795 kJ/kg. These values are obtained from steam tables which are described later. This is an important concept for marine engineers as we use this concept to recover fresh water from sea water. The sea water is introduced to a machine where its pressure is reduced to such an extent that it boils at as low a temperature as 45°C. This means that 'waste heat' from the main engine can be used as an energy source which improves the overall thermal efficiency of the plant.

When heat is transferred to or from a substance which changes only its temperature, and there is no physical change of state, it is sometimes referred to as *sensible heat*. This distinguishes it from latent heat which changes the physical state of the substance without change of temperature.

We shall see later that, for a constant pressure process, the heat energy transferred to a substance is termed *enthalpy*; then, latent heat of fusion is termed enthalpy of fusion, and latent heat of evaporation is termed *enthalpy of evaporation*, and so on.

Example 2.5. Calculate the heat required to be given to 2 kg of ice at −15°C to change it into steam at atmospheric pressure, taking the following values:

Specific heat of ice = 204 kJ/kg K

Latent heat of fusion = 335 kJ/kg

$$\text{Specific heat of water} = 4.2 \text{ kJ/kg K}$$

$$\text{Latent heat of evaporation} = 2256.7 \text{ kJ/kg}$$

The steps in the calculation consist of the following:

Heat to raise the temperature of the ice from $-15°C$ to its melting point of $0°C$, that is, a temperature rise of $15°C$:

$$\text{Sensible heat} = m \times c \times \text{Temp. rise}$$
$$= 2 \times 204 \times 15$$
$$= 61.2 \text{ kJ}$$

Heat to change the ice at $0°C$ into water at $0°C$:

$$\text{Latent heat} = 2 \times 335$$
$$= 670 \text{ kJ}$$

Heat to raise the temperature of the water from $0°C$ to its boiling point of $100°C$, that is, a temperature rise of $100°C$:

$$\text{Sensible heat} = 2 \times 4 - 2 \times 100$$
$$= 840 \text{ kJ}$$

Heat to evaporate the water at $100°C$ into steam at $100°C$:

$$\text{Latent heat} = 2 \times 2256.7$$
$$= 4513.4 \text{ kJ}$$
$$\text{Total heat} = 61.2 + 670 + 840 + 4513.4$$
$$= 6084.6 \text{ kJ Ans.}$$

Working is simplified by finding the total heat transfer required per unit mass and then finally multiplying by the total mass. Thus:

$$Q = 2 \; (2.04 \times 15 + 335 + 4.2 \times 100 + 2256.7)$$
$$= 2 \; (30.6 + 335 + 420 + 2256.7)$$
$$= 2 \times 3042.3$$
$$= 6084.6 \text{ kJ}$$

Change of temperature, change of quantity of heat energy and so on are often written as $\Delta\theta$, ΔQ, and so on.

Test Examples 2

1. A water brake, coupled to an engine flywheel on test, absorbs 70 kW of power. Find the heat generated at the brake per minute and the mass flow of fresh water through the brake, in kg/min, if the temperature increase of the water is 10°C. Assume all the heat generated is carried away by the cooling water.

2. The effective radius of the pads in a single collar thrust block is 230 mm and the total load on the thrust block is 240 kN when the shaft is running at 93 rev/min. Taking the coefficient of friction between thrust collar and pads to be 0.025, find:

   (i)   the power lost due to friction

   (ii) the heat generated per hour

   (iii) the mass flow of oil in kilograms per hour through the block, assuming all the heat is carried away by the oil, allowing an oil temperature rise of 20°C and taking the specific heat of the oil to be 2 kJ/kg K.

3. Calculate the temperature of flue gases if 1.8 kg of copper of specific heat 0.395 kJ/kg K is suspended in the flue until it reached the temperature of the gases, and then dropped into 2.27 kg of water at 20°C. If the resultant temperature of the copper and water was 37.2°C, find the temperature of the flue gases.

4. In an experiment to find the specific heat of iron, 2.15 kg of iron cuttings at 100°C are dropped into a vessel containing 2.3 l of water at 17°C and the resultant temperature of the mixture is 24.4°C. If the water equivalent of the vessel is 0.18 kg, determine the specific heat of the iron.

5. 0.5 kg of ice at −5°C is put into a vessel containing 1.8 kg of water at 17°C and mixed together, the result being a mixture of ice and water at 0°C. Calculate the final masses of ice and water, taking the water equivalent of the vessel to be 0.148 kg, specific heat of ice 2.04 kJ/kg K and latent heat of fusion 335 kJ/kg.

3

THERMAL
EXPANSION

Expansion of Metals

The effect of increasing the temperature of metals will generally cause their dimensions to increase. Most metals expand when they are heated and contract when they are cooled, and the amount of expansion per degree rise of temperature differs with different metals. Some alloys are manufactured to have a minimum amount of expansion over a considerable working temperature range; these are usually for special purposes such as measuring instruments and gauges. However, this concept is being used in the construction of components such as pistons, which gives manufacturers the ability to improve the accuracy over a large thermal range. Although the expansion of a heated metal happens in all directions and there is an increase in all dimensions, it is sometimes only relevant to consider the expansion in one direction.

LINEAR EXPANSION. When a linear dimension is under consideration, the amount that a metal will expand lengthwise is expressed by its *coefficient of linear expansion*. This is the increase in length per unit length per degree increase in temperature. For example, if the coefficient of linear expansion of copper is given as 1.7×10^{-5}/°C (which is 0.000017 per degree Celsius), it means that each metre of length will increase in length by 1.7×10^{-5} metre when heated through one degree Celsius. This coefficient may be represented by the symbol α. Hence, representing the original length by l, and the temperature rise by $(\theta_2 - \theta_1)$, then the

$$\text{Increase in length} = \alpha \times l \times (\theta_2 - \theta_1)$$

The new length of the metal will then be:

$$\text{New length} = \text{Original length} + \text{Increase in length}$$
$$= l + \alpha l\left(\theta_2 - \theta_1\right)$$
$$= l\left\{1 + \alpha\left(\theta_2 - \theta_1\right)\right\}$$

Design and operational engineers need to understand this natural property of metals when building or repairing all the different parts of a ship. If the dimensions are not correct then the thermal expansion of metals will cause damage if allowances are not built into the design.

One of the most challenging of design environments is within a diesel engine or steam/gas turbine. The difference in temperature over a small distance is extraordinary. Recent advances in material science have allowed engine manufacturers to increase the performance of engines due to the stability of the dimensions of the internal components, over a high temperature range. This development enables the engine to stay more efficient than engines in the past, over a longer period of time.

Example 3.1. A main steam pipe is 6.5 m long when fitted at a temperature of 15°C. Calculate how much allowance should be made for its increase in length if it is subjected to a steam temperature of 300°C, taking the coefficient of linear expansion of the material to be 1.2×10^{-5}/°C.

$$\text{Increase in length} = \alpha \times l \times \left(\theta_2 - \theta_1\right)$$
$$= 1.2 \times 10^{-5} \times 6.5 \times (300 - 15)$$
$$= 0.02223 \text{ m}$$
$$= 22.23 \text{ mm} \quad \text{Ans.}$$

Example 3.2. A brass liner is 270 mm in diameter when the temperature is 17°C. Take the coefficient of linear expansion of the brass to be 1.9×10^{-5}/°C, and find the temperature to which the liner should be heated in order to increase the diameter by 2 mm.

Diameter is a linear dimension and therefore the same rule can be applied as for length. *Note*: the diameter and increase in diameter must be expressed in the same unit.

$$\text{Increase in diameter} = \alpha \times d \times \left(\theta_2 - \theta_1\right)$$
$$2 = 1.9 \times 10^{-5} \times 270 \times \left(\theta_2 - \theta_1\right)$$
$$\text{Increase in temperature} = \frac{2 \times 10^5}{1.9 \times 270} = 389.7°C$$

$$\text{Required temperature} = 17 + 389.7$$
$$= 406.7 °C \quad \text{Ans.}$$

SUPERFICIAL EXPANSION. This refers to an increase in area of an object as it is heated. The *coefficient of superficial expansion* is the increase in area per unit area per degree increase in temperature. Therefore, if A represents original area, and $(\theta_2 - \theta_1)$ the increase in temperature, then:

$$\text{Increase in area} = \text{Coeff. of superficial expansion} \times A \times (\theta_2 - \theta_1)$$

Consider an area of metal of unit length and unit breadth (Figure 3.1) and let this be heated through one degree. The length and breadth will each increase by an amount equal to a, the coefficient of linear expansion.

$$\text{Original area} = 1 \times 1 = 1$$
$$\text{New length and new breadth} = 1 + \alpha$$
$$\text{New area} = \left(1 + \alpha\right)^2 = 1 + 2\alpha + \alpha^2$$
$$\text{Increase in area} = \text{New area} - \text{Original area}$$
$$= 1 + 2\alpha + \alpha^2 - 1$$
$$= 2\alpha + \alpha^2$$

The coefficient of linear expansion, a, is a very small quantity for any metal (for example, it is about 1.2×10^{-5} for steel); therefore α^2 being the second order of smallness is also a very small quantity indeed and is completely negligible as a quantity to be added for most practical purposes. We can therefore take the increase to be $2a$. As this is the increase in area per unit area for one degree increase in temperature, it is the value of the coefficient of superficial expansion. Hence:

$$\text{Coeff. of superficial expansion} = 2 \times \text{Coeff. of linear expansion}$$

Therefore:

$$\text{Increase in area} = 2a \times A \times (\theta_2 - \theta_1)$$

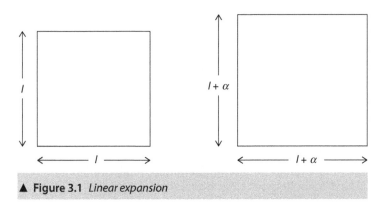

▲ **Figure 3.1** *Linear expansion*

CUBICAL EXPANSION. This refers to the increase in volume. The *coefficient of cubical* (or *volumetric*) *expansion* is the increase in volume per unit volume per degree increase in temperature. Therefore, if V represents the original volume, and $(\theta_2 - \theta_1)$ the increase in temperature, then:

$$\text{Increase in volume} = \text{Coeff. of cubical expansion} \times V \times (\theta_2 - \theta_1)$$

Consider a block of metal of unit length, unit breadth and unit thickness (Figure 3.2), and let this be heated through one degree. The length, breadth and thickness will each increase by an amount equal to a, the coefficient of linear expansion.

$$\text{Original volume} = 1 \times 1 \times 1 = 1$$
$$\text{New volume} = (1 + \alpha)^3$$
$$= 1 + 3\alpha + 3\alpha^2 + \alpha^3$$
$$\text{Increase in volume} = \text{New volume} - \text{Original volume}$$
$$= 1 + 3\alpha + 3\alpha^2 + \alpha^3 - 1$$
$$= 3\alpha + 3\alpha^2 + \alpha^3$$

As α^2 and α^3 are the second and third order of smallness, respectively, they are negligible quantities for addition, and hence the increase may be taken as 3a. As this is the increase in volume per unit volume per degree increase in temperature, it is also the value of the coefficient of cubical expansion.

$$\text{Coeff. of cubical expansion} = 3 \times \text{Coeff. of linear expansion}$$

Therefore:

$$\text{Increase in volume} = 3a \times V \times (\theta_2 - \theta_1)$$

Note that *linear* refers to any linear dimension such as length, breadth, thickness, diameter, circumference and so on, and the expression for linear expansion can be applied to any of these dimensions. It is true for both internal and external dimensions.

The expression for superficial expansion covers any area of the solid, cross-sectional area, surface area and so on and holds good for both internal and external areas.

The expression for cubical expansion is also applicable for internal volumes of a hollow vessel.

Thinking about and allowing for expansion is very important to the marine engineer. This aspect will have been considered at the design stage, especially where piping will be subjected to differing temperatures over its working cycle.

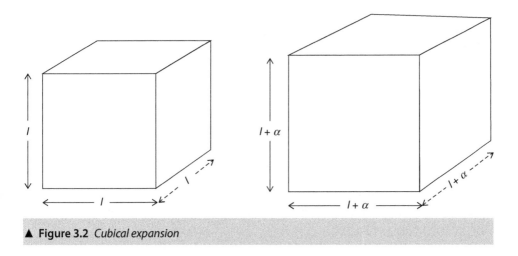

▲ **Figure 3.2** *Cubical expansion*

Example 3.3. A metal sphere is exactly 25 mm diameter at 20°C. Find the increase in diameter, increase in surface area and increase in volume when heated to 260°C, if the coefficient of linear expansion of the metal is 1.8×10^{-5}/°C.

$$\text{Increase in temperature} = 260 - 20 = 240 \, °C$$
$$\text{Increase in diameter} = \alpha \times d \times (\theta_2 - \theta_1)$$
$$= 1.8 \times 10^{-5} \times 25 \times 240$$
$$= 0.108 \text{ mm} \quad \text{Ans.(i)}$$

$$\text{Surface area of sphere} = \pi d^2$$
$$\text{Increase in area} = 2\alpha \times A \times (\theta_2 - \theta_1)$$
$$= 2 \times 1.8 \times 10^{-5} \times \pi \times 25^2 \times 240$$
$$= 16.96 \text{ mm}^2 \quad \text{Ans.(ii)}$$

$$\text{Volume of sphere} = \frac{\pi}{6} d^2$$
$$\text{Increase in volume} = 3\alpha \times V \times (\theta_2 - \theta_1)$$
$$= 3 \times 1.8 \times 10^{-5} \times \frac{\pi}{6} \times 25^3 \times 240$$
$$= 106 \text{ mm}^3 \quad \text{Ans.(iii)}$$

Expansion of Liquids

Liquids have no definite shape of their own, therefore no linear dimensions, and hence the coefficient of cubical expansion of a liquid is an independent quantity. The coefficient of cubical expansion is usually represented by β.

$$\text{Increase in volume} = \beta \times V \times (\theta_2 - \theta_1)$$

Example 3.4. 2500 litres of oil are heated through 50°C. If the coefficient of cubical expansion of this oil is 0.0008/°C, find the increase in volume in cubic metres.

$$2500 \text{ litres} = 2.5 \text{ m}^3$$
$$\text{Increase in volume} = \beta \times V \times \left(\theta_2 - \theta_1\right)$$
$$= 0.0008 \times 2.5 \times 50$$
$$= 0.1 \text{ m}^3 \quad \text{Ans.}$$

APPARENT CUBICAL EXPANSION. The tank or vessel which contains a liquid will also expand when heated. It is therefore useful to know the expansion of the liquid relative to its container so that the correct allowance can be made for changes in temperature.

The apparent or relative increase in volume of a liquid is the difference between the volumetric expansion of the liquid and the volumetric expansion of its container. If both have the same initial volume and are raised through the same range of temperature, then, letting suffix L represent the liquid and suffix C the container:

Apparent increase in volume of the liquid
$$= \text{Vol. increase of liquid} - \text{Vol. increase of container}$$
$$= \beta_L \times V \times (\theta_2 - \theta_1) - \beta_C \times V \times (\theta_2 - \theta_1)$$
$$= (\beta_L - \beta_C) \times V \times (\theta_2 - \theta_1)$$

The difference between the coefficients of cubical expansion of the liquid and its container can therefore be termed the *apparent* coefficient of cubical expansion of the liquid.

Restricted Thermal Expansion

If the natural thermal expansion of a metal is restricted, the metal will be strained from its natural length and the material will be stressed. To enable the effects of restricted expansion to be seen here, it is necessary to revise the relationships between stress, strain and modulus of elasticity.

STRESS. This is the internal resistance set up in a material when an external force is applied. It is expressed as the force carried by the material per unit area of its cross-section.

$$\text{Stress [N/m}^2] = \frac{\text{Force [N]}}{\text{Area [m}^2]}$$

High values of stress are expressed in multiples of the force unit, such as kN/m² ($=10^3$ N/m²) and MN/m² ($=10^6$ N/m²). In some industries, stress may be expressed in hectobars (hbar): 1 hectobar $= 10^7$ N/m².

STRAIN. This is the change of shape that takes place in a material due to it being stressed. Linear strain is the change of length per unit length. Thus:

$$\text{Linear strain} = \frac{\text{Change of length}}{\text{Original length}}$$

MODULUS OF ELASTICITY. The modulus of elasticity of a material is the constant obtained by dividing the stress set up in it by the strain it endures under that stress, provided the elastic limit is not exceeded.

$$\text{Modules of elasticity } (E) = \frac{\text{Stress}}{\text{Strain}}$$

Example 3.5. A solid steel stay of 50 mm diameter and 300 mm length at a temperature of 25°C is firmly secured at each end so that expansion is fully restricted. Find the stress set up in the stay and the equivalent total axial force when it is heated to 150°C, taking the coefficient of linear expansion of steel to be 1.2×10^{-5}/°C and its modulus of elasticity to be 206 GN/m².

If the stay was perfectly free to expand without restriction,

$$\begin{aligned}
\text{Free expansion} &= \alpha \times l \times (\theta_2 - \theta_1) \\
&= 1.2 \times 10^{-5} \times 300 \times (150 - 25) \\
&= 0.45 \text{ mm}
\end{aligned}$$

Free and unrestrained length would then be:

$$300 + 0.45 = 300.45 \text{ mm}$$

If prevented from expanding, the effect is that the stay is compressed from its natural unstrained length of 300.45 mm to 300 mm. (Please see formula for strain in the next paragraph.)

It is usual, however, to make a slight approximation at this stage by taking the original cold length of 300 mm instead of the heated length of 300.45 mm. The difference in the final result is negligible, and it proves much more convenient in solving complicated problems. It will also be seen that by making this slight approximation the length of the material will not be required as the strain can be obtained directly from:

$$\text{Strain} = \frac{\text{Change of length}}{\text{Original length}}$$
$$= \frac{\alpha \times l \times (\theta_2 - \theta_1)}{l} = \alpha\,(\theta_2 - \theta_1)$$

Hence, the strain for this stay can be taken as:

$$\text{Strain} = \alpha\left(\theta_2 - \theta_1\right)$$
$$= 1.2 \times 10^{-5} \times 125 = 0.0015$$

$$\text{which is the same as } \frac{0.45}{300} = 0.0015$$

$$E\ [\text{N/m}^2] = \frac{\text{Stress}\ [\text{N/m}^2]}{\text{Strain}}$$
$$\therefore \text{Stress} = \text{Strain} \times E$$
$$= 0.0015 \times 206 \times 10^9$$
$$= 3.09 \times 10^8\ \text{N/m}^2$$
$$= 309\ \text{MN/m}^2 \text{ or } 30.9 \text{ hbar} \quad \text{Ans.(i)}$$

$$\text{Stress}\ [\text{N/m}^2] = \frac{\text{Force}\ [\text{N}]}{\text{Area}\ [\text{m}^2]}$$
$$\therefore \text{Total axial force} = \text{Stress} \times \text{Area}$$
$$= 3.09 \times 108 \times 0.7854 \times 0.052$$
$$= 6.068 \times 10^5\ \text{N}$$
$$= 606.8\ \text{kN} \quad \text{Ans.(ii)}$$

Test Examples 3

1. A steam pipe is 3.85 m long when fitted at a temperature of 18°C. Find the increase in length if free to expand, when carrying steam at a temperature of 260°C, taking the coefficient of linear expansion of the pipe material to be 1.25×10^{-5}/°C.

2. A solid cast iron sphere is 150 mm in diameter. If 2110 kJ of heat energy is transferred to it, find the increase in diameter, taking the following values for cast iron: density = 7.21 g/cm³, specific heat = 0.54 kJ/kg K, coefficient of linear expansion = 1.12×10^{-5}/°C.

3. A bi-metal control device is made up of a thin flat strip of aluminium and a thin flat strip of steel of the same dimensions, connected together in parallel and separated from each other by two brass distance pieces 2.5 mm long, their centres being 50 mm apart. Find the radius of curvature of the strips when heated through 200°C, taking the following values for the coefficients of linear expansion:

$$\text{Aluminium } \alpha = 2.5 \times 10^{-5} \text{ /°C}$$
$$\text{Steel } \alpha = 1.2 \times 10^{-5} \text{ /°C}$$
$$\text{Brass } \alpha = 2.0 \times 10^{-5} \text{ /°C}$$

4. The pipeline of a hydraulic system consists of a total length of steel pipe of 13.7 m and internal diameter 30 mm. If the coefficient of linear expansion of the steel is 1.2×10^{-5}/°C and coefficient of cubical expansion of the oil in the pipe is 9×10^{-4}/°C, calculate the volumetric allowance in litres to be made for oil overflow from the pipe when the temperature rises by 27°C.

5. A straight length of steam pipe is to be fitted between two fixed points with no allowance for expansion. If the compressive stress in the pipe is to be limited to 35 hbar (=350 MN/m²) calculate the initial tensile stress to be exerted on the pipe when fitted cold at 17°C to allow for a steam temperature of 220°C. Take the coefficient of linear expansion of the pipe material to be 1.12×10^{-5}/°C and the modulus of elasticity to be 206 GN/m².

4

HEAT TRANSFER

Heat is transferred from one system to another by one of the three methods known as *conduction, convection*, and *radiation*, or by a combination of these. This area of marine engineering is becoming increasingly important as restricting the 'loss' of heat from a system is a way of improving the overall thermal efficiency of the plant.

Conduction

Conduction is the flow of heat energy through a body, or from one body to another where the two 'bodies' are in contact with each other, due to difference in temperature. The *natural* flow of heat *always* takes place from a region of high temperature to a region of lower temperature.

Generally speaking, metals are good conductors of heat. Air, together with materials such as polyester, cork, glass and wool, are examples of very bad conductors; these very bad conductors are given the term 'insulators', and they can be used to minimise heat loss. Special polymer based compositions are used to lag boilers, pipes, casings and other hot surfaces in a ship's machinery space to reduce loss of heat energy to the colder outside surroundings.

Cork and fibre glass are common insulating materials to pack the hollow walls of refrigerating chambers to reduce heat flow into the cold chambers from the warmer outside surroundings.

The quantity of heat conducted through a material, in a given time, depends upon the thermal conductivity of the material, and is:

- proportional to the surface area exposed to the source of heat
- proportional to the temperature difference between the hot and cold ends
- inversely proportional to the distance or thickness through which the heat is conducted. Therefore, arranged as a formula we have:

$$\text{Quantity of heat varies as } \frac{\text{Area} \times \text{Time} \times \text{Temp. difference}}{\text{Thickness}}$$

The thermal conductivity depends upon the nature of the material and its ability to conduct heat. This varies for different materials and sometimes varies slightly for the same material depending upon the temperature range.

The *thermal conductivity* (*k*) of a material expresses the quantity of heat energy conducted through a unit of area in unit of time for a unit of temperature between two opposite faces of a material that is a unit distance apart.

Considering the heat flow through a flat wall, taking the following symbols and basic units, and referring to Figure 4.1:

Q = Quantity of heat energy conducted, in joules [J]

A = Area through which heat flows, in square metres [m²]

t = Time of heat flow, in seconds [s]

$T_1 - T_2$ = Temperature difference between the two faces [K]

S = Thickness of wall, in metres [m]

Then, the units of k, the thermal conductivity, are Jm/m²s K. For convenience this is usually shortened by cancelling m into m² and substituting W watts in the place of J/s.

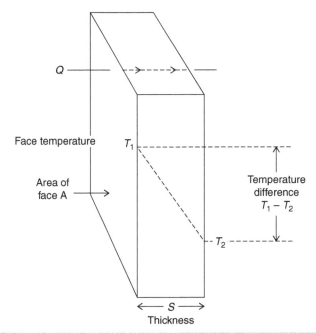

▲ **Figure 4.1** *Heat flow through a wall of uniform composition*

$$\frac{Jm}{m^2 s\ K} = \frac{J}{ms\ K} = \frac{W}{mK} = W/m\ K$$

This gives us the quantity of heat energy transferred by conduction as:

$$Q = \frac{kAt(T_1 - T_2)}{S}$$

Example 4.1. Calculate the heat transfer per hour through a solid brick wall 6 m long, 2.9 m high and 225 mm thick, when the outer surface is at 5°C and the inner surface 17°C, the thermal conductivity of the brick being 0.6 W/m K.

$$Q\ [J] = \frac{k\ [W/m\ K] \times A\ [m^2] \times t\ [s] \times (T_1 - T_2)\ [K]}{S\ [m]}$$
$$= \frac{0.6 \times 6 \times 2.9 \times 3600 \times (17 - 5)}{0.225}$$
$$= 2.004 \times 10^6\ J$$
$$= 2.004\ MJ\ or\ 2004\ kJ\quad Ans.$$

Note: T_1 is $17 + 273 = 290$ K and T_2 is $5 + 273 = 278$ K; their difference, $T_1 - T_2$, is the same as 17°C − 5°C.

COMPOSITE WALL OR PARTITION. Consider the transfer of heat energy by conduction through a wall made up of a number of layers of different materials; in addition, take as an example three slabs of different thicknesses, as shown in Figure 4.2.

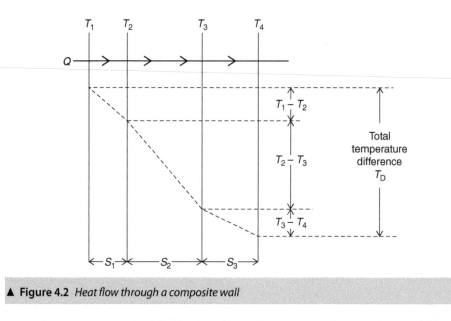

▲ **Figure 4.2** *Heat flow through a composite wall*

For each thickness the calculation is as follows:

$$Q = \frac{kAt \times \text{Temp. diff.}}{S} \qquad \therefore \text{Temp. diff.} = \frac{QS}{kAt}$$

Therefore, the total drop in temperature across all three thicknesses is made up of three individual formulae:

$$T_1 - T_4 = \frac{Q_1 S_1}{k_1 A_1 t_1} + \frac{Q_2 S_2}{k_2 A_2 t_2} + \frac{Q_3 S_3}{k_3 A_3 t_3}$$

The same quantity of heat energy is transferred across each layer through the same area in the same time; therefore Q, A and t are common:

$$T_1 - T_4 = \frac{Q}{At}\left\{\frac{S_1}{k_1} + \frac{S_2}{k_2} + \frac{S_3}{k_3}\right\}$$

For any number of layers, let T_D represent the total temperature drop, that is:

$$T_D = (T_1 - T_2) + (T_2 - T_3) + (T_3 - T_4) + \cdots$$

Using the summation sign (Sigma) Σ for the sum of the quantities inside the brackets, we get:

$$\Sigma\left\{\frac{S}{k}\right\} = \frac{S_1}{k_1} + \frac{S_2}{k_2} + \frac{S_3}{k_3} + \cdots$$

Then:

$$T_D = \frac{Q}{At}\Sigma\left\{\frac{S}{k}\right\} \quad \text{or} \quad Q = \frac{AtT_D}{\Sigma\left\{\dfrac{S}{k}\right\}}$$

Example 4.2. One insulated wall of a cold-storage compartment is 8 m long by 2.5 m high and consists of an outer steel plate 18 mm thick and an inner wood wall 22.5 mm thick. The steel and wood are 90 mm apart to form a cavity which is filled with cork. If the temperature drop across the extreme faces of the composite wall is 15°C, calculate the heat transfer per hour through the wall and the temperature drop across the thickness of the cork. Take the thermal conductivity for steel, cork and wood to be 45, 0.045 and 0.18 W/m K, respectively.

For the composite wall:

$$T_1 - T_4 = \frac{Q}{At}\left\{\frac{S_1}{k_1} + \frac{S_2}{k_2} + \frac{S_3}{k_3}\right\}$$

where $T_1 - T_4 = 15$ K,

$$A = 8 \times 2.5 = 20 \text{ m}^2$$
$$t = 3600 \text{ s}$$
$$\Sigma\left\{\frac{S}{k}\right\} = \frac{0.018}{45} + \frac{0.09}{0.045} + \frac{0.0225}{0.18}$$
$$= 0.0004 + 2 + 0.125$$
$$= 2.1254$$
$$\therefore 15 = \frac{Q}{20 \times 3600} \times 2.1254$$
$$Q = \frac{15 \times 20 \times 3600}{2.1254}$$
$$= 508.2 \times 105 \text{ J}$$
$$= 508.2 \text{ kJ} \quad \text{Ans. (i)}$$

Temperature drop across the cork:

$$= \frac{QS}{Atk}$$
$$= \frac{5.082 \times 10^5 \times 0.09}{20 \times 3600 \times 0.045} = 14.11 \text{ K}$$
$$= 14.11°\text{C} \quad \text{Ans. (ii)}$$

CYLINDRICAL WALL. A cylindrical wall (in marine engineering this could be a cooling water pipe, boiler tube or heat exchanger tube) could be considered as being made up of many thin cylindrical elements all with the same coefficient of thermal conductivity.

In Figure 4.3 if dT is the small temperature difference radially across the cylindrical element whose thickness is dr, length L and area A.

$$\text{Then } Q = kAt\left(-\frac{dT}{dr}\right)$$
$$A = 2\pi rL$$
$$\text{Hence } Q = k2\pi rL\left(-\frac{dT}{dr}\right)$$

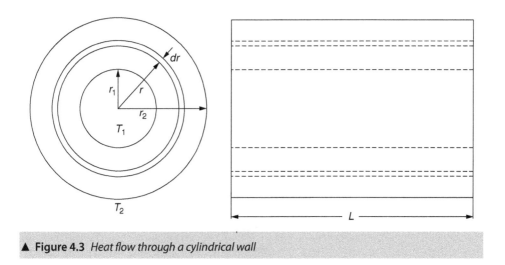

▲ **Figure 4.3** *Heat flow through a cylindrical wall*

Using integration by the use of the calculus to account for the complete cylinder thickness:

$$Q\int_{r_1}^{r_2} \frac{dr}{r} = -k \times 2\pi Lt \int_{T_1}^{T_2} dT$$

$$Q \ln\left(\frac{r_2}{r_1}\right) = -2\pi kLt(T_2 - T_1)$$

$$Q = \frac{2\pi kLt(T_1 - T_2)}{\ln\left(\frac{r_2}{r_1}\right)}$$

Note: ln *x*, the natural logarithm of *x*, is often written as $\log_e x$.

Example 4.3. A steam pipe has an external diameter of 85 mm and is 25 m long. It carries 1250 kg of steam per hour at a pressure of 20 bar. The steam enters the pipe 0.98 dry and it is to leave the outlet end of the pipe with a minimum dryness fraction of 0.96. The pipe is to be lagged with a material having a thermal conductivity of 0.2 W/m K and outer surface temperature of 26°C. Any temperature drop across the wall of the pipe is to be neglected and the rate of heat transfer through the wall of a thick cylinder is given by:

$$Q = \frac{2\pi k\theta}{\ln\left(\frac{D_2}{D_1}\right)} \text{ per metre length}$$

where:

k = The thermal conductivity of cylinder material

θ = Temperature difference

D_2 = External diameter

D_1 = Internal diameter

Determine the thickness of the lagging required.

$$\text{Heat loss} = m \times hfg(x_i - x_o)$$

$$= \frac{1250 \times 1890 \times (0.98 - 0.96)}{3600}$$

$$= 13.125 \text{ kW}$$

$$Q = \frac{2\pi k\theta}{\ln\left(\dfrac{D_2}{D_1}\right)}$$

$$\frac{13125}{25} = \frac{2\pi \times 0.2 \times (212.4 - 26)}{\ln\left(\dfrac{D_2}{0.085}\right)}$$

$$\ln\left(\frac{D_2}{0.085}\right) = \frac{0.4\pi \times 186.4}{525}$$

$$\lg\left(\frac{D_2}{0.085}\right) = \frac{0.4462}{2.3026}$$

$$D_2 = 0.0085 \times 1.563$$

$$= 0.1329 \text{ m}$$

$$\text{Thickness} = \frac{132.9 - 85}{2}$$

$$= 23.9 \text{ mm} \quad \text{Ans.}$$

Notes: **(i)** At 20 bar t_s = 212.4°C.

(ii) $\lg x$, the common (base 10) logarithm of x, is often written as log x. (See *Reeds Vol 1*, Chapter 1, for more details about logs.)

(iii) Ratio of diameters is the same as ratio of radii.

Convection

Convection is the method of transferring heat through a fluid (liquid or gas) by the movement of the heated particles within the fluid. The natural tendency is for hotter particles to move upwards. Marine engineers can use their knowledge of this natural state to good effect by, for example, analysing the flow of the liquids through a heater or cooler.

Further examples are explained on this page. For example, Figure 4.4(a) shows a vessel with an inclined tube connected at the bottom. When this contains water, and heat is applied to the tube, the heated particles of the water become less dense and rise, denser particles move to take their place and thus convection currents are set in motion, resulting in all the water in the vessel and tube becoming heated almost uniformly due to the continuous circulation of the water. This is the operating principle of the water-tube boiler – if the heat source is strong, then the circulation currents will also be strong.

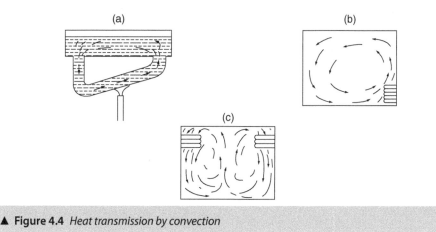

▲ **Figure 4.4** *Heat transmission by convection*

Figure 4.4(b) shows air in a room heated by convection, the fire, radiator or other heat source being placed at the bottom of the room.

Figure 4.4(c) illustrates how the air in a room is cooled by convection, the coolers (such as refrigerated pipes) being situated near the top of the room.

The examples given above illustrate the natural circulation of the fluid which can be referred to as *free convection*. When the motion of the fluid is produced mechanically, as by the use of a pump or a fan, it is referred to as *forced circulation*.

Radiation

Radiation is the transfer of heat energy from one body to another through space by electromagnetic waves. The heat travel is in straight lines in all directions and is at the same velocity as light, that is, at nearly 300,000 kilometres per second. Most radiated heat is associated with the infra-red part of the electromagnetic spectrum.

Some of the radiant heat falling upon a body is reflected in the same manner as light is reflected, and the remainder is absorbed or transmitted through the material. Dark and rough surfaces are good at absorbing radiant heat whereas bright, light and polished surfaces reflect most of the heat and therefore the amount of absorption is small. A body which is poor at absorbing heat is also a poor radiator, and a body that is good is also a good radiator. A body that is perfect at absorbing and radiating heat energy is called a 'perfect' black body. Perfect black bodies do not naturally exist but they can be simulated in the laboratory, and the first person to show this was Max Planck.

The *emissivity* of a radiating body is the ratio of the heat emitted by that body compared with the heat emitted by a perfect black body of the same surface area and temperature in the same time. Emissivity may be represented by ε, and its value for the ideal radiator (a perfect black body) is therefore unity (1).

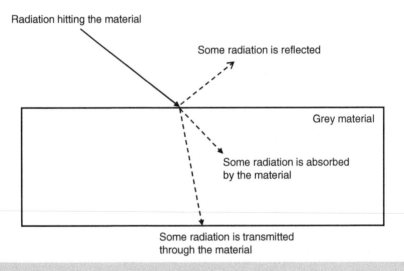

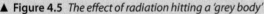

▲ **Figure 4.5** *The effect of radiation hitting a 'grey body'*

Bodies, other than perfect black bodies, are given the generic term 'grey bodies', and they will have an emissivity coefficient of less than 1 (Figure 4.5).

STEFAN–BOLTZMANN LAW. This law states that the heat energy radiated by a perfect radiator is proportional to the fourth power of its absolute temperature. Hence Q = quantity of heat energy radiated, A = surface area radiating heat, t = time of radiation, T = absolute temperature, then:

$$Q = AtT^4 \times \text{a constant}$$

The value of the constant depends upon the units employed, and for the basic SI units of area in square metres, time in seconds and absolute temperature in degrees Kelvin, the value of the constant to give kilojoules of heat energy is 5.67×10^{-11} kJ/m²s K⁴. Please note that J/m²s K = W/m² K and students will often see the Stefan–Boltzmann constant expressed to 4 decimal places as 5.6703×10^{-8} W/m² K.

Thus, the quantity of heat energy radiated from a hot body of absolute temperature T_1 to its surroundings at absolute temperature T_2 is:

$$Q = 5.67 \times 10^{-11} \times \varepsilon At(T_1^4 - T_2^4)$$

Example 4.4. The temperature of the flame in a boiler furnace is 1277°C and the temperature of its surrounds is 277°C. Calculate the maximum theoretical quantity of heat energy radiated per minute per square metre to the surrounding surface area.

$$T_1 = 1550 \text{ K}$$
$$T_2 = 550 \text{ K}$$
$$Q = 5.67 \times 10^{-11} \times \varepsilon At \ (T_1^4 - T_2^4)$$
$$= 5.67 \times 10^{-11} \times 1 \times 60 \times (1550^4 - 550^4)$$
$$= 5.67 \times 10^{-11} \times 1 \times 60 \times 2.705 \times 2.1 \times 10^{12}$$
$$= 1.933 \times 10^4 \text{ kJ or 19.33 MJ} \quad \text{Ans.}$$

Combined Modes

The transfer of heat from one fluid to another through a dividing wall is an example of a practical application in many engineering applications. Consider the transfer of heat from a fluid to a flat plate, through the plate, and from the plate to another fluid, as illustrated in Figure 4.6. On each side of the plate, a thin film of almost stagnant fluid clings to the surface, the heat transfer through the film is by conduction, convection and radiation.

The quantity of heat transfer through the film of fluid per unit area of surface, in unit time, for unit temperature drop across the thickness of the film is expressed by the *surface heat transfer coefficient* (h), and this depends to a large extent upon the velocity of the fluid and the condition of the surfaces.

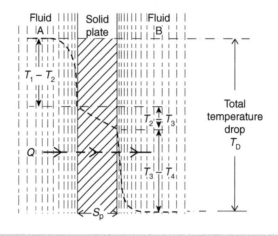

▲ **Figure 4.6** *Transfer of heat from one fluid to another through a metal wall*

Since the quantity of heat conducted in a given time is proportional to the surface area and the temperature drop, then the units of h are joules per square metre of area per second per degree = J/m^2s K = W/m^2 K. Hence:

$$Q\,[J] = h[W/m^2\ K] \times A[m^2] \times t\,[s] \times (T_1 - T_2)[K]$$

Referring to Figure 4.6,

Heat transfer through film of fluid A:

$$Q = h_A At\ (T_1 - T_2) \qquad \therefore T_1 - T_2 = \frac{Q}{h_A At}$$

Heat transfer through solid plate:

$$Q = \frac{h_p At\ (T_2 - T_3)}{S_p} \qquad \therefore T_2 - T_3 = \frac{QS_p}{h_p At}$$

Heat transfer through film of fluid B:

$$Q = h_B At\ (T_3 - T_4) \qquad \therefore T_3 - T_4 = \frac{Q}{h_B At}$$

Total drop in temperature:

$$T_D = (T_1 - T_2) + (T_2 - T_3) + (T_3 - T_4)$$

$$= \frac{Q}{At}\left\{\frac{1}{h_A} + \frac{S_P}{k_P} + \frac{1}{h_B}\right\}$$

The quantity inside the brackets may be represented by $1/U$, where U is called the *overall heat transfer coefficient*. Then:

$$\frac{1}{U} = \frac{1}{h_A} + \frac{S_P}{k_P} + \frac{1}{h_B}$$

Hence:

$$T_D = \frac{Q}{UAt} \quad \text{or,} \quad Q = UAtT_D$$

Example 4.5. A cubical tank of 2 m sides is constructed of metal plate 12 mm thick and contains water at 75°C. The surrounding air temperature is 16°C. Calculate (i) the overall heat transfer coefficient from water to air, and (ii) the heat loss through each side of the tank per minute. Take the thermal conductivity of the metal to be 48 W/m K, the heat transfer coefficient of the water to be 2.5 kW/m² K and the heat transfer coefficient of the air to be 16 W/m² K.

$$\frac{1}{U} = \frac{1}{h_W} + \frac{S_P}{k_P} + \frac{1}{h_A}$$

$$= \frac{1}{2.5 \times 10^3} + \frac{0.125}{48} + \frac{1}{16}$$

$$= 0.0004 + 0.00025 + 0.0625$$

$$= 0.06315$$

$$U = 1/0.06315 = 15.84 \text{ W/m}^2 \text{ K} \quad \text{Ans.(i)}$$

$$Q = UAtT_D$$

$$= 15.84 \times 2^2 \times 60 \times (75 - 16)$$

$$= 2.243 \times 10^5$$

$$= 224.3 \text{ kJ} \quad \text{Ans.(ii)}$$

HEAT EXCHANGERS. Common examples, in marine plant, of heat exchangers in which heat is transferred from one fluid to another are: plate type heat exchangers used as diesel engine fresh-water coolers, tube type heat exchangers used as lubricating oil coolers and main engine charge air coolers.

To determine the amount of heat transferred in a heat exchanger it is necessary to know the overall heat transfer coefficient U for the wall and fluid boundary layers and also the *logarithmic mean temperature difference*. Which is the temperature difference between the fluids.

$$\text{Heat transferred } Q = UAt\,\theta_m \ [\text{J}]$$

For both heat exchangers shown in Figure 4.7:

$$\theta_m = \frac{\theta_1 - \theta_2}{\ln\left(\dfrac{\theta_1}{\theta_2}\right)}$$

where

θ_1 = The temperature difference between hot and cold fluid at inlet

θ_2 = The temperature difference between hot and cold fluid at outlet

θ_m = Mean temperature difference between the fluids, usually called the *logarithmic mean temperature difference*

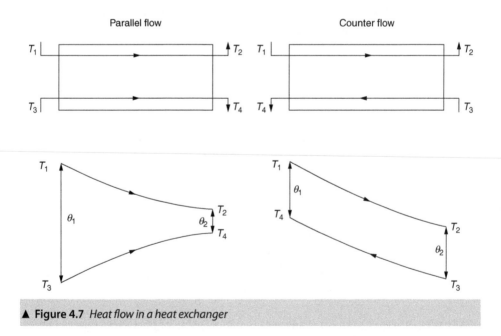

▲ **Figure 4.7** *Heat flow in a heat exchanger*

Example 4.6. A counter-flow cooler is required to cool 0.5 kg/s of oil from 50°C to 20°C with a cooling water inlet temperature at 10°C and mass flow rate of 0.375 kg/s. There are 200 thin walled tubes of diameter 12.5 mm. Calculate the length of the tube required.

Overall heat transfer coefficient: 70 W/m² K

Specific heat capacities: oil 1400 J/kg K, water 4200 J/kg K

Let T_4 be water outlet temperature.

$$Q = 0.5 \times 1400\ (50 - 20) = 0\text{--}375 \times 4200\ (T_4 - 10) = 21000\ \text{W}$$
$$T_4 = 13.33°\text{C}$$
$$= \frac{\theta_1 - \theta_2}{\ln\left(\dfrac{\theta_1}{\theta_2}\right)}$$
$$= \frac{(T_1 - T_4) - (T_2 - T_3)}{\ln\left(\dfrac{T_1 - T_4}{T_2 - T_3}\right)}$$
$$= \frac{(50 - 13.33) - (20 - 10)}{\ln\left(\dfrac{50 - 13.33}{20 - 10}\right)}$$
$$= \frac{26.67}{1.229}$$
$$= 20.53°\text{C}$$
$$Q = UAt\ \theta_m$$
$$21000 = 70(\pi \times 0.0125 \times l \times 200) \times 1 \times 20.53$$
$$l = 1.86\ \text{m} \quad \text{Ans.}$$

Test Examples 4

1. Calculate the quantity of heat conducted per minute through an aluminum alloy circular disc, 127 mm diameter and 19 mm thick, when the temperature drop across the thickness of the plate is 5°C. Take the thermal conductivity of the alloy to be 150 W/m K.

2. A cold storage compartment is 4.5 m long by 4 m wide by 2.5 m high. The four walls, ceiling and floor are covered to a thickness of 150 mm with insulating material which has a thermal conductivity of 5.8 x 10⁻² W/m K. Calculate the quantity of heat leaking through the insulation per hour when the outside and inside face temperatures of the material are 15°C and −5°C, respectively.

3. One side of a refrigerated cold chamber is 6 m long by 3.7 m high and consists of 168 mm thickness of cork between outer and inner walls of wood. The outer wood wall is 30 mm thick and its outside face temperature is 20°C, the inner wood wall is 35 mm thick and its inside face temperature is −3°C. Taking the thermal conductivity of cork and wood to be 0.042 and 0.2 W/m K, respectively, calculate (i) the heat transfer per second per square metre of surface area, (ii) the total heat transfer through the chamber side per hour and (iii) the interface temperatures.

4. A flat circular plate is 500 mm diameter. Calculate the theoretical quantity of heat radiated per hour when its temperature is 215°C and the temperature of its surrounds is 45°C. Take the value of the radiation constant to be 5.67 x 10^{-11} kJ/m²s K⁴.

5. Hot gases at 280°C flow on one side of a metal plate of 10 mm thickness and air at 35°C flows on the other side. The surface heat transfer coefficient of the gases is 31.5 W/m² K and that of air is 32 W/m² K. The thermal conductivity of the metal plate is 50 W/m K. Calculate (i) the overall heat transfer coefficient, and (ii) the heat transfer from gases to air per minute per square metre of plate area.

6. The wall of a cold room consists of a layer of cork sandwiched between outer and inner walls of wood, the wood walls being each 30 mm thick. The inside atmosphere of the room is maintained at −20°C when the external atmospheric temperature is 25°C, and the heat loss through the wall is 42 W/m². Taking the thermal conductivity of wood and cork to be 0.2 W/m K and 0.05 W/m K, respectively, and the surface heat transfer coefficient between each exposed wood surface and their respective atmospheres to be 15 W/m² K, calculate (i) the temperatures of the exposed surfaces, (ii) the temperatures of the interfaces and (iii) the thickness of the cork.

7. The steam drum of a water-tube boiler has hemispherical ends, the diameter is 1.22 m and the overall length is 6 m. Under steaming conditions the temperature of the shell before lagging was 230°C and the temperature of the surrounds was 51°C. The temperature of the cleading after lagging was 69°C and the surrounds 27°C. Assuming 75% of the total shell area to be lagged and taking the radiation constant as 5.67 × 10^{-11} kJ/m²s K⁴, estimate the saving in heat energy per hour due to lagging.

8. A pipe 200 mm outside diameter and 20 m length is covered with a layer, 70 mm thick, of insulation having a thermal conductivity of 0.05 W/m K and a surface heat transfer coefficient of 10 W/m² K at the outer surface. If the temperature of the pipe is 350°C and the ambient temperature is 15°C calculate: (i) the external surface temperature of the lagging and (ii) the heat flow rate from the pipe.

The heat flow rate through the lagging per metre length of pipe is $2\pi kT/\ln(D/d)$, where

k is the thermal conductivity of the insulation

T is the temperature difference across the insulation

D and d are the external and internal diameters of the insulation, respectively.

9. In an inert gas system the boiler exhaust is cooled from 410°C to 130°C in a parallel flow heat exchanger. Gas flow rate is 0.4 kg/s, cooling water flow rate is 0.5 kg/s and cooling water inlet temperature 10°C. Overall heat transfer coefficient from the gas to the water is 140 W/m² K. Determine the cooling surface area required. Take c_p for exhaust gas to be 1130 J/kg K and c_p for water to be 4190 J/kg K.

Note: logarithmic mean temperature difference is a $\theta_m = \theta_1 - \theta_2 \Big/ \ln\left(\dfrac{\theta_1}{\theta_2}\right)$, where

θ_1 = temperature difference between hot and cold fluid at inlet

θ_2 = temperature difference between hot and cold fluid at outlet.

10. Steam at 007 bar condenses on the outside of a thin tube of 0.025 m diameter and 2.75 m length. Cooling water of density 1000 kg/m³, c_p 4.18 kJ/kg K, flows through the tube at 0.6 m/s rising in temperature from 12°C to 24°C. The surface heat transfer coefficient between steam and tube is 17 000 W/m² K. Determine:

(i) the overall heat transfer coefficient and

(ii) the surface heat transfer coefficient between water and tube.

5

LAWS OF

PERFECT GASES

When a substance is progressing through an evaporation process it can exist as a gas or vapour, and one of its most important characteristics is its elastic property. For example, if a specific volume of a liquid is put into a vessel of larger volume, the liquid will only partially fill the vessel, taking up no more or less volume than it did before. However, when a gas enters a vessel, it immediately fills up every part of that vessel irrespective of the new vessel's size.

During the evaporation process a substance will slowly transform from a solid to a liquid. As the heating progresses the liquid will start to change into a gas. During that change the gas will be able to hold some of the moisture in suspension. The mixture of gas and water droplets is called 'a vapour'.

There are occasions when the heating of a substance is so rapid that a solid transforms into a gas, in effect bypassing the 'liquid' phase, the action is called 'sublimation'.

Another observation to note is that, within practical limits, liquids cannot be compressed or expanded, but gases can be compressed into smaller volumes and expanded to larger volumes.

When learning about the 'gas laws' it is usual to consider a perfect or ideal gas which is a gas that strictly follows Boyle's and Charles' laws of gases. The difference between an 'ideal gas' and an actual gas is that an actual gas does not 'always' behave according to the two laws of gases under all conditions of temperature and pressure.

However, over the range of temperatures and pressures found in practical circumstances, a real gas does behave in a similar way to an ideal gas.

Consider a given mass of a perfect gas enclosed in a cylinder by a gas-tight movable piston. When the piston is pushed inward, the gas is compressed to a small volume, when pulled outward the gas is expanded to a larger volume. However, not only is there a change in volume but the pressure and temperature also change.

These three quantities, pressure, volume and temperature, are related to each other. To determine their relationship, it is usual to perform experiments where each one of these quantities in turn is kept constant and the change in relationship with the other two is observed.

Scientists and students completing experiments and observations to test these basic laws must have the pressure, the volume and the temperature, all expressed in their absolute values. As we have seen that means being measured from absolute zero, and not measured from some artificial level such as the freezing point of water. Absolute values were explained in Chapter 1, and so a brief reminder will suffice.

ABSOLUTE PRESSURE (p). This is the pressure measured above a perfect vacuum, obtained by adding the atmospheric pressure to the gauge pressure. 'Absolute' need not follow the given pressure; it is to be taken as such unless the pressure is distinctly marked 'gauge' to indicate that it is a pressure gauge reading.

VOLUME (V). The volume of the gas is equal to the full volume of the vessel containing it. Note that in the case of the cylinder of an air compressor or reciprocating engine, the total volume of air or gas includes the volume of the clearance space between the piston at its top dead centre and the cylinder cover, as well as the piston swept volume.

ABSOLUTE TEMPERATURE (T). This is the temperature in degrees kelvin measured above the absolute zero of temperature. 0 [°C] = +273 K.

Boyle's Law

All the time that a substance is in its gaseous state, the mixture behaves in a predictable and measurable way. In the past scientists have carefully completed experiments under tightly controlled conditions in order to understand these properties.

The most important, and relevant for marine engineers, are the properties and behaviour that were found by Robert Boyle, Jacques Charles and Amedeo Avogadro. Boyle's law states that the absolute pressure of a fixed mass of a perfect gas varies inversely to its volume if the temperature remains unchanged.

$$p \propto \frac{1}{V}, \text{ therefore } p \times V = \text{a constant}$$

This means that $p_1 \times V_1 = p_2 \times V_2$

To illustrate this, imagine 2 m³ of gas at a pressure of 100 kN/m² (=10⁵ N/m² = 1 bar) contained in a cylinder with a gas-tight movable piston, as illustrated in Figure 5.1. When the piston is pushed inward the pressure will increase as the gas is compressed to a smaller volume and, provided the temperature remains unchanged, the product of pressure and volume will be a constant quantity for all positions of the piston. From the known initial conditions, the constant can be calculated:

$$p_1 \times V_1 = \text{constant}$$

$$100 \times 2 = 200$$

and the pressure at any other volume can be determined.

When the volume is 1.5 m³:

$$p_2 \times 1.5 = 200$$
$$p_2 = 133.3 \text{ kN/m}^2$$

When the volume is 1 m³:

$$p_3 \times 1 = 200$$
$$p_3 = 200 \text{ kN/m}^2$$

When the volume is 0.5 m³:

$$p_4 \times 0.5 = 200$$
$$p_4 = 400 \text{ kN/m}^2$$

and so on.

The variation of pressure with change of volume is shown in the graph below the cylinder in Figure 5.1. The graph produced by joining up the plotted points is called a rectangular hyperbola; as a result we refer to compression or expansion where $pV = $ constant as hyperbolic compression or hyperbolic expansion. When the temperature is constant throughout the process, as in this example, the operation may also be termed 'isothermal'.

Note that the ordinates of the vertical measurements represent pressure, and the ordinates of the horizontal measurements, represent volume. This means that since the result of a pressure multiplied by a corresponding volume is constant, then all the rectangles drawn from each of the axes with their corners touching the curve, shown in Figure 5.1 (bottom), will be of equal area.

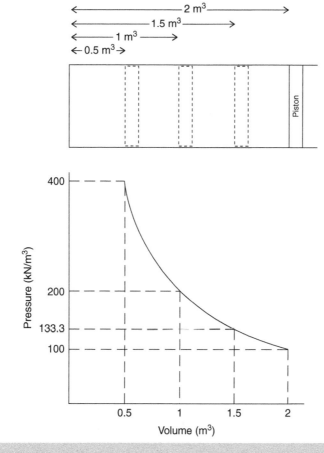

▲ **Figure 5.1** *A piston moving in a gas tight cylinder*

Example 5.1. 3.5 m³ of air at a pressure of 20 kN/m² gauge is compressed at constant temperature to a pressure of 425 kN/m² gauge. Taking the atmospheric pressure to be 100 kN/m², calculate the final volume of air.

Initial absolute pressure = 20 + 100 = 120 kN/m²

Final absolute pressure = 425 + 100 = 525 kN/m²

$$p_1 V_1 = p_2 V_2$$
$$120 \times 3.5 = 525 \times V_2$$
$$V_2 = \frac{120 \times 3.5}{525} = 0.8 \text{ m}^3 \quad \text{Ans.}$$

Charles' Law

Charles' law states that the volume of a fixed mass of a perfect gas varies directly as its absolute temperature, if the pressure remains unchanged; also, the absolute pressure varies directly as the absolute temperature, if the volume remains unchanged.

From the above statement we have:

For constant pressure, $V \infty T$ $\quad \therefore \dfrac{V}{T} = $ constant

$$\text{Hence, } \frac{V_1}{T_1} = \frac{V_2}{T_2} \qquad \text{or} \qquad \frac{V_1}{V_2} = \frac{T_1}{T_2}$$

Example 5.2. The pressure of the air in a starting air vessel is 40 bar (= 40 × 10^5 N/m^2 or 4 MN/m^2) and the temperature is 24°C. If a fire in the vicinity causes the temperature to rise to 65°C, find the pressure of the air. Neglect any increase in volume of the vessel.

As the term 'gauge' does not follow the given pressure, it is assumed that this is the initial absolute pressure.

Initial absolute temperature = 24°C + 273 = 297 K.

Final absolute temperature = 65°C + 273 = 338 K.

$$\text{Hence, } \frac{p_1}{T_1} = \frac{p_2}{T_2} \qquad \text{or} \qquad \frac{p_1}{p_2} = \frac{T_1}{T_2}$$

$$\frac{p_1}{T_1} = \frac{p_2}{T_2} \qquad \therefore p_2 = \frac{p_1 T_2}{T_1}$$

= 45.52 × 10^5 N/m^2 or 4.552 MN/m^2 Ans.

Combination of Boyle's and Charles' Laws

Each one of the gas laws states how one quantity varies with another if the third quantity remains unchanged, but if the three quantities change simultaneously, it is necessary to combine these laws in order to determine the final conditions of the gas.

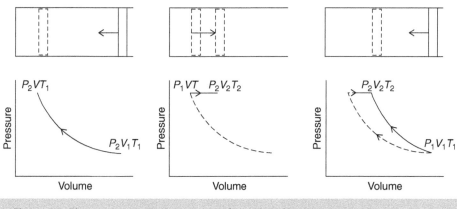

▲ **Figure 5.2** *The compression and expansion of gas when the piston is moved*

Figure 5.2 again represents a cylinder with a piston. The system is gas-tight, which is taken to mean that the mass of gas within the cylinder always remains the same. Starting the process let the gas be compressed from its initial state of pressure p_1, volume V_1 and temperature T_1 to its final state of p_2, V_2 and T_2; however, to arrive at the final state it passes through two stages. The first stage will satisfy Boyle's law and the second will satisfy Charles' law.

Imagine the piston pushed inward to compress the gas until it reaches the final pressure of p_2 and let its volume then be represented by V. Normally the temperature would tend to increase due to the work done in compressing the gas, but if any heat so generated is removed during compression then its temperature remains unchanged at T_1. And following Boyle's law, the values are:

$$p_1V_1 = p_2V$$ (i)

Now apply heat to raise the temperature from T_1 to T_2 and at the same time draw the piston outward to prevent a rise of pressure and keep it constant at p_2. The volume will increase in direct proportion to the increase in absolute temperature according to Charles' law:

$$\frac{V_2}{V} = \frac{T_2}{T_1}$$ (ii)

By substituting the value of V from (ii) into (i), this intermediate quantity V will be eliminated and the final equation arrived at:

$$\text{From (ii) } V = \frac{V_2 T_1}{T_2}$$

Substituting into (i)

$$p_1V_1 = p_2 \times \frac{V_2T_1}{T_2}$$

$$\therefore \frac{p_1V_1}{T_1} = \frac{p_2V_2}{T_2}$$

This combined law of Boyle's and Charles' is true for a given mass of any perfect gas subjected to any form of compression or expansion.

Example 5.3. 0.5 m³ of a perfect gas at a pressure of 0.95 bar (= 95 kN/m²) and temperature 17°C are compressed to a volume of 0.125 m³ and the final pressure is 5.6 bar (= 560 kN/m²). Calculate the final temperature.

Initial absolute temperature = 290 K

$$\frac{p_1V_1}{T_1} = \frac{p_2V_2}{T_2}$$

$$\frac{95 \times 0.5}{290} = \frac{560 \times 0.125}{T_2}$$

$$T_2 = \frac{560 \times 0.125 \times 290}{95 \times 0.5}$$

$$= 427.4 \text{ K}$$

$$= 154.4°C \text{ Ans.}$$

The Ideal Gas (Characteristic) Equation

Since pV/T is a constant, its value can be determined for a given mass of any perfect gas. To establish a basis on which to work, the constant is calculated on the specific volume (v) of the gas, which is the volume in cubic metres occupied by a mass of 1 kilogram. The resultant constant obtained in this way is called the *gas constant*, and it is represented by the symbol R and is different for different gases:

$$\frac{pv}{T} = R \quad \text{or} \quad pv = RT$$

The volume occupied by m kg of mass being represented by V then is:

$$pV = mRT$$

This generic expression is called the *characteristic equation of a perfect gas.*

Taking air as an example, experiments show that at standard atmospheric pressure and temperature, that is, at 101.325 kN/m² and 0°C, the specific volume is 0.7734 m³/kg. Using these values to find R:

$$R = \frac{pv}{T}$$

$$= \frac{101.325 \ [kN/m^2] \times 0.7734 \ [m^3/kg]}{273 \ [K]}$$

$$= 0.287 \ kJ/kg \ K$$

Note the units for R. Repeating the above with the units only:

$$R = \frac{kN/m^2 \times m^3/kg}{K} = \frac{kN}{m^2} \times \frac{m^3}{kg} \times \frac{1}{K}$$

$$= \frac{kNm}{kg \ K} = kNm/kg \ K = kJ/kg \ K$$

Example 5.4. An air compressor delivers 0.2 m³ of air at a pressure of 850 kN/m² and 31°C into an air reservoir. Taking the gas constant for air as 0.287 kJ/kg K, calculate the mass of air delivered.

$$pV = mRT$$

$$m = \frac{pV}{RT}$$

$$= \frac{850 \times 0.2}{0.287 \times 304} = 1.948 \ kg \quad Ans.$$

Universal Gas Constant

The *kilogram-mol* of a substance is a mass of that substance numerically equal to its molecular weight. The kilogram-mol is the derived SI unit, but it may simply be abbreviated to *mol*.

As examples, the molecular weight of oxygen is 32, and therefore the mass of 1 mol of oxygen is 32 kg. The molecular weight of hydrogen is 2, and hence 1 mol of hydrogen is a mass of 2 kg, and so on.

AVOGADRO'S LAW. This law states that for the same values of temperature and pressure, equal volumes of all gases contain the same number of molecules.

For example, consider two gases, one of mass m_1 having a molecular weight of M_1 and containing n_1 molecules, the other of mass m_2, of molecular weight M_2 and containing n_2 molecules. Then:

$$\text{Ratio of their masses} = \frac{m_1}{m_2} = \frac{n_1 M_1}{n_2 M_2}$$

This means that for equal volumes of the gases at the same temperature and pressure, the gases contain an equal number of molecules, therefore $n_1 = n_2$. They cancel each other out and we are left with the ratio of their masses being the same as the ratio of their molecular weight, which is:

$$\frac{m_1}{m_2} = \frac{M_1}{M_2}$$

From the characteristic gas equation $pV = mRT$, therefore $m = pV/RT$ and substituting this into the ratio above we have the equation:

$$\frac{p_1 V_1 / R_1 T_1}{p_2 V_2 / R_2 T_2} = \frac{M_1}{M_2} \quad \therefore \quad \frac{p_1 V_1 / R_2 T_2}{p_2 V_2 / R_1 T_1} = \frac{M_1}{M_2}$$

pV and T cancel because they are equal. Therefore we have:

$$\frac{R_2}{R_1} = \frac{M_1}{M_2} \quad \text{or} \quad R_1 M_1 = R_2 M_2$$

Therefore the product of the gas constant R and the molecular weight M is the same for all gases. This product is termed the *universal gas constant*, and it is represented by R_0 and its value has been found by experiment to be 8.314 kJ/mol K.

$$R_0 = RM$$

This means that the gas constant R for a specific gas can be found if its molecular weight is known.

For example, the molecular weight of nitrogen is 28, and therefore the gas constant for nitrogen is:

$$R = \frac{R_0}{M} = \frac{8.314}{28} = 0.2969 \text{ kJ/kg K}$$

Dalton's Law of Partial Pressures

This law states that the pressure exerted in a vessel by a mixture of gases is equal to the sum of the pressures that each separate gas would exert if it alone occupied the whole volume of the vessel.

The pressure exerted by each gas is termed a partial pressure:

Total pressure of the mixture

	Partial press. due to gas1	+	Partial press. due to gas2	+	Partial press. due to gas3	+ ...
=						

$$p = p_1 + p_2 + p_3 + \cdots$$

From the characteristic gas equation $pV = mRT$, substituting for the pressure of each gas, and also for the mixture:

$$\frac{mRT}{V} = \frac{m_1 R_1 T_1}{V_1} + \frac{m_2 R_2 T_2}{V_2} + \frac{m_3 R_3 T_3}{V_3} + \cdots$$

Since it can be considered as though each gas on its own occupied the whole space, then the volume V is common throughout. The temperature is also common; therefore if V and T are also common to all terms, they will cancel out, leaving:

$$\frac{mRT}{V} = \frac{T}{V}\left\{ m_1 R_1 + m_2 R_2 + m_3 R_3 + \cdots \right\}$$

where m is the total mass of the mixture and R is the gas constant of the mixture.

Example 5.5. The analysis by mass of a sample of air shows 23.14% oxygen, 75.53% nitrogen, 1.28% argon and 0.05% carbon dioxide. Estimate the gas constant for air (to the nearest four figures), taking the molecular weights of O_2, N_2, Ar and CO_2 to be 32, 28, 40 and 44, respectively, and the universal gas constant $R_0 = 8.314$ kJ/mol K.

$$R = \frac{R_0}{M}$$

Considering 1 kg of air:

$$\text{Oxygen } R_1 = \frac{8.314}{32} = 0.2598$$

$$\text{Nitrogen } R_2 = \frac{8.314}{28} = 0.2969$$

$$\text{Argon } R_3 = \frac{8.314}{40} = 0.2078$$

$$\text{Carbon dioxide } R_4 = \frac{8.314}{44} = 0.1889$$

Therefore making up the gas constant for air, we have the percentage mass of each gas multiplied by its own gas constant. For example:

$$M_4 R_4 = 0.0005 \times 0.1889 = 0.00009445$$

The sum of all four comes to $\Sigma mR = 0.28712058$

$$R \text{ air} = \frac{0.2871}{1} = 0.2871 \text{ kJ/kg K} \quad \text{Ans.}$$

PARTIAL VOLUMES. If each gas of a mixture in a closed vessel at the full volume of V and partial pressure p_1 (p_2, etc.) is considered as being compressed until its pressure is the full pressure p and its volume is its partial volume V_1 (V_2, ...), then, since its temperature remains the same:

$$p_1 \times V = p \times V_1 \text{ and } p_2 \times V = p \times V_2 \cdots$$

Therefore $(p_1 + p_2 + \text{etc.}) \times V = p \times (V_1 + V_2 + \cdots)$

Hence, the total volume of the mixture is equal to the sum of the partial volumes of the gases at the same total pressure and temperature.

$$\text{Also, } \frac{p_1}{p_2} = \frac{V_1}{V_2}$$

Therefore, the ratio of the partial volumes is equal to the ratio of the partial pressures.

Dalton's law can also be applied by marine engineers to determine the quantity of air leakage into steam condensers. An example is given in Chapter 10 after the study of properties of steam.

Specific Heats of Gases

The specific heat (c) of a substance is defined as the quantity of heat energy required to be transferred to unit mass of that substance to raise its temperature by one degree. Hence, the quantity (Q) of heat in kilojoules required to be given to a mass of m kilograms of a substance to raise its temperature from to T_2 is:

$$Q[kJ] = m[kg] \times c[kJ/kg\ K] \times (T_2 - T_1)\ [K]$$

Since the characteristics of gases vary considerably at different temperatures and pressures, heat energy may be transferred under an infinite number of conditions; therefore, the specific heat can have an infinite number of different values. Consider, however, two important conditions under which heat may be transferred: (i) while the *volume* of the gas remains constant and (ii) while the *pressure* of the gas remains constant.

The specific heat of a gas at constant volume is represented by c_v.

The specific heat of a gas at constant pressure is represented by c_p.

C_p is a higher value than c_v because when the gas is receiving heat it must be allowed to expand in volume to prevent a rise in pressure and, while expanding, the gas is expending energy in doing external work. Hence extra heat energy must be supplied equivalent to the external work done. (See also page 22 for information about the specific heat capacity of water.)

Example 5.6. A quantity of air of mass 0.23 kg, pressure 100 kN/m², volume 0.1934 m³ and temperature 20°C is enclosed in a cylinder with a gas-tight movable piston, and heat energy is transferred to the air to raise its temperature to 142°C.

(i) If the piston is prevented from moving during heat transfer so that the volume of the air remains unchanged, calculate (a) the heat supplied, taking the specific heat at constant volume $c_v = 0.718$ kJ/kg K, and (b) the final pressure.

(ii) If the piston moves to allow the air to expand in volume at such a rate as to keep the pressure constant, calculate (a) the heat supplied, taking the specific heat at constant pressure $c_p = 1.005$ kJ/kg K, and (b) the final volume.

Initial absolute temperature = 293 K

Final absolute temperature = 415 K

Temperature rise = 142 − 20 = 122°C = 122 K

$$(i) \ Q = m \times c_v \times \left(T_2 - T_1\right)$$
$$= 0.23 \times 0.718 \times 122$$
$$= 20.15 \ \text{kJ} \quad \text{Ans.} \ (i)(a)$$

$$\frac{p_1 V_1}{T_1} = \frac{p_2 V_2}{T_2}$$

The volume is constant, and therefore $V_2 = V_1$ and cancels:

$$= \frac{100 \times 415}{293} = 141.6 \ \text{kN/m}^2 \quad \text{Ans.} \ (i)(b)$$

$$p_2 = \frac{p_1 T_2}{T_1} \ \text{(which is Charles' law)}$$

$$(ii) \ Q = m \times c_p \times \left(T_2 - T_1\right)$$
$$= 0.23 \times 1.005 \times 122$$
$$= 28.19 \ \text{kJ} \quad \text{Ans.} \ (ii)(a)$$

$$\frac{p_1 V_1}{T_1} = \frac{p_2 V_2}{T_2}$$

The pressure is constant and therefore $p_1 = p_2$ and cancels:

$$V_2 = \frac{V_1 T_2}{T_1} \ \text{(which is Charles' law)}$$

$$= \frac{0.1934 \times 415}{293} = 0.2738 \ \text{m}^3 \quad \text{Ans.} \ (ii)(b)$$

Note the difference between the quantity of heat energy supplied to the air in the two cases:

$$28.19 - 20.15 = 8.04 \ \text{kJ}$$

This amount of heat energy was expended in moving the piston.

Referring to Figure 5.3, the piston is pushed forward by the gas at a constant pressure of, say p [kN/m²]. Let A [m²] represent the area of the piston; then the total force on the piston is $p \times A$ [kN]. If the piston moves S [m], the work done, being the product of force and distance, is $p \times A \times 5$ [kNm = kJ]. The product of the area A and the distance moved S is the volume moved (or swept) through by the piston. Hence:

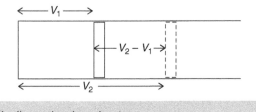

▲ **Figure 5.3** *Symbols allocated to show the piston movement*

Work done $= p\,(V_2 - V_1)$

In the last example, when the pressure was constant at 100 kN/m² the volume of the air increased from 0.1934 m³ to 0.2738 m³, and so the piston's movement or swept volume is the difference between these two volumes.

$$\text{Work done} = p\left(V_2 - V_1\right)$$
$$= 100\,\left(0.2738 - 0.1934\right)$$
$$= 100 \times 0.0804$$
$$= 804 \text{ kJ}$$

This shows that the extra heat energy given to the air at constant pressure, compared with that at constant volume, is the heat energy expended in doing external work by pushing the piston forward.

Energy Equation (Closed Systems)

The *internal energy* (*U*) of a gas is the energy contained within it as 'stored up work' by virtue of the internal movement of its molecules. Water collected in a header tank does not appear to have any energy. However, in reality it is a mass of high-speed particles and its 'raised' position, both of which are examples of stored energy.

JOULE'S LAW. This law states that the internal energy of a gas depends only upon its temperature and is independent of changes in pressure and volume.

By the principle of the conservation of energy, that is, that energy can neither be created nor destroyed, it follows that the total heat energy (*Q*) transferred to a gas will be the sum of the increase in its internal energy ($U_2 - U_1$) and any work (*W*) that is done by the gas during the transfer of heat energy to it. Thus:

Heat energy transferred	=	Increase in internal	+	External work done by the
to the gas		energy of the gas		gas

$$Q = (U_2 - U_1) + W$$

This is known as the *First Law of Thermodynamics*.

However, any of these three terms may be negative, which then represents the following conditions:

(i) heat rejected (extracted),

(ii) decrease of internal energy,

(iii) work done on the gas (compression).

They are also non-flow (across a system boundary) or reversible, processes for perfect gases.

RELATIONSHIP BETWEEN SPECIFIC HEATS. Referring again to the last example, with Figure 5.3, and applying the energy equation in the first case, heat energy is supplied to the gas to raise its temperature from T_1 to T_2 at *constant volume*. The heat supplied is $mc_v(T_2 - T_1)$, but no work is done. This is because, as the volume is constant, the piston does not move.

Heat supplied	=	Increase in internal energy	+	External work done

$$mc_v(T_2 - T_1) = (U_2 - U_1) + 0 \qquad \text{(i)}$$

In the second case, heat energy is supplied to raise the temperature through the same range, from T_1 to T_2 at *constant pressure*. The heat supplied is $mc_p(T_2 - T_1)$ and external work done is equal to $p(V_2 - V_1)$. Expressing the work done in terms of temperature by substituting the value of V from the characteristic gas equation, and inserting it into the energy equation, the resultant equation is:

$$p(V_2 - V_1) = p\left\{ \frac{mRT_2}{p} - \frac{mRT_1}{p} \right\} = mR(T_2 - T_1)$$

Heat supplied	=	Increase in internal energy	+	External work done

$$mc_p(T_2 - T_1) = (U_2 - U_1) + mR(T_2 - T_1)$$

$$(U_2 - U_1) = mc_p(T_2 - T_1) - mR(T_2 - T_1) \qquad \text{(ii)}$$

Since the temperature change is the same in each case, the change in internal energy is also the same; hence, from equations (i) and (ii) above:

$$mc_v(T_2 - T_1) = mc_p(T_2 - T_1) - mR(T_2 - T_1)$$

m and $(T_2 - T_1)$ are common to all terms and therefore cancel each other:

$$c_v = c_p - R$$

Rearranging the equation, then $R = c_p - c_v$.

Inserting the values for air as previously given:

$$R = 1\,005 - 0.718 = 0.287 \text{ kJ/kg K}$$

RATIO OF SPECIFIC HEATS. Another important relationship between the different specific heats is the ratio c_p/c_v. The symbol representing this ratio is γ:

$$\gamma = \frac{c_p}{c_v}$$

and for air this ratio is:

$$\gamma = \frac{1.005}{0.718} = 1.4$$

Taking this further:

$$\gamma = \frac{c_p}{c_v} \qquad \therefore c_p = \gamma c_v$$

Substituting this into the relationship:

$$R = c_p - c_v$$
$$= \gamma c_v - c_v$$

Therefore $R = c_v(\gamma - 1)$

$$\text{or} \quad c_v = \frac{R}{\gamma - 1}$$

These expressions will be found useful in later calculations.

Enthalpy

The enthalpy (*H*) of any fluid (liquid, vapour, gas) in a system is a term given to the total heat content of that system. It represents a grouping of terms (in open systems for reversible steady-flow processes) which are the sum of two kinds of energy transfer, that is, internal energy (*U*) and work transfer (*pV*).

$$H = U + pV \text{ (J)}$$

$$h = u + pv \text{ (J/kg) Specific enthalpy}$$

Energy Equation (Open Systems)

Consider the mass of fluid entering a system as m_1 and the volume to be V_1.

Therefore, work done against the resistance $= p_1 V_1$.

Consider mass of fluid leaving the systems as m_2, and the final volume to be V_2.

The work done against that resistance $= p_2 V_2$.

If $m_1 = m_2$ (for a steady flow process where the mass of fluid within a system is constant),

then $p_1 V_1 = p_2 V_2$ and this represents work, or energy, transfer through the system.

Referring to Figure 5.4, where the mass of fluid within a boundary is constant, that is, the mass entering equals the mass leaving.

Let h_1 and h_2 = specific enthalpy of fluid entering and leaving the system, respectively (kJ/kg).

Let C_1 and C_2 = velocity of fluid entering and leaving the system, respectively (m/s).

Let q = quantity of heat per unit mass crossing the boundary into the system (kJ/kg).

Let w = external work done per unit mass by the fluid (kJ/kg).

Assume no change in potential energy, then:

$$h_1 + \text{Kinetic energy}_1 + q = h_2 + \text{Kinetic energy}_2 + w$$

$$\text{i.e. } h_1 + 1/2C_1^2 + q = h_2 + 1/2C_2^2 + w$$

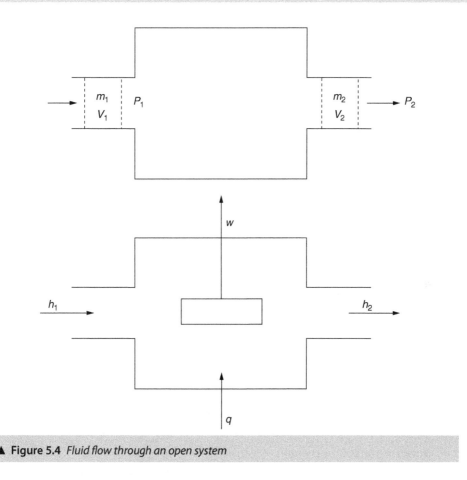

▲ **Figure 5.4** *Fluid flow through an open system*

This is known as the *steady flow energy equation*.

Example 5.7. The working fluid passes through a gas turbine at a steady rate of 10 kg/s. It enters with a velocity of 100 m/s and specific enthalpy of 2000 kJ/kg and leaves at 50 m/s with a specific enthalpy of 1500 kJ/kg. If the heat lost to surroundings, as the fluid passes through the turbine, is 40 kJ/kg, calculate the power developed.

Steady flow energy equation:

$$h_1 + 1/2C_1^2 + q = h_2 + 1/2C_2^2 + w$$

Therefore, $2000 \times 10^3 + \frac{1}{2} \times 100^2 - 40 \times 10^3 = 1500 \times 10^3 + \frac{1}{2} \times 50^2 + w$

(*Note: q* is negative as it is heat crossing the boundary out of the system.)

$$460 \times 10^3 + \frac{1}{2}(100^2 - 50^2) = w$$

$$4.6375 \times 10^5 = w$$

Therefore $w = 4.64 \times 10^5$ J

$$\text{Power} = 4.64 \times 10^5 \text{ J/s}$$

$$= 4.64 \text{ MW}$$

Note: For perfect gases,

$$h_2 - h_1 = (u_2 + p_2 v_2) - (u_1 + p_1 v_1)$$

$$= c_v(T_2 - T_1) + R(T_2 - T_1)$$

$$= c_p(T_2 - T_1)$$

Test Examples 5

1. Assuming compression according to the law $pV =$ constant:
   (i) Calculate the final volume when 1 m³ of gas at 120 kN/m² is compressed to a pressure of 960 kN/m².
   (ii) Calculate the initial volume of gas at a pressure of 1.05 bar which will occupy a volume of 5.6 m³ when it is compressed to a pressure of 42 bar.

2. 0.2 m³ of gas at a pressure of 1350 kN/m² and temperature 177°C is expanded in a cylinder to a volume of 0.9 m³ and pressure 250 kN/m². Calculate the final temperature.

3. An air receiver containing 2 m³ of air at 10 bar, 20°C has a relief valve set to operate at 20 bar. If 8% of the air were to leak out, calculate the temperature at which the relief valve would operate. *Note*: for air $R = 287$ J/kg K.

4. An air reservoir contains 20 kg of air at 3200 kN/m² gauge and 16°C. Calculate the new pressure and heat energy transfer if the air is heated to 35°C. Neglect any expansion of the reservoir; take R for air $= 0.287$ kJ/kg K, specific heat at constant volume $c_v = 0.718$ kJ/kg K, and atmospheric pressure $= 100$ kN/m².

5. The dimensions of a large room are 12 m by 16.5 m by 4 m. The air is completely changed once every 30 minutes and the temperature is maintained at 21°C. If the temperature of the outside atmosphere is 30°C, calculate the quantity of heat required to be extracted from the supply air per hour, and the equivalent power, taking the density of air at atmospheric pressure and 0°C as 1.293 kg/m³ and the specific heat at constant pressure as 1.005 kJ/kg K.

6. In a steady flow process the working fluid enters and leaves a horizontal system with negligible velocity. The temperature drop from inlet to outlet is 480°C and the heat losses from the system are 10 kJ/kg of fluid. Determine the power output from the system for a fluid flow of 1.7 kg/s. For the fluid $c_p = 900$ J/kg K.

7. Heat energy is transferred to 1.36 kg of air which causes its temperature to increase from 40°C to 468°C. Calculate, for the two separate cases of heat transfer at (i) constant volume, (ii) constant pressure:

   (a) the quantity of heat energy transferred,

   (b) the external work done,

   (c) the increase in internal energy.

   Take c_v and c_p as 0.718 and 1 005 kJ/kg K, respectively.

8. A closed vessel of 500 cm³ capacity contains a sample of flue gas at 1015 bar and 20°C. If the analysis of the gas by volume is 10% carbon dioxide, 8% oxygen and 82% nitrogen, calculate the partial pressure and mass of each constituent in the sample.

   R for CO_2, O_2 and $N_2 = 0.189$, 0.26 and 0.297 kJ/kg K, respectively.

9. A gas is discharged from a horizontal convergent nozzle at a steady rate of 1 kg/s. Conditions at inlet are 10 bar and 200°C and at exit 5 bar and 100°C. The change in specific internal energy passing through the nozzle is 80 kJ/kg and heat lost to the surrounds is negligible. If the gas enters the nozzle at 50 m/s, determine the exit velocity.

   For the gas M is 30 and R_0 may be taken to be 8.314 kJ/mol K.

10. A vessel of volume 0.4 m³ contains 0.45 kg of carbon monoxide and 1 kg of air at 15°C. Calculate:

    (i) the total pressure in the vessel,
    (ii) the partial pressure of the nitrogen.

    Note: Atomic mass relationships: carbon 12, oxygen 16, nitrogen 14.

    Universal gas constant = 8.314 kJ/mol K.

    Air is 23% oxygen by mass.

6

EXPANSION AND COMPRESSION OF PERFECT GASES

Compression of a Gas in a Closed System

When a gas is compressed, for example, in the cylinder of an internal combustion engine, by the inward movement of a gas-tight piston (Figure 6.1), then the pressure of the gas increases, as the volume decreases. The work done *on* the gas to compress it appears as heat energy in the gas and the temperature tends to rise. This effect can be seen with a tyre inflator; in pumping up the tyre the discharge end of the inflator gets hot due to the compression of the air. The reverse can also be observed when a carbon dioxide (CO_2) fire extinguisher is operated, and the high pressure gas is released to atmospheric pressure. The reduction in pressure reduces the temperature of the gas and the characteristic 'mist' can be observed as the moisture in the surrounding air condenses.

ISOTHERMAL COMPRESSION. Imagine an experiment where the piston is pushed inward slowly to compress a gas and, at the same time, the heat generated is taken away via the cylinder walls by a cooling water-jacket or other means. This will avoid any rise in temperature and the compression of the gas will have occurred *at constant temperature*. Then the process would be referred to as *isothermal compression*, and the relationship between pressure and volume would follow Boyle's law as stated in the previous chapter which is:

$$pV = \text{constant therefore } p_1 V_1 = p_2 V_2$$

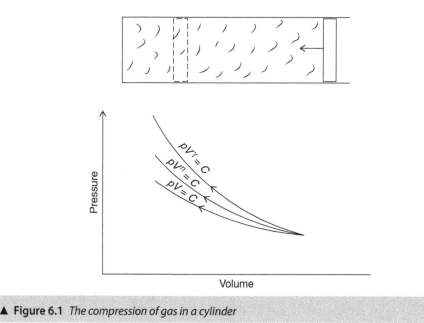

▲ **Figure 6.1** *The compression of gas in a cylinder*

ADIABATIC COMPRESSION. Now imagine that the piston is pushed inward quickly so that there is insufficient time for any heat energy to be transferred from the gas to the cylinder walls. All the work done in compressing the gas then appears as stored-up heat energy. The temperature at the end of compression will therefore be high and, for the same ratio of compression as with the isothermal compression, the pressure will consequently be higher. This form of compression, where no heat energy transfer takes place between the gas and an external source is known as *adiabatic compression*. The relationship between pressure and volume for adiabatic compression is:

$$pV^\gamma = \text{constant} \quad \therefore \quad p_1 V_1^\gamma = p_2 V_2^\gamma$$

where γ (gamma) is the ratio of the specific heat of the gas at constant pressure to the specific heat at constant volume, and thus:

$$\gamma = \frac{C_p}{C_v}$$

POLYTROPIC COMPRESSION. In the real world, neither isothermal nor adiabatic processes can be achieved perfectly. Some heat energy is always lost from the gas through the cylinder walls, especially if the cylinder is water cooled; however, the amount of heat 'lost' is never as much as the amount of heat generated by the compression. Consequently, the compression curve, representing the relationship between pressure and volume, lies somewhere between the two theoretical cases of isothermal and adiabatic compression.

Such compression, where a partial amount of heat energy exchange takes place between the gas and an outside source during the process, is termed *polytropic compression* and the compression curve follows the law:

$$pV^n = \text{constant} \quad \therefore p_1 V_1^n = p_2 V_2^n$$

Thus, the law 'pV^n = constant' may be taken as the general case to cover all forms of compression from isothermal to adiabatic wherein the value of n for isothermal compression is unity and for adiabatic compression $n = \gamma$; therefore for polytropic compression n generally lies somewhere between 1 and γ.

Expansion of a Gas in a Closed System

When a gas is expanded in a cylinder (Figure 6.2) the pressure falls and the volume increases as the piston is pushed outward by the energy in the gas.

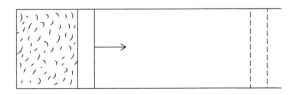

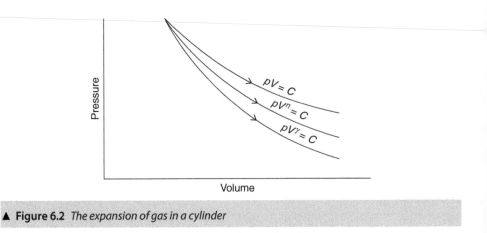

▲ **Figure 6.2** *The expansion of gas in a cylinder*

This is exactly the opposite of compression. Work is done *by* the gas in pushing the piston outward and there is a tendency for the temperature to fall due to the heat energy in the gas being converted into mechanical energy. Therefore, to expand the gas *isothermally*, heat energy must be transferred to the gas from an external source during the expansion in order to maintain its temperature constant. The expansion would then follow Boyle's law, pV = constant.

The gas would expand *adiabatically* if no heat energy transfer, to or from the gas, occurs during the expansion. In this case the external work done, in pushing the piston forward, is achieved entirely by reducing the stored-up heat energy. Therefore, the temperature of the gas will fall during the expansion. As for adiabatic compression, the law for adiabatic expansion is pV^γ = constant.

During *polytropic* expansion, a partial amount of heat energy will be transferred to the gas from an outside source but not sufficient to maintain a uniform temperature during the expansion. The law for polytropic expansion is 'pV^n = constant' as it is for polytropic compression.

With reference to Figures 6.1 and 6.2, note that the adiabatic curve is the steepest, the isothermal curve is the least steep and the polytropic curve lies between the two. Thus the higher the index of the law of expansion or compression, the steeper will be the curve.

It must also be noted that for any mode of expansion or compression in a closed system, the combination of Boyle's and Charles' laws and the characteristic gas equation given in the previous chapter are always true:

$$\frac{pV}{T} = \text{constant} \qquad \frac{p_1V_1}{T_1} = \frac{p_2V_2}{T_2}$$

$$\therefore pV = mRT$$

Example 6.1. 0.25 m³ of air at 90 kN/m² and 10°C is compressed in an engine cylinder to a volume of 0.05 m³, the law of compression being $pV^{1.4}$= constant. Calculate (i) the final pressure, (ii) the final temperature and (iii) the mass of air in the cylinder, taking the characteristic gas constant for air $R = 0.287$ kJ/kg K.

$$p_1V_1^{1.4} = p_2V_2^{1.4}$$
$$90 \times 0.25^{1.4} = p2 \times 0.05^{1.4}$$
$$p_2 = \frac{90 \times 0.25^{1.4}}{0.05^{1.4}}$$
$$= 90 \times 5^{1.4} = 856.7 \text{ kN/m}^2 \quad \text{Ans. (i)}$$

$$\frac{p_1 V_1}{T_1} = \frac{p_2 V_2}{T_2}$$

$$= 265.8°C \quad \text{Ans. (ii)}$$

$$p_1 V_1 = mRT_1,$$

$$\frac{90 \times 0.25}{283} = \frac{856.7 \times 0.05}{T_2}$$

$$T_2 = \frac{283 \times 856.7 \times 0.05}{90 \times 0.25} = 538.8 \text{ K}$$

$$= 265.8°C$$

$$p_1 V_1 = mRT_1$$

$$m = \frac{90 \times 0.25}{0.287 \times 283} = 0.277 \text{ kg} \quad \text{Ans. (iii)}$$

Example 6.2. 0.07 m³ of gas at 4.14 MN/m² is expanded in an engine cylinder and the pressure at the end of expansion is 310 kN/m². If expansion follows the law $pV^{1.35} = $ constant, find the final volume.

$$p_1 V_1^{1.35} = p_2 V_2^{1.35}$$

$$4140 \times 0.07^{1.35} = 310 \times V_2^{1.35}$$

$$V_2 = 0.07 \times \sqrt[1.35]{\frac{4140}{310}} = 0.4774 \text{ m}^3 \quad \text{Ans.}$$

Example 6.3. 0.014 m³ of gas at 3.15 MN/m² is expanded in a closed system to a volume of 0.154 m³ and the final pressure is 120 kN/m². If the expansion takes place according to the law $pVn = $ constant, find the value of n.

$$p_1 V_1^n = p_2 V_2^n$$

$$3150 \times 0.014^n = 120 \times 0.154^n$$

$$\frac{3150}{120} = \left\{\frac{0.154}{0.014}\right\}^n$$

$$26.25 = 11^n$$

$$n = 1.363 \quad \text{Ans.}$$

Determination of *n* from a Graph

Students will appreciate that with modern electronic sensors, feeding information into a computer, about the compression and expansion within an operating engine can now be measured accurately in real time. This information can be used to inform a control system about the most appropriate settings for the engine controls, or it can be used to inform engineers about the 'condition' of the engine.

Where engines are fitted with this technology the recording and control process can happen many times a second. However, it can also be appreciated that without the most up-to-date electronics fitted it would be difficult to obtain two pairs of sufficiently accurate values of pressure and volume from a running engine to enable the law of expansion or compression to be determined, as in the previous example. One practical method of finding a fairly close approximation of the law is as follows:

(i) measure a series of connected values of p and V from the curve of a mechanical indicator diagram; (see *Reeds Vol. 12*, Chapter 3);

(ii) reduce the equation $pV^n = C$ to a straight line logarithmic equation;

(iii) draw a straight line graph as near as possible through the plotted points of log p and log V to eliminate slight errors of measurement; and

(iv) determine the law of this graph to obtain the value of n. Thus:

$$p \times V^n = C$$

$$\log p + n \log V = \log C$$

$$\log p = \log C - n \log V$$

This is the same form of equation as,

$$y = a - bx$$

which represents a straight line graph.

The terms log p and log V are the two variables comparable with y and x, respectively, and log C and n are constants comparable with a and b, respectively. The constant n (like the constant b) represents the slope of the straight line graph and, being a negative value, the line will slope downwards from top left to bottom right.

Studying this procedure will help students to appreciate the basic way in which the combustion process can be monitored. This will guide their understanding of the physics involved, which will help in any fault diagnostic process that may have to be undertaken in the future.

Example 6.4. The following related values of the pressure p in kN/m² and the volume V in m³ were measured from the compression curve of an internal combustion engine indicator diagram (see pages 97, 101 & 103 [Engine Indicator Diagram]). Assuming that p and V are connected by the law $pV^n = C$, find the value of n.

p	3450	2350	1725	680	270	130
V	0.0085	0.0113	0.0142	0.0283	0.0566	0.0991

The scales of the graph can be of any convenient choice. In this case both pressure and volume can be expressed in more convenient units, the pressure in bars (1 bar = 10^5 N/m^2 = 10^2 kN/m^2) to proportionally reduce the high figures, and the volume in litres (10^3 l = 1 m^3) to express all volumes above unity, thereby avoiding negative logs and so reducing labour. Hence the following tabulated values of p and V are in bars and litres, respectively, with their corresponding logarithms to obtain graph plotting points from the respective pairs of log p and log V (base 10 used here).

$$\frac{3150}{120} = \left\{\frac{0.154}{0.014}\right\}^n$$

p [bar]	Log p	V [litres]	Log V
34.5	1.5378	8.5	0.9294
23.5	1.3711	11.3	1.0531
17.25	1.2368	14.2	1.1523
6.8	0.8325	28.3	1.4518
2.7	0.4314	56.6	1.7528
1.3	0.1139	99.1	1.9961

The graph is then plotted as shown in Figure 6.3. Note that it is not necessary to commence at zero origin when only the slope of the line (value of n) is required. A larger graph on the available squared paper can be drawn by starting and finishing to suit the minimum and maximum values to be plotted.

Choosing any two points on the line such as those shown:

$$n = \frac{\text{Decrease of log } p}{\text{Decrease of log } V}$$

$$= \frac{1.5 - 0.15}{1.95 - 0.95}$$

$$= \frac{1.35}{1} = 1.35$$

$$\log p = \log C - 1.35 \log V$$

$$p = C \times V^{-1.35}$$

$$pV^{1.35} = C \quad \text{Ans.}$$

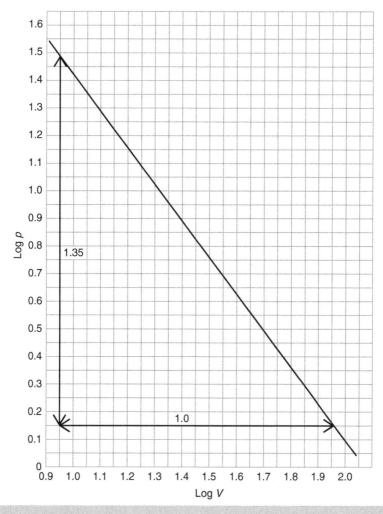

▲ **Figure 6.3** *Logarithmic values of pressure and volume plotted on one graph*

Ratios of Expansion and Compression

The ratio of expansion of gas in a cylinder is the ratio of the volume at the end of expansion to the volume at the beginning of expansion. It is usually denoted by *r*.

$$\text{Ratio of expression} = r = \frac{\text{Final volume}}{\text{Initial volume}}$$

The ratio of compression is the ratio of the volume of the gas at the beginning of compression to the volume at the end of compression. This can also be denoted by r.

$$\text{Ratio of compression} = r = \frac{\text{Initial volume}}{\text{Final volume}}$$

It will be seen that in each of the above ratios, it is the larger volume divided by the smaller, therefore, the ratio of expansion and ratio of compression are always greater than unity.

Relationships between Temperature and Volume, and between Temperature and Pressure, when $pV^n = C$

As stated previously, the equations:

$$p_1V_1^n = p_2V_2^n \qquad \text{and} \qquad \frac{p_1V_1}{T_1} = \frac{p_2V_2}{T_2}$$

are always true for any kind of expansion or compression of a perfect gas in a closed system. Some problems arise, however, where neither p_1 nor p_2 are known/given, and the unknown temperature or volume has to be solved by substituting the value of one of the pressures from one equation into the other. Similarly, where neither volume is known/given, substitution has to be made for one of the volumes to obtain the unknown temperature or pressure.

Substitution can be made in one of the above equations to eliminate either pressure or volume and so derive relationships for a direct solution as shown.

$$p_1V_1^n = p_2V_2^n \qquad \therefore \; p_1 = \frac{p_2V_2^n}{V_1^n}$$

Substituting this value of p_1 into the combined law equation:

$$\frac{p_1V_1}{T_1} = \frac{p_2V_2}{T_2}$$

$$\frac{p_2 \times V_2^n \times V_1}{T_1 \times V_1^n} = \frac{p_2V_2}{T_2}$$

$$T_1 \times V_1^n \times p_2 \times V_2 = T_2 \times p_2 \times V_2^n \times V_1$$

$$\frac{T_1}{T_2} = \frac{p_2 \times V_2^n \times V_1}{p_2 \times V_2 \times V_1^n}$$

p_2 cancels:

dividing V_2^n by $V_2 = V_2^n \div V_2 = V_2^{n-1}$

dividing V_1^n by $V_1 = V_1^n \div V_1 = V_1^{n-1}$

$$\frac{T_1}{T_2} = \frac{V_2^{n-1}}{V_1^{n-1}}$$

$$\frac{T_1}{T_2} = \left\{ \frac{V_2}{V_1} \right\}^{n-1} \qquad\qquad \text{(i)}$$

Again from $p_1 V_1^n = p_2 V_2^n$ $V_1^n = \dfrac{p_2 V_2^n}{p_1}$

$$\therefore V_1 = p_2^{1/n} \times \frac{V_2}{p_1^{1/n}}$$

Substituting this value of V_1 into the combined law equation:

$$\frac{p_1 V_1}{T_1} = \frac{p_2 V_2}{T_2}$$

$$\frac{p_1 \times p_2^{1/n} \times V_2}{T_2 \times p_1^{1/n}} = \frac{p_2 \times V_2}{T_2}$$

$$T_1 \times p_1^{1/n} \times p_2 \times V_2 = T_2 \times p_1 \times p_2^{1/n} \times V_2$$

$$\frac{T_1}{T_2} = \frac{p_1 \times p_2^{1/n} \times V_2}{p_1^{1/n} \times p_2 \times V_2}$$

V_2 cancels:

$$\frac{T_1}{T_2} = \frac{p_1^{1-1/n}}{p_2^{1-1/n}}$$

$$\frac{T_1}{T_2} = \left\{ \frac{p_1}{p_2} \right\}^{1-1/n}$$

$$\frac{T_1}{T_2} = \left\{ \frac{p_1}{p_2} \right\}^{\frac{n-1}{n}} \qquad\qquad \text{(ii)}$$

From (i) and (ii) we have the very useful relationship:

$$\frac{T_1}{T_2} = \left\{\frac{V_2}{V_1}\right\}^{n-1} = \left\{\frac{P_1}{P_2}\right\}^{\frac{n-1}{n}}$$

For an adiabatic process, the adiabatic index γ is substituted for the polytropic index n.

Example 6.5. Air is expanded adiabatically from a pressure of 800 kN/m² to 128 kN/m². If the final temperature is 57°C, calculate the temperature at the beginning of expansion, taking γ = 1.4.

$$\frac{T_1}{T_2} = \left\{\frac{P_1}{P_2}\right\}^{\frac{\gamma-1}{\gamma}}$$

$$\frac{T_1}{330} = \left\{\frac{800}{128}\right\}^{2/\gamma}$$

$$T_1 = 330 \times 6.25^{2/7}$$

$$= 557.1 \text{ K}$$

$$= 284.1°C \quad \text{Ans.}$$

Example 6.6. The ratio of compression in a petrol engine is 9:1. Find the temperature of the gas at the end of compression if the temperature at the beginning is 24°C, assuming compression to follow the law pVn = constant, where n = 1.36.

$$\frac{T_1}{T_2} = \left\{\frac{V_2}{V_1}\right\}^{n-1} \quad \text{or} \quad \frac{T_2}{T_1} = \left\{\frac{V_1}{V_2}\right\}^{n-1}$$

$$T_2 = 297 \times 9^{0.36} = 655.1 \text{ K}$$

$$= 382.1°C \quad \text{Ans.}$$

Example 6.7. The volume and temperature of a gas at the beginning of expansion are 0.0056 m³ and 183°C, respectively; at the end of expansion the values are 0.0238 m³ and 22°C. Assuming expansion follows the law pV^n = C, find the value of n.

$$\frac{T_1}{T_2} = \left\{\frac{V_2}{V_1}\right\}^{n-1}$$

$$\frac{T_1}{330} = \left\{\frac{800}{128}\right\}^{2/\gamma}$$

$$1.546 = 4.25^{n-1}$$

$$n = 1.301 \quad \text{Ans.}$$

Work (Energy Transfer)

Work is the result of energy imparting a force to an object. The action is measured in the SI units of 'Watts'. Initially consider a gas expanding, due to the heat energy generated by combustion, at *constant pressure* in a cylinder fitted with a gas-tight piston (refer also to Figure 5.3 together with notes in Chapter 5).

$$\text{Work done } [\text{kJ} = \text{kN m}] = \text{Force } [\text{kN}] \times \text{Distance } [\text{m}]$$

The total force [kN] on the piston is the product of the pressure p [kN/m²] of the gas acting on the area A [m²] of the piston and, if the piston then moves through a distance of S metres as the gas expands, the calculation for the work done is:

$$\text{Work done} = P \times A \times S$$

The product of the piston area A and the distance it moves S is the volume swept, or moved through, by the piston. This is also equal to the increase in volume of the gas in the cylinder. If V_1 is the volume of the gas at the beginning of the expansion, and V_2 is the volume at the end of expansion, the swept volume ($A \times S$) is equal to $V_2 - V_1$ in cubic metres; therefore:

$$\text{Work done} = p(V_2 - V_1)$$

Figure 6.4 (left drawing) shows the pressure–volume diagram representing work done at constant pressure. The graph is a straight horizontal line and the area under the graph is a rectangle. The area of a rectangle is height × length which, in this case, is $p \times (V_2 - V_1)$. Therefore *the area under the pressure–volume line must represent work done.*

Now consider the cases where pressure falls during the expansion of the gas (Figure 6.4 centre and right drawings). The formula calculating the area under the polytropic curve representing the general relationship between pressure and volume, that is, $pVn = C$, can only be derived satisfactorily with the use of differential calculus. The expression is illustrated in Figure 6.4 (centre drawing) and is derived in the following example.

Example 6.8. A quantity of gas undergoes a non-flow process from an initial pressure of p_1 and volume V_1 to a final pressure p_2 and volume V_2.

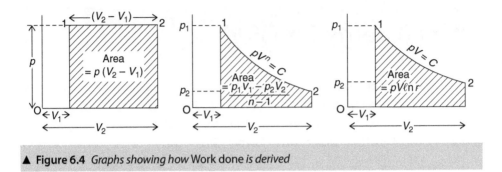

▲ **Figure 6.4** *Graphs showing how* Work done *is derived*

(a) It is given that the area beneath the pV curve is expressed by the formula:

$$\int_1^2 pdV$$

Now show that the total area beneath the pV curve may be written as:

$$\frac{p_2V_2 - p_1V_1}{1-n}$$

(b) During the process a quantity of heat q per unit mass is transferred. Show also that q is given by the equation:

$$q = \left(c_V + \frac{R}{1-n}\right)(T_2 - T_1)$$

$$pV^n = C$$

$$\text{Area} = \int_1^2 pdV$$

$$= C\int_{V_1}^{V_2} V^{-n}dV$$

$$= pV^n\left[\frac{V^{-n+1}}{-n+1}\right]_{V_1}^{V_2}$$

$$= \frac{pV^n(V_2^{1-n} - V_1^{1-n})}{1-n}$$

$$\text{Area} = \frac{p_2V_2 - p_1V_1}{1-n} \quad \text{Ans. (a)}$$

Heat transfer = Change of internal energy + External work

$$q = c_V(T_2 - T_1) + \frac{p_2 V_2 - p_1 V_1}{1-n}$$

$$pv = RT$$

$$q = c_V(T_2 - T_1) + \frac{R(T_2 - T_1)}{1-n}$$

$$q = \left(c_V + \frac{R}{1-n}\right)(T_2 - T_1) \quad \text{Ans. (b)}$$

$$\text{Work done during polytropic expansion} = \frac{p_1 V_1 - p_2 V_2}{n-1}$$

This is the general expression for work done and for adiabatic expansion, n is replaced by γ. For isothermal expansion however, since the value of n is 1, and as $p_1 V_1 = p_2 V_2$ then substitution in this expression for work will produce $0 \div 0$ which is indeterminate. A different expression is therefore necessary to obtain work during isothermal expansion (see Figure 6.4, right-hand drawing):

$$pV = C$$

$$p = \frac{C}{V}$$

$$\text{Area} = \int_{V_1}^{V_2} p\, dV$$

$$= C\int_{V_1}^{V_2} \frac{dV}{V}$$

$$= C(\ln V_2 - \ln V_1)$$

$$= pV \ln\left(\frac{V_2}{V_1}\right)$$

Work done during isothermal expansion $= pV \ln r$, where r is the ratio of expansion.

The above expressions give the work done *by* the gas during expansion. The same expressions give the work done *on* the gas during compression.

In the case of expansion the initial condition of $p_1 V_1$ will exceed the final condition of $p_2 V_2$ and the expression for work will produce a *positive* result, indicating that work is done *by* the gas in pushing the piston forward. Conversely, for compression, the initial condition of $p_1 V_1$ will be less than the final condition of $p_2 V_2$ and therefore a *negative* result will be obtained, indicating that work is done *on* the gas by the piston.

It is important to bear in mind that in calculating work, the units must be consistent. For example, to express work in kilojoules, the pressure must be in kN/m² and the volume in m³. Thus,

$$kN/m^2 \times m^3 = kN\,m = kJ$$

Work is a transfer of energy to and/or from a system; therefore, the above expression can be referred to as *work transfer* **from** the closed system, within the boundary of the cylinder, for example to the external mechanism, or it can be described as *work transfer* **to** the closed system within the boundary of the cylinder from the external mechanism.

In the first case, when the gas expands, work is being transferred from the energy in the gas to the piston, which, in turn, transmits the work through the connecting mechanism to the crank shaft, and the work transfer in this case is referred to as being positive. In the second case, when the gas is compressed, work is being transferred from the crank shaft, through the connecting mechanism and piston to the gas, thereby increasing the energy in the gas, and this work transfer is called negative transfer.

Further, since $pV = mRT$, the expressions for work may be stated in terms of mRT instead of pV:

Polytropic expansion is expressed as:

$$\text{Work} = \frac{p_1 V_1 - p_2 V_2}{n-1} = \frac{mR(T_1 - T_2)}{n-1}$$

Isothermal expansion is expressed as:

$$\text{Work} = pV \ln r = mRT \ln r$$

Example 6.9. 0.04 m³ of gas at a pressure of 1482 kN/m² is expanded isothermally until the volume is 0.09 m³. Calculate the work done during the expansion.

$$\text{Work done} = pV \ln r$$
$$r = \frac{\text{Final volume}}{\text{Initial volume}} = \frac{0.09}{0.04} = 2.25$$
$$\text{Work done} = 1482 \times 0.04 \times 0.81809$$
$$= 48.08 \text{ kJ} \quad \text{Ans.}$$

Example 6.10. A 7.08 l of air at a pressure of 13.79 bar and temperature 335°C are expanded according to the law $pV^{1.32} = \text{constant}$, and the final pressure is 1.206 bar. Calculate

(i) the volume at the end of expansion,

(ii) the work transfer from the air,

(iii) the temperature at the end of expansion,

(iv) the mass of air in the system, taking $R = 0.287$ kJ/kg K.

$$P_1V_1^{1.32} = P_2V_2^{1.32}$$
$$1379 \times 0.00708^{1.32} = 120.6 \times V_2^{1.32}$$
$$V_2 = 0.00708 \times \sqrt[1.32]{\frac{1379}{120.6}}$$
$$= 0.04484 \text{ m}^3 \text{ or } 44.84 \text{ litres} \quad \text{Ans. (i)}$$

Note that if the units of pressure and volume are the same on each side of the equation, the units cancel each other out. This means that any convenient and/or 'relevant' units can be used. The above could therefore be worked in bars of pressure and litres of volume or in whatever units are given in the question or found in the data. However, as already indicated, it is essential to work in fundamental SI units in such expressions, as used in parts (ii) and (iv) of this problem. It is preferable to use fundamental units throughout by expressing the pressure in kN/m² and the volume in m³.

$$\text{Work} = \frac{p_1V_1 - p_2V_2}{n - 1}$$
$$= \frac{1379 \times 0.00708 - 120.6 \times 0.04484}{1.32 - 1}$$
$$= 13.61 \text{ kJ} \quad \text{Ans. (ii)}$$

$$\frac{p_1V_1}{T_1} = \frac{p_2V_2}{T_2}$$
$$\frac{1379 \times 0.00708}{608} = \frac{120.6 \times 0.04484}{T_2}$$
$$T_2 = \frac{608 \times 120.6 \times 0.04484}{1379 \times 0.00708}$$
$$T_2 = 336.6 \text{ K}$$
$$= 63.6°C \quad \text{Ans. (iii)}$$

$$P_1V_1 = mRT_1$$
$$m = \frac{1379 \times 0.00708}{0.287 \times 608}$$
$$= 0.05595 \text{ kg} \quad \text{Ans. (iv)}$$

Example 6.11. A perfect gas is compressed in a cylinder according to the law $pV^{1.3}$ = constant. The initial condition of the gas is 1.05 bar, 0.34 m³ and 17°C. If the final pressure is 6.32 bar, calculate (i) the mass of gas in the cylinder, (ii) the final volume, (iii) the final temperature, (iv) the work done to compress the gas, (v) the change in internal energy and (vi) the transfer of heat between the gas and cylinder walls.

Take $c_v = 0.7175$ kJ/kg K and $R = 0.287$ kJ/kg K.

$$P_1V_1 = mRT_1$$

$$m = \frac{105 \times 0.34}{0.287 \times 290}$$

$$= 0.4289 \text{ kg} \quad \text{Ans. (i)}$$

$$P_1V_1^{1.3} = P_2V_2^{1.3}$$
$$1.05 \times 0.34^{1.3} = 6.32 \times V_2^{1.3}$$

$$V_2 = 0.34 \times \sqrt[1.3]{\frac{1.05}{6.32}}$$

$$= 0.08549 \text{ m}^3 \quad \text{Ans. (ii)}$$

$$\frac{P_1V_1}{T_1} = \frac{P_2V_2}{T_2}$$

$$\frac{1.05 \times 0.34}{290} = \frac{6.32 \times 0.08549}{T_2}$$

$$T_2 = \frac{290 \times 6.32 \times 0.08549}{1.05 \times 0.34}$$

$$T_2 = 438.8 \text{ K}$$

$$= 165.8°C \quad \text{Ans. (iii)}$$

Alternatively, the final temperature could be obtained from

$$\frac{T_1}{T_2} = \left\{\frac{P_1}{P_2}\right\}^{\frac{n-1}{n}}$$

$$\text{Work done} = \frac{P_1V_1 - P_2V_2}{n-1}$$

$$= \frac{105 \times 0.34 - 632 \times 0.08549}{1.3 - 1}$$

$$= -61.1 \text{ kJ}$$

Alternatively, the work done could be obtained from:

$$\frac{mR(T_1 - T_2)}{n-1}$$

Note that the minus sign indicates that work is done *on* the gas.

Work to compress gas = 61.1 kJ Ans. (iv)

Increase in internal energy:

$$U_2 - U_1 = mc_v(T_2 - T_1)$$
$$= 0.4289 \times 0.7175 \times (438.8 - 290)$$
$$= 45.78 \text{ kJ} \quad \text{Ans.} (v)$$

Heat supplied to the gas	=	Increase in internal energy	+	Work done by the gas

$$= 45.78 - 61.1$$
$$= -15.32 \text{ kJ}$$

The minus sign means that heat is rejected by the gas during compression, that is, this amount of heat energy is transferred *from* the gas *to* the cylinder wall surrounds.

Transfer of heat = 15.32 kJ Ans. (vi)

The Relationship between Heat Energy Supplied and Work Done

Thinking about a heat engine we first consider polytropic expansion of a gas in a cylinder from initial conditions, which are represented by state-point 1 to the final

Heat supplied to the gas	=	Increase in internal energy	+	Work done by the gas

conditions of state-point 2, and then apply the energy equation:

$$= mc_v(T_2 - T_1) + \frac{mR(T_1 - T_2)}{n-1}$$

During the expansion of the gas, work is carried out *by* the gas and therefore the expression is a positive quantity, which means that a positive quantity of heat energy is supplied from the cylinder walls to the gas.

In the compression of a gas, work is done *on* the gas which is negative work done by the gas, the result of the above expression is negative heat supplied, meaning that heat energy is transferred from the gas to the cylinder walls.

The relations between the properties of a perfect gas in its initial and final states are the same for both reversible (ideal, frictionless), steady-flow (open) and the non-flow (closed) processes as detailed in this chapter.

Test Examples 6

1. Gas is expanded in an engine cylinder, following the law pV^n = constant, where the value of n is 1.3. The initial pressure is 2550 kN/m² and the final pressure is 210 kN/m². If the volume at the end of expansion is 0.75 m³, calculate the volume at the beginning of expansion.

2. The ratio of compression in a petrol engine is 8.6:1. At the beginning of compression the pressure of the gas is 98 kN/m² and the temperature is 28°C. Find the pressure and temperature at the end of compression, assuming it follows the law $pV^{1.36}$ = constant.

3. Gas is expanded in an engine cylinder according to the law $pV^n = C$. At the beginning of expansion the pressure and volume are 1750 kN/m² and 0.05 m³, respectively, and at the end of expansion the respective values are 122.5 kN/m² and 0.375 m³. Calculate the value of n.

4. The temperature and pressure of the air at the beginning of compression in a compressor cylinder are 20°C and 101.3 kN/m², and the pressure at the end of compression is 1420 kN/m². If the law of compression is $pV^{1.35}$ = constant, find the temperature at the end of compression.

5. 0.014 m³ of a gas at 66°C is expanded adiabatically in a closed system and the temperature at the end of expansion is 2°C. Taking the specific heats of the gas at constant pressure and constant volume as 1.005 and 0.718 kJ/kg K, respectively, calculate the volume at the end of the expansion.

6. Air is compressed in a diesel engine from 1.17 bar to 36.55 bar. If the temperatures at the beginning and end of compression are 32°C and 500°C, respectively, find the law of compression assuming it is polytropic.

7. 1 kg of air at 20 bar, 200°C is expanded to 10 bar, 125°C by a process which is represented by a straight line on the pV diagram.

   Calculate for the air:

   (i) the work transfer,

   (ii) the change in internal energy,

(iii) the heat transfer,

(iv) the change in enthalpy.

Note: For air $R = 287$ J/kg K and $c_p = 1005$ J/kg K.

8. A closed and insulated vessel contains 1 kg of air at 213°C and 1 bar. A second vessel, which is also insulated, has a volume of 0.2 m³ and contains air at 6 bar and 412°C. The two vessels are then connected by a pipe of negligible volume. Calculate:

   (i) the final pressure of air in the vessels,

   (ii) the final temperature of the air in the vessels.

   Note: For air $R = 287$ J/kg K.

9. A cylinder fitted with a piston contains 0.1 m³ of air at 1 bar, 15°C. Heat is supplied to the air until the temperature reaches 500°C while the piston is fixed. The piston is then released and the air expands according to the law $pV^{1.5} = C$ until the pressure is 1 bar. Calculate:

   (i) the final temperature of the air,

   (ii) the work done during expansion,

   (iii) the heat transferred during each process.

   Note: For the gas $c_v = 718$ J/kg K and $R = 287$ J/kg K.

10. A gas is expanded in a cylinder behind a gas-tight piston. At the beginning of expansion the pressure is 36 bar, volume 0.125 m³ and temperature 510°C. At the end of expansion the volume is 1.5 m³ and temperature 40°C. Taking $R = 0.284$ kJ/kg K and $c_v = 0.71$ kJ/kg K, calculate (i) the pressure at the end of expansion, (ii) the index of expansion, (iii) the mass of gas in the cylinder, (iv) change of internal energy, (v) work done by the gas and (vi) heat transfer during expansion.

7

IC ENGINES: ELEMENTARY PRINCIPLES

Internal combustion (IC) engines are given this name due to the combustion of the fuel taking place inside the engine. The rapid oxidation of fuel (combustion) inside an engine's cylinder releases heat which is absorbed by the gaseous mixture that is already in the cylinder. The temperature of the mixture is therefore increased with a consequent increase in pressure and/or change in volume as the piston moves due to the heat energy imparted to it. The linear motion of the piston is converted into a rotary motion in the crankshaft by the interaction of the connecting rod and crank or by the piston rod, crosshead and crank in the case of the large two-stroke marine engines.

With IC engines the method of igniting the fuel shows a fundamental difference between the diesel engine and the 'petrol' engine. In diesel engines the air in the cylinder is compressed to a high pressure and therefore, following the gas laws, the air reaches a high temperature. When the fuel is injected into this high temperature it ignites after a short delay (see the section on combustion and heat release – page 231).

When the ignition of the fuel is only caused by the heat from the compression of the air, the engine is classed as a compression–ignition (CI) engine. Please note that in modern engines some of the exhaust gas might be recirculated and introduced back into the cylinder along with the fresh charge of air. This is known as exhaust gas recirculation (EGR) and is a process designed to reduce the more harmful contents of the final exhaust gas exiting the vessel's funnel.

In petrol engines however, the fuel is usually taken in with the charge of air. The charge is compressed and then ignited by an electrically induced spark. These internal combustion engines are designated as spark ignition (SI) engines.

Students will be able to work out that the air/fuel mix in the SI engine will not be compressed to the same level as the CI engine since the same temperature level is not required. The reason for this is that the auto-ignition temperature of the fuel must not be reached in the SI engine before the spark occurs. It should also be noted that under cold conditions the CI engine might need some assistance with starting. For marine engineers this point is important, especially in connection with the operation of lifeboat engines.

Engines keep running due to their ability to combust fuel and extract the energy turning it into useful work, on a continual basis. To enable this to happen the engine must draw in a charge of air containing oxygen or an air/fuel mix, combine the fuel and the oxygen together in each combustion chamber and then set light to the mixture. Further to this the engine must also remove the waste in preparation to start the process again. Students need to start by learning the basic concept which is usually described as the completion of four phases in one cycle of operation. These are induction, compression, ignition and exhaust. This cycle is then repeated in a running engine.

The Four-Stroke Diesel Engine

As described on page 18, one stroke is the term given to one movement of a piston from one end of its travel to the other. If the piston is arranged vertically then the top of the travel is called the top dead centre (TDC) and the bottom of its travel is called the bottom dead centre (BDC). Therefore, four strokes would be four movements of the piston, two from TDC to BDC and two from BDC to TDC. The piston is connected to the crankshaft, in a trunk type engine, via the 'connecting rod', and the connecting rod turns the crankshaft through 180° for one stroke of the piston.

In this type of engine it takes four strokes of the piston (i.e., two revolutions of the crank) to complete one working cycle of operations, and hence the name *four-stroke* cycle.

Figure 7.1 illustrates how each of these four strokes happen in one cylinder. On the top of the cylinder is the 'cylinder head' and as it is a diesel engine, there is a fuel valve (or injector) which lifts to admit oil fuel (under pressure) into the cylinder. Then, there is an inlet valve that opens to allow air into the operating cylinder, and the exhaust valve through which the exhaust gases are expelled from the cylinder at the end of each cycle. There are two more valves which are not shown here because they do not

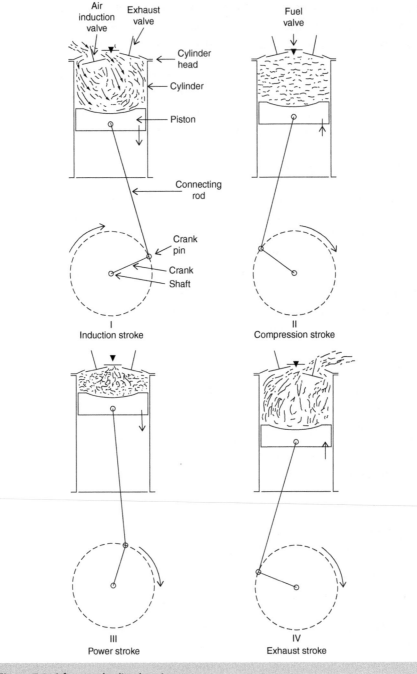

▲ **Figure 7.1** *A four-stroke diesel engine*

operate during the normal working cycle; one is the relief valve which opens against the compression of its spring if the pressure in the cylinder rises too high. The other valve is the air-starting valve which is used on the majority of marine diesel engines, to admit high pressure air into the cylinder to move the piston and start the engine.

Figure 7.2 shows two drawings that are very useful in understanding the operation of the IC engine. They relate to the four-stroke engine. The top drawing shows the position of operation of the operating valves (in degrees) in relation to the position of the crankshaft. It also shows the position of the start and finish of fuel injection.

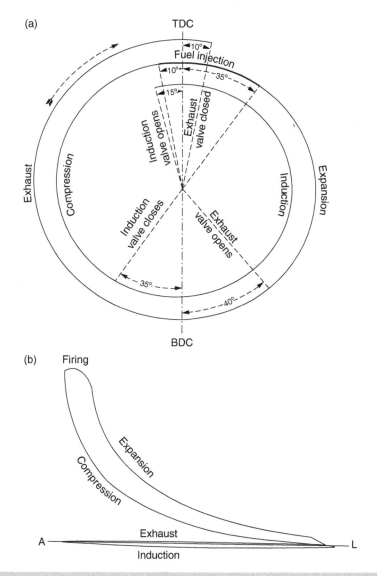

▲ **Figure 7.2** *(a) Timing diagram for a four-stroke diesel engine. (b) Four-stroke diesel indicator diagram*

The Two-Stroke Diesel Engine

The two-stroke diesel engine is so named because it takes two strokes of the piston to complete one working cycle. Every downward stroke of the piston is a power stroke and every upward stroke is a compression stroke. The exhaust of the burned gases from the first combustion and the fresh charge for the next combustion are taken in during the late period of the downward stroke and the early part of the upward stroke.

In the older and less efficient designs the exhaust gases pass through a set of ports in the lower part of the cylinder and the air is admitted through a similar set of ports. The ports are covered and uncovered by the piston itself which must be designed with a long piston skirt so that the ports are covered when the piston is at the top of its stroke.

As there is no complete stroke to draw the air into the cylinder, the air must be pumped in at a low pressure from a scavenge pump or a mechanical blower of some type. With modern engines this means supplying the compressed air by means of a turbocharger. The inlet air supplied is referred to as scavenge-air and the ports in the cylinder through which the air is admitted are termed scavenge ports. It is the function of this air to sweep around the cylinder and so 'scavenge' or clean out the cylinder by pushing the remains of the exhaust gases out, leaving a clean charge of air to be compressed (see Figure 7.3).

Modern Two-Stroke Designs

The modern two-stroke (crosshead) engines all operate on the uni-flow scavenge design. In this design the inlet or scavenge ports at the bottom of the cylinder liner are retained. The exhaust ports, however, are replaced by a mechanically operated poppet valve similar to the four-stroke engine valves.

This arrangement has the considerable advantage of the incoming (scavenge) air only having to push the exhaust in one direction and for a shorter distance (as the exhaust is removed from the top; see Figure 7.3).

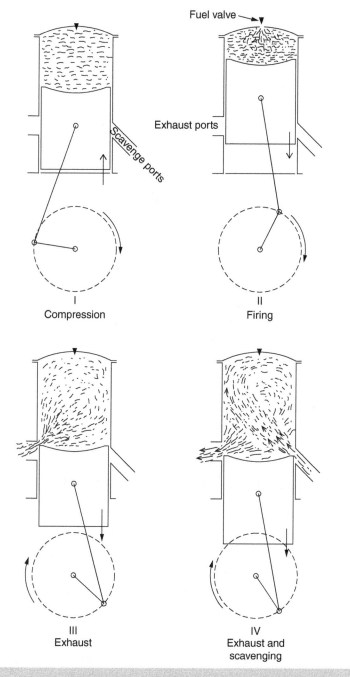

Fuel valve

Exhaust ports

Scavenge ports

I
Compression

II
Firing

III
Exhaust

IV
Exhaust and
scavenging

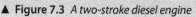

▲ **Figure 7.3** *A two-stroke diesel engine*

Petrol Engines

Engines which run with petrol as the fuel might be termed 'light oil' engines but are usually referred to as petrol or gasoline engines. The main difference between the original petrol engine and the diesel engine is that the petrol engine takes in a charge of air and petrol vapour, and this explosive mixture is compressed and ignited, at the correct moment, by an electric spark. In the diesel engine, however, the cylinder is charged with air only, and it is only the charge air being compressed that provides the temperature for ignition. Students will need to note that the most up-to-date engines using exhaust gas recirculation will have some exhaust gas mixed in with the fresh charge of air.

When the air is compressed in a diesel engine there is no possibility of firing before the fuel is injected and therefore the temperature can be raised to a very high value. In a petrol engine, however, an explosive mixture of petrol and air is compressed and there is danger of the mixture firing spontaneously due to the heat of compression alone and before the electric spark occurs; therefore the ratio of compression must be limited to prevent this. The ratio of compression in diesel engines can be higher, such as 20:1 and upwards whereas the ratio of compression in petrol engines is much less, in the region of 12:1.

The modern petrol engine, however, is usually 'fuel injected', meaning that the fuel and the air are mixed by injecting the fuel into the air stream just before it enters the cylinder. For more information about 'combustion', see Chapter 13.

Mean Effective Pressure and Power

We have seen in the last chapter that the area under the pV diagram represents work. The indicator diagrams shown in Figures 7.2 and 7.4 are examples of practical pV diagrams that have been taken from operating engines by means of an *engine indicator* (Figure 7.5), and the areas of these indicator diagrams represent the work done per cycle.

Figure 7.5 shows an apparatus called an engine indicator. This equipment can be fitted to an indicator cock, which is a fitting to each cylinder head of a marine diesel engine, by means of the screw coupling. The engine indicator can be used on engines with rotational speeds of up to 300 rev/min. The indicator chord is attached to a mechanism on the engine that is operated by the camshaft but replicates the movement of the crankshaft. As the indicator cock is opened so the full and fluctuating pressure generated in the cylinder will operate the piston and piston rod that is inside the cylinder of

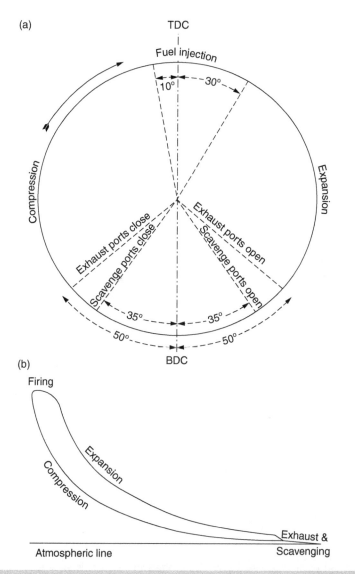

(a)

TDC

Fuel injection

30°

10°

Compression

Expansion

Exhaust ports close

Scavenge ports close

Exhaust ports open

Scavenge ports open

35°

35°

50°

50°

BDC

(b)

Firing

Expansion

Compression

Exhaust & Scavenging

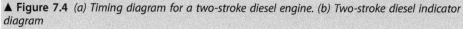

Atmospheric line

▲ **Figure 7.4** *(a) Timing diagram for a two-stroke diesel engine. (b) Two-stroke diesel indicator diagram*

the apparatus. A spring of the correct strength, shown just above the pencil lever in Figure 7.5, is chosen to balance the gas pressure for a particular engine. The spring can be changed when used on an engine with a different pressure range within its cylinder.

The cylinder pressure will, via the piston, rod and frame, operate the pencil lever in time with its rise and fall. The indicator cord will rotate the drum in time with the crankshaft. Students will appreciate that this simple mechanical apparatus will now be able to produce a pressure/volume graph representative of the combustion process, as these two readings come together.

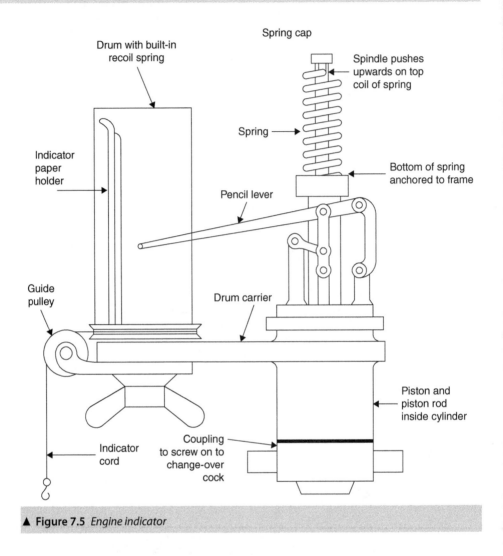

Drum with built-in recoil spring

Spring cap

Spindle pushes upwards on top coil of spring

Indicator paper holder

Spring

Pencil lever

Bottom of spring anchored to frame

Guide pulley

Drum carrier

Piston and piston rod inside cylinder

Indicator cord

Coupling to screw on to change-over cock

▲ **Figure 7.5** *Engine indicator*

The mechanical equipment will have its limitations, which is why it is only used on slower speed engines. However, the very important principle, of combining these two continuously varying readings is illustrated well with this equipment.

Any faults or developing faults with the combustion process within each individual cylinder can be identified by using this technique/equipment.

Therefore, if modern materials, electronics and manufacturing techniques are used to design equipment that works on the same principle, the output can be used as the basis for computerised control systems. Students will now be able to appreciate the

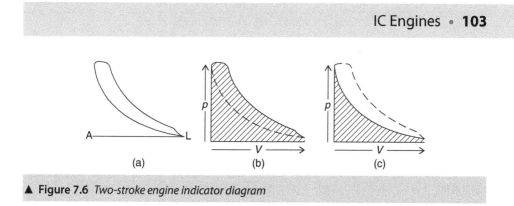

▲ **Figure 7.6** *Two-stroke engine indicator diagram*

operating mechanism that is driving the combustion control of modern diesel engines. More information can be found about this subject in *Reeds Vol. 12*, Chapter 1.

MEAN EFFECTIVE PRESSURE. Consider the two-stroke diesel engine indicator diagram shown in Figure 7.6. The positive work done in one cycle of operations *by* the gas during the burning period and expansion of the gas is shown by the shaded area of Figure 7.6(b).

The work done *on* the air during the compression period, representing negative work done by the engine, is shown by the shaded area of Figure 7.6(c). Hence the net useful work done in one cycle is the difference between positive and negative work and is represented by the actual diagram of Figure 7.6(a).

If the area of the indicator diagram (Figure 7.6a) is divided by its length, the average height is obtained, which, to scale, is the average or mean pressure effectively pushing the piston forward and transmitting useful energy to the crank during one cycle. When this pressure is expressed in N/m² or a suitable multiple of the basic pressure unit, it is termed the *indicated mean effective pressure.*

The area of the paper diagram is usually measured by an instrument called a planimeter. If the area is measured in mm² then dividing this area by the length of the graph, in mm, gives an answer that represents the mean height in mm. The mean height in mm can now be multiplied by the pressure scale of the indicator spring in N/m²/mm to obtain the indicated mean effective pressure in N/m². The usual convenient multiples of N/m² for such pressures are kN/m² and bars, and the spring may be graduated in either of these units.

If a planimeter is not available, the mean height of the indicator diagram may be obtained by the application of Simpson's rule or the mid-ordinate rule. See *Reeds Vol. 1*, Chapter 9.

POWER. This is a measure of the rate of doing work. It is recorded as the quantity of work done in a given time and the basic unit of power is the watt [W]. This is equal to the rate of one joule of work being done every second. In symbols it becomes:

$$1 \text{ W} = 1 \text{ J/s} = 1 \text{ Nm/s}$$

The watt is a small unit and only suitable for expressing the output of low power machines. For normal power ratings of marine engines, mechanical, electrical or hydraulic, the kilowatt [kW] is usually a more convenient size to use, and the power from large engines may be expressed in megawatts [MW].

The power recorded for the same engine could give different values depending upon where and how the measurement takes place. For example, the power measured at the engine's flywheel will be less than the calculated power from each cylinder added together. The reason for this is that some power will be required to operate the internal components of the engine and to overcome frictional losses (see page 106), and therefore the power output will be less than the power generated in each cylinder. The power measured at different points on the engine are given different names as show in the rest of this chapter.

INDICATED POWER. This is the term given to the power generated inside each cylinder and will represent the total amount of power available from the engine. The calculation for finding the indicated power of an engine is shown below and can be remembered by the pneumonic PLAN, where:

p_m = Mean effective pressure [N/m²]

L = Length of stroke [m]

A = Area of piston [m²]

n = Number of power strokes per second

Then:

Average force [N] on piston = $p_m \times A$ newtons

Work done [J] in one power stroke = $p_m \times L \times A$ newton-metres = joules

Work per second [J/s = W] = $p_m \times L \times A \times n$ watts of power

Therefore, the indicated power = $p_m LAn$.

This is the power indicated in one cylinder and the total power of a multi-cylinder engine is that one power multiplied by the number of cylinders, as long as the mean effective pressure is the same for all cylinders.

Students will note that when the mean effective pressure is in N/m² the power obtained by the above calculation will be in watts. If the mean effective pressure in kN/m² is used in the calculation then the result will be the power in kW, and this is usually more convenient for marine applications.

The value of n, the number of power strokes per second, also depends upon the working cycle of the engine (two-stroke or four-stroke), its rotational speed and, in the past, whether it was a single-acting or double-acting engine.

Modern engines are almost universally single-acting engines, which is where the cycle of operations takes place only on the top side of the piston.

In the four-stroke cycle, there is one power stroke in every four strokes, that is, one power stroke in every two revolutions. Hence:

$$n = \text{rev/s} \div 2$$

In the two-stroke cycle, there is one power stroke in every two strokes, that is, one power stroke in every revolution. Hence:

$$n = \text{rev/s}$$

Example 7.1. The area of an indicator diagram taken off one cylinder of a four-cylinder, four-stroke, single-acting internal combustion engine is 378 mm², the length is 70 mm, and the indicator spring scale is 1 mm = 2 bar. The diameter of the cylinders is 250 mm, stroke 300 mm and rotational speed 5 rev/s. Calculate the indicated power of the engine assuming all cylinders develop equal power.

$$\text{Mean height of diagram} = \text{Area} \div \text{Length}$$
$$= 378 \div 70 = 54 \text{ mm}$$

$$\text{Indicated } p_m = \text{Mean height} \times \text{Spring scale}$$
$$= 5.4 \times 2 = 10.8 \text{ bar}$$
$$10.8 \text{ bar} \times 10^2 = 1080 \text{ kN/m}^2$$
$$n = \text{rev/s} \div 2$$
$$= 5 \div 2 = 2.5$$

$$\text{Indicated power} = p_m LAn$$
$$= 1080 \times 0.7854 \times 0.25^2 \times 0.3 \times 2.5$$
$$= 39.74 \text{ kW}$$

Total power for four cylinders
$$= 4 \times 39.74 = 158.96 \text{ kW} \quad \text{Ans}$$

Example 7.2. The diameter of the cylinders of a six-cylinder, single-acting, two-stroke diesel engine is 635 mm and the stroke is 1010 mm. Indicator diagrams taken off the engine when running at 2.2 rev/s give an average area of 563 mm², the length of the diagram being 80 mm and the scale of the indicator spring 1 mm = 160 kN/m². Calculate the indicated power.

Indicated mean effective pressure

$$= \frac{\text{Area of diagram}}{\text{Length of diagram}} \times \text{Spring scale}$$
$$= \frac{563}{80} \times 160 = 1126 \text{ kN/m}^2$$

For a single-acting, two-stroke engine:

$$n = \text{rev/s} = 2.2$$

For a six-cylinder engine:

$$\text{Indicated power} = p_m LAn \times 6$$
$$= 1126 \times 0.7854 \times 0.635^2 \times 1.01 \times 2.2 \times 6$$
$$= 4754 \text{ kW} \quad \text{Ans.}$$

Brake Power and Mechanical Efficiency

Power is required to overcome the engine's internal frictional resistances at the various rubbing surfaces of the engine. This could be at the piston rings, the crosshead slide and bearing, the crank and crankshaft bearings; therefore, only part of the *indicated power* (ip) developed in the cylinders is transmitted as useful power at the engine output shaft. The power absorbed in overcoming friction is termed the *friction power* (fp). The power available at the shaft is termed *shaft power* (sp) or, as this is measured (at the time of manufacture) by means of a brake, it is also called *brake power* (bp).

Brake power = Indicated power – Friction power

The *mechanical efficiency* is the ratio of the brake power to the indicated power:

$$\text{Mechanical efficiency} = \frac{\text{Brake power}}{\text{Indicated power}}$$

Since the brake power is always less than the indicated power, the above expresses the mechanical efficiency as a fraction less than unity. It is common practice to state the efficiency as a percentage, by multiplying the fraction by 100.

Brake power is measured by applying a resisting torque as a brake on the shaft, the heat generated by the friction at the brake being transferred to and carried away by circulating water.

Let F = resisting force of brake, in newtons, applied at a radius of R metres when the rotational speed is in revolutions per second. Then:

$$\text{Work absorbed per revolution} \left[\text{Nm} = \text{J} \right]$$
$$= \text{Force} \left[\text{N} \right] \times \text{Circumference} \left[\text{m} \right]$$
$$= F \times 2\pi R$$

$$\text{Work absorbed per second} \left[\text{J/s} \right] = \text{Power absorbed} \left[\text{W} \right]$$
$$= F \times 2\pi R \times \text{rev/s}$$
$$F \times R = \text{Torque in Nm} = T$$
$$\text{i.e. Brake power} = T \times 2\pi \times \text{rev/s}$$
$$2\pi \times \text{rev/s} = \text{Angular velocity in radians/second}$$
$$= \omega$$
$$\text{i.e. Brake power} = T\omega$$

Common types of brakes for measuring brake power are, for small engines, a loaded rope or steel band around a flywheel on the shaft and, for large engines, a hydraulic dynamometer.

Figure 7.7 illustrates a simple rope brake. The rope passes over the flywheel with one end of the rope anchored to the engine base and the other end hanging freely and loaded against the direction of rotation of the flywheel, the amount of loading depending upon the desired speed.

If w = weight of the load in newtons, and S = reading of spring balance in newtons, then the effective tangential braking force on the flywheel rim is $(W - S)$ newtons. If R = effective radius in metres from centre of shaft to centre of rope, then the braking torque in newton-metres is:

$$T = (W - S) \times R$$

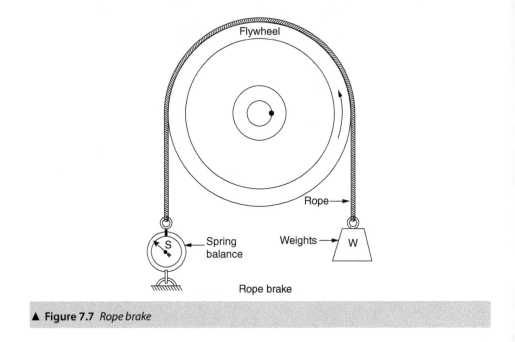

Flywheel

Rope→

Spring balance

Weights → W

S

Rope brake

▲ Figure 7.7 *Rope brake*

Example 7.3. In a single-cylinder, four-stroke, single-acting gas engine, the cylinder diameter is 180 mm and the stroke 350 mm. When running at 250 rev/min the mean area of the indicator diagrams taken off the engine is 355 mm², length of diagram 75 mm, scale of the indicator spring 90 kN/m² per mm and the number of explosions was counted to be 114/min. Calculate (i) the indicated power. If the effective radius of the rope brake on the flywheel is 600 mm, load on free end of rope 425 N and reading of spring balance 72 N, calculate (ii) the brake power and (iii) the mechanical efficiency.

Indicated mean effective pressure

$$= \text{Mean height of diagram} \times \text{Spring scale}$$

$$= \frac{\text{Area of diagram}}{\text{Length of diagram}} \times \text{Spring scale}$$

$$= \frac{355}{75} \times 90 = 426 \text{ kN/m}^2$$

Indicated power $= p_m LAn$

$$= 426 \times 0.7854 \times 0.18^2 \times 0.35 \times \frac{114}{60}$$

$$= 7.21 \text{ kW} \quad \text{Ans. (i)}$$

$$\text{Braking torque} = (W - S) \times R$$
$$= (425 - 72) \times 0.6 = 211.8 \text{ Nm}$$
$$\text{Brake power} = T\omega$$
$$= 211.8 \times \frac{250 \times 2\pi}{60}$$
$$= 5546 \text{ W} = 5.546 \text{ kW} \quad \text{Ans.(ii)}$$

$$\text{Mech. efficiency} = \frac{\text{Brake power}}{\text{Indicated power}}$$
$$= \frac{5.546}{7.21}$$
$$= 0.7693 \text{ or } 76.93\% \quad \text{Ans.(iii)}$$

Fault Finding – Morse Test

In multi-cylinder IC engines where all the cylinders are of the same cubic capacity, a reasonable estimate of the indicated power developed in each cylinder can be made by measuring the full output of the engine and then 'cutting out' each cylinder in turn and measuring the reduction. This is known as the *Morse test*. The test is most useful in small high-speed engines where indicator diagrams cannot be taken satisfactorily by the standard mechanical indicator.

The test consists of measuring the brake power at the shaft when all cylinders are firing and then measuring the brake power of the remaining cylinders when each one is 'cut out' in turn. Cutting out the power of each cylinder is done in petrol engines by shorting the sparking plug, and in diesel engines by by-passing the cylinder fuel supply. The speed of the engine and the petrol throttle or fuel pump setting is kept constant during the test so that friction and pumping losses are approximately constant. Accomplishing this task on modern engines might prove complicated and could be tackled by, for example, asking the electronic control unit (ECU) to cut out each cylinder in turn. This, of course, would require specialist equipment such as a 'diagnostic computer' that could interface with the ECU.

Taking a four-cylinder engine as an example:

With all four cylinders working,

Total bp = Total ip – Total fp

= Sum of ips of the four cylinders – Sum of the fps of the four cylinders

When the power of one cylinder is cut out,

Total bp = Sum of ips of the three cylinders − Sum of the fps of the four cylinders

Hence, we can see from the above that when one cylinder is cut out, the loss of *brake* power at the shaft is the loss of the *indicated* power of that cylinder which is not firing.

Example 7.4. During a Morse test on a four-cylinder, four-stroke petrol engine, the throttle was set in a fixed position and the speed maintained constant at 35 rev/s by adjusting the brake, and the following powers in kW were measured at the brake,

With all cylinders working, bp developed = 57

With sparking plug of no. 1 cyl shorted, bp = 38.5

. . .no. 2 cyl . . .bp = 37

. . .no. 3 cyl . . .bp = 37.5

. . .no. 4 cyl . . .bp = 38

Estimate the indicated power of the engine and the mechanical efficiency.

Ip of no. 1 cyl = 57 − 38.5 = 18.5

. . .no. 2 cyl = 57 − 37 = 20

. . .no. 3 cyl = 57 − 37.5 = 19.5

$$\text{Mech. efficiency} = \frac{\text{Brake power}}{\text{Indicated power}}$$

$$= \frac{5.546}{7.21}$$

Thermal Efficiency

When ships have just been built it is usual to test all the systems and machinery at 'sea trials' to establish an 'as built' set of performance data. During these trials it is usually most convenient to base the engine performance calculations on a running time of 1 h. Also, to enable comparisons to be made on the quantity of fuel oil to run the engine under different conditions, or comparisons of one engine with another, the fuel consumption is expressed per unit power developed.

The fuel consumed in unit time per unit power developed is termed the *specific fuel consumption* and commonly stated in the units kilograms of fuel per kilowatt-hour [kg/kWh].

The thermal efficiency of an engine is the relationship between the quantity of heat energy converted into work and the quantity of heat energy supplied to the engine in the fuel:

$$\text{Thermal efficiency} = \frac{\text{Heat energy converted into work}}{\text{Heat energy supplied}}$$

In IC engines the heat is supplied directly into the cylinders by the burning of the injected fuel. The heat energy given off during complete combustion of unit mass of the fuel is termed the *calorific value* and may be expressed in kilojoules of heat energy given off during the burning of 1 kg of fuel, giving a value in [kJ/kg]. However, the calorific value of fuel oil ranges from about 40 000 to 44 000 kJ/kg and therefore the more usual expression is in megajoules per kilogram [MJ/kg], that is, 40 to 44 MJ/kg.

Hence the heat supplied in megajoules is the product of the mass of fuel burned in kilograms and its calorific value in megajoules per kilogram. Therefore, on a basis of 1 kWh:

$$\text{Thermal effic.} = \frac{\text{Heat energy equivalent of 1 kWh [MJ/kWh]}}{\text{Spec. fuel cons. [kg/kWh]} \times \text{cal. value [MJ/kg]}}$$

The heat energy equivalent of 1 kWh is:

$$
\begin{aligned}
\text{Energy} &= \text{Power} \times \text{Time} \\
&= 1000\,[\text{W}] \times 3600\,[\text{s}] \\
&= 3.6 \times 10^6\,\text{J or } 3.6 \times 10^3\,\text{kJ or } 3.6\,\text{MJ/kg}
\end{aligned}
$$

Thermal efficiency may be based on the heat energy supplied to develop 1 kW of indicated power in the cylinders, or the heat energy supplied to obtain 1 kW of brake power at the shaft. In the former, the specific fuel consumption (indicated) is expressed as the kilogram of fuel per indicated kilowatt-hour [kg/ind. kWh] and the efficiency is the *indicated thermal efficiency*. In the latter, the specific fuel consumption (brake) is expressed as the kilogram of fuel per brake kilowatt-hour [kg/brake kWh] and the efficiency is the *brake thermal efficiency*.

Thus, on the basis of 1 kWh:

$$\text{Indicated thermal effic.} = \frac{3.6 \ [\text{MJ/kW h}]}{\text{kg fuel/ind. kW h} \times \text{cal. value [MJ/kg]}}$$

$$\text{Brake thermal effic.} = \frac{3.6 \ [\text{MJ/kW h}]}{\text{kg fuel/brake. kW h} \times \text{cal. value [MJ/kg]}}$$

The brake thermal efficiency is also the indicated thermal efficiency multiplied by the mechanical efficiency. The symbol to represent efficiency is η.

Example 7.5. The following data were taken during a 1-h trial run on a single-cylinder, single-acting, four-stroke diesel engine of cylinder diameter 175 mm and stroke 225 mm, the speed being constant at 1000 rev/min:

Indicated mean effective pressure = 5.5 bar

Effective diameter of rope brake = 1066 mm

Load on brake = 400 N

Reading of spring balance = 27 N

Fuel consumed = 5.7 kg

Calorific value of fuel = 44.2 MJ/kg

Calculate the indicated power, brake power, specific fuel consumption per indicated kWh and per brake kWh, mechanical efficiency, indicated thermal efficiency and brake thermal efficiency.

$$ip = p_m LAn$$

$$= 5.5 \times 10^2 \times 0.7854 \times 0.175^2 \times 0.225 \times \frac{1000}{60 \times 2}$$

$$= 24.8 \text{ kW} \quad \text{Ans.} \quad (\text{i})$$

$$bp = T\omega$$

$$= (400 - 27) \times \frac{1.066}{2} \times \frac{1000 \times 2\pi}{60}$$

$$= 2.082 \times 10^4 \text{ W} = 20.82 \text{ kW} \quad \text{Ans.} \quad (\text{ii})$$

Spec. fuel cons (indicated)

$$= \frac{5.7}{24.8} = 0.2298 \text{ kg/ind. kW h} \quad \text{Ans.} \quad \text{(iii)}$$

Spec. fuel cons (brake)

$$= \frac{5.7}{20.82} = 0.2738 \text{ kg/brake kW h} \quad \text{Ans.} \quad \text{(iv)}$$

$$\text{Mech. effic.} = \frac{\text{Brake power}}{\text{Indicated power}}$$
$$= \frac{20.82}{24.8}$$
$$= 0.8395 \text{ or } 83.95\% \quad \text{Ans.} \left(\text{v} \right)$$

$$\text{Ind. therm. effic.} = \frac{3.6 \text{ [MJ/kWh]}}{\text{kg fuel/ind. kW h} \times \text{cal. value [MJ/kg]}}$$
$$= \frac{3.6}{0.2298 \times 44.2}$$
$$= 0.3544 \text{ or } 35.44\% \quad \text{Ans.} \left(\text{vi} \right)$$

$$\text{Brake therm. effic.} = \frac{3.6 \text{ [MJ/kWh]}}{\text{kg fuel/brake kW h} \times \text{cal. value [MJ/kg]}}$$
$$= \frac{3.6}{0.2738 \times 44.2}$$
$$= 0.2975 \text{ or } 29.75\% \quad \text{Ans.} \left(\text{vii} \right)$$

Alternatively, Brake therm. effic. = Ind. therm. effic. × Mech. effic.

$$= 0.3544 \times 0.8395 = 0.2975 \text{ (as above).}$$

Heat Balance

Students will now be able to see that of the total heat energy supplied to an engine, only a proportion is converted directly into useful work (38% to 40% at best). The heaviest losses are those due to the heat energy transferred to and carried away by the cooling water, and the heat energy remaining in the gas which is released from the cylinders and into the exhaust trunking. A clear picture of the distribution of heat is shown by constructing a heat balance chart, based on taking the heat supplied by burning the fuel as 100%.

The International Maritime Organization (IMO) requires ships to be designed to be the most fuel efficient as possible. The measure of this fuel efficiency is called the Energy Efficiency Design Index (EEDI). One of the most obvious places to start this process is with the design of the machinery plant. The direct conversion of heat from the combustion of fuel into useful work is limited by the physical constraints of IC engine design. For example, some of the heat will naturally transfer to the cylinder walls of the engine and be taken up by the cooling water. It is just not possible with current technology to stop this happening.

In addition, given the time constraints of each cycle, some of the heat will be retained in the exhaust gas. Therefore, the 'waste heat' left over from the initial combustion process will need to be 'recovered' as much as possible to improve the EEDI rating.

A simple heat balance is shown in Figure 7.8. There are some factors not considered in drawing up this balance, but as a first analysis this serves to give a useful indication of the heat distribution for the IC engine. The high thermal efficiency and low fuel consumption obtained by modern diesel engines are superior to any other form of ship's propulsion in use at present. However, the losses and efficiency gains are due to the following features:

1. The development of waste heat recovery systems gives the marine plant an efficiency gain as this is heat that would otherwise be lost to the environment.

2. The recent efficiency increases of exhaust gas driven turbo-chargers not only contribute to high mechanical efficiency, by taking no mechanical power from the engine, but they take a smaller percentage of the exhaust gas output to drive the charge air compressor. This means that more gas is left over to drive turbo-generators, exhaust gas boilers and other waste heat recovery systems.

3. Cooling loss includes an element of heat energy due to generated friction.

4. Propellers do not usually have propulsive efficiencies exceeding 70% which reduces brake power according to the output power.

5. In the previous remarks no account has been taken of the increasing common practice of using a recovery system for heat normally lost in coolant systems.

Analysis of the simplified heat balance shown in Figure 7.8 reveals two important observations.

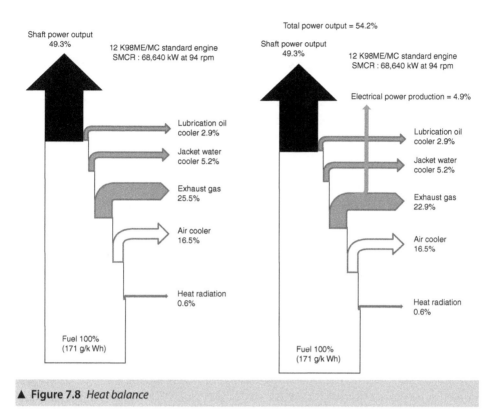

▲ **Figure 7.8** *Heat balance*

1. The difference between the indicated power and brake power is not only the power absorbed by the friction losses as some power is required to drive engine components such as camshafts, pumps, etc., which means a reduced potential for brake power.

2. Friction also results in heat generation which is dissipated by the various fluid cooling media, that is, oil and water, and hence the cooling analysis in a heat balance equation will include the frictional heat effect as an estimation.

Clearance and Stroke Volume

In the design of an engine a physical clearance is built between the lower face of the cylinder cover and the top of the piston, when the piston is at the top of its stroke (TDC). This gap maintains the correct compression ratio (CR) and avoids mechanical contact. The distance is measured and expressed as the distance between those two parts. In reciprocating air compressors this clearance is called the *bump clearance*.

The *clearance volume* is the volume of the enclosed space above the piston, when it is at the top of its stroke. This includes all cavities up to the valve faces when the valves are closed. In IC engines the clearance volume is also the space required to accommodate sufficient air for the complete combustion of the fuel and to limit the rise of temperature during burning. It used to be designed as near a spherical space as practicable, in many cases by concave piston tops and concave cylinder covers. Modern engine combustion space design now concentrates on promoting a swirl and piston crown that tends to have a raised section in the centre.

The *stroke volume*, or *swept volume*, is the volume swept out by the piston as it moves through one complete stroke. It is, therefore, equal to the product of the cross-sectional area of the cylinder and the length of the stroke.

Since the ratio of compression is the ratio of the volume at the beginning of compression to the volume at the end of compression, then, referring to Figure 7.9:

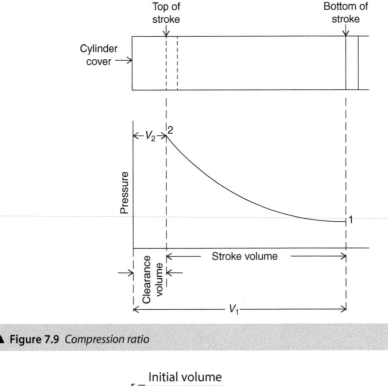

▲ **Figure 7.9** *Compression ratio*

$$r = \frac{\text{Initial volume}}{\text{Final volume}}$$
$$= \frac{V_1}{V_2} = \frac{\text{Clearance vol.} + \text{Stroke vol.}}{\text{Clearance vol.}}$$

Then, the magnitude of the clearance volume affects the ratio of compression (and expansion) and can be adjusted by shims under the foot of the connecting rod or plates in the clearance space. The next generation of variable speed engines will benefit from the development of variable compression ratio technology.

Dividing the stroke volume by the cross-sectional area of the cylinder gives the length of the stroke. Similarly, dividing the clearance volume by the cross-sectional area of the cylinder gives the clearance in terms of length. Hence, for convenience, if the stroke is expressed in millimetres, the clearance may be expressed as a length in millimetres.

Alternatively, the clearance may be expressed as 'a fraction of the stroke' or 'a percentage of the stroke', for instance, if the stroke is 200 mm and the clearance length is 20 mm, the clearance could be expressed as 'one-tenth of the stroke', or, '10% of the stroke'.

Example 7.6. The stroke of an engine is 450 mm. The pressure at the beginning of compression is 1.01 bar and at the end of compression it is 11.1 bar. Assuming compression follows the law $pV^{1.36} =$ a constant, calculate the clearance between the piston and cylinder cover at the end of compression in millimetres of length.

$$\text{Let clearance} = c$$
$$V_1 = \text{stroke} + \text{clearance} = 450 + c$$
$$V_2 = \text{clearance} = c$$
$$p_1 V_1^{1.36} = p_2 V_2^{1.36}$$
$$\left\{ \frac{V_1}{V_2} \right\}^{1.36} = \frac{p_2}{p_1}$$
$$\frac{V_1}{V_2} = {}^{1.36}\sqrt{\frac{p_2}{p_1}}$$
$$\frac{450 + c}{c} = {}^{1.36}\sqrt{\frac{11.1}{1.01}}$$
$$\frac{450 + c}{c} = 5.826$$
$$450 + c = 5.826c$$
$$450 = 4.826c$$
$$c = 93.26 \text{ mm} \quad \text{Ans.}$$

Test Examples 7

1. The area of an indicator diagram taken off a four-cylinder, single-acting, four-stroke, internal combustion engine when running at 5.5 rev/s is 390 mm², the length is 70 mm and the scale of the indicator spring is 1 mm = 1.6 bar. The diameter of the cylinders is 150 mm and the stroke is 200 mm. Calculate the indicated power of the engine assuming all cylinders develop equal power.

2. The cylinder diameters of an eight-cylinder, single-acting, four-stroke diesel engine are 750 mm and the stroke is 1125 mm. The indicated mean effective pressure in the cylinders is 1172 kN/m² when the engine is running at 110 rev/min. Calculate the indicated power and the brake power if the mechanical efficiency is 86%.

3. Calculate the cylinder diameters and stroke of a six-cylinder, single-acting, two-stroke diesel engine to develop a brake power of 2250 kW at a rotational speed of 2 rev/s when the indicated mean effective pressure in each cylinder is 10 bar. Assume a mechanical efficiency of 84% and the length of the stroke 25% greater than the diameter of the cylinders.

4. The flywheel of a rope brake is 1.22 m diameter and the rope is 24 mm diameter. When the engine is running at 250 rev/min the load on the brake is 480 N on one end of the rope and 84 N on the other end. Calculate the brake power. If the rise in temperature of the brake cooling water is 18 K, calculate the quantity of water flowing through the brake in litres per hour assuming that the water carries away 90% of the heat generated at the brake. Take the specific heat of the water to be 4.2 kJ/kg K.

5. During a Morse test on a four-cylinder petrol engine, the speed was kept constant at 24.5 rev/s by adjusting the brake and the following readings were taken:

   With all cylinders firing, torque at brake = 193.8 Nm

   . . . no. 1 cyl cut out = 130.8 . . .

   . . . no. 2 cyl cut out = 130.2 . . .

   . . .. no. 3 cyl cut out = 129.9 . . .

   . . . no. 4 cyl cut out = 131.1 . . .

   Calculate the bp, ip and mechanical efficiency.

6. When developing a certain power, the specific fuel consumption of an internal combustion engine is 0.255 kg/kWh (brake) and the mechanical efficiency is 86%. Calculate (i) the indicated thermal efficiency and (ii) the brake thermal efficiency, taking the calorific value of the fuel as 43.5 MJ/kg. If 35 kg of air is supplied per

kilogram of fuel, the air inlet being at 26°C and exhaust at 393°C, find (iii) the heat energy carried away in the exhaust gases as a percentage of the heat supplied, taking the specific heat of the gases to be 1.005 kJ/kg K.

7. A diesel engine uses 27 tonne of fuel per day when developing 4960 kW indicated power and 4060 kW brake power. Of the total heat supplied to the engine, 31.7% is carried away by the cooling water and radiation, and 30.8% in the exhaust gases. Calculate the indicated thermal efficiency, mechanical efficiency, overall efficiency, specific fuel consumption (indicated) and the calorific value of the fuel.

8. The stroke of a petrol engine is 87.5 mm and the clearance is equal to 12.5 mm. A compression plate is now fitted which has the effect of reducing the clearance to 10 mm. Assuming the compression period to be the whole stroke, the pressure at the beginning of compression as 0.97 bar and the law of compression $pV^{1.35} = C$, calculate the pressure at the end of compression before and after the compression plate is fitted.

9. The mean effective pressure measured from the indicator diagram taken off a single-cylinder, four-stroke gas engine was 3.93 bar when running at 5 rev/s and developing a brake power of 4.33 kW. The number of explosions per minute was 123 and the gas consumption 3.1 m³/h. The diameter of the cylinder is 180 mm, stroke 300 mm and calorific value of the gas 17.6 MJ/m³. Calculate the indicated power, mechanical efficiency, indicated thermal efficiency and brake thermal efficiency.

10. A two-stroke cycle compression–ignition engine has a stroke of 1.5 m, mean piston speed 6 m/s and brake mean effective pressure of 7 bar. A similar engine with the same stroke/bore rate of 2:1 is to develop 370 kW. Both engines have the same power-to-swept-volume ratio and operate at the same speed. For the engine which is to develop 370 kW, calculate:

   (i)   the cylinder bore;

   (ii) brake mean effective pressure.

8

IDEAL CYCLES

Theoretical cycles are reversible with either isentropic (frictionless adiabatic) or isothermal processes and they can be applied to both gas and vapour power cycles. The efficiency of such cycles is called the ideal cycle (thermal) efficiency. By introducing process efficiencies, for example, using polytropics (considering the small scale or micro environment) to modify the isentropic (end to end of the process) of actual compression/expansion. Using the final isentropic efficiencies, it is then possible to estimate the actual cycle efficiencies. The ratio of actual-to-ideal cycle efficiency is called the efficiency ratio.

Consider now gas power cycles or systems which can be classified into two main groups. The internal combustion (IC) reciprocating engines use non-flow systems and gas turbines use steady-flow.

Performance can be assessed using air as the working fluid, that is, air standard cycles and air standard efficiency.

In any cycle:

$$\text{Heat converted into work} = \text{Heat supplied} - \text{Heat rejected}$$

And therefore:

$$
\begin{aligned}
\text{Thermal efficiency} &= \frac{\text{Heat converted into work}}{\text{Heat supplied}} \\
&= \frac{\text{Heat supplied} - \text{Heat rejected}}{\text{Heat supplied}} \\
&= 1 - \frac{\text{Heat rejected}}{\text{Heat supplied}}
\end{aligned}
$$

Note: The heat converted into work, that is, work done, is represented by the area of the pV diagram and mean effective pressure (m.e.p.) can be calculated by dividing by the length equivalent – using correct units, that is, kN/m^2 and m^3 (see page 103).

Constant Volume Cycle

This is also known as the Otto cycle and is the basis on which petrol, paraffin and gas engines usually work. Designating in sequence, the four cardinal state points of the cycle as 1, 2, 3 and 4, respectively (Figure 8.1), the cycle of operations commences with a volume of air V_1 at pressure p_1 and temperature T_1. The piston moves inward and the air is compressed adiabatically to a volume V_2 and the pressure and temperature rise to p_2 and T_2, respectively.

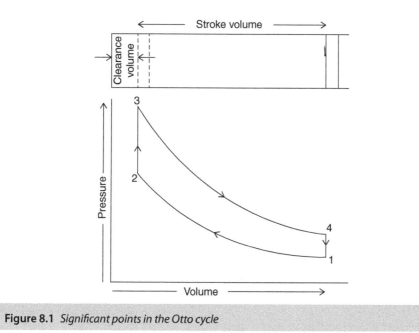

▲ **Figure 8.1** *Significant points in the Otto cycle*

Heat energy is now given from some outside source, and it is assumed that the air receives this heat instantaneously and so there is no time for any change of volume to occur. The pressure and temperature consequently rise to p_3 and T_3 while the volume remains unchanged and, therefore, V_3 is equal to V_2. Adiabatic expansion of the air now takes place while the piston is pushed outward on its power stroke, the volume increasing to V_4, which is the same as the initial volume V_1, and the pressure and temperature during expansion falling to p_4 and T_4. Finally, the cycle is completed by the air rejecting heat (theoretically instantaneously) at constant volume, to an outside source, which causes the pressure and temperature to fall to their initial values of p_1 and T_1.

It should be noted that, since the compression and expansion of the air is adiabatic, then there is no exchange of heat during these operations. This means that all the heat supplied takes place at constant volume between the state points 2 and 3, and all the heat rejected takes place at constant volume between the state points 4 and 1.

$$\text{Heat supplied or rejected} = \text{Mass} \times \text{spec. ht.} \times \text{Temp. change}$$

Hence:

$$\text{Heat supplied} = m \times c_V \times (T_3 - T_2)$$
$$\text{Heat rejected} = m \times c_V \times (T_4 - T_1)$$

Therefore:

$$\text{Ideal thermal efficiency} = 1 - \frac{\text{Heat rejected}}{\text{Heat supplied}}$$
$$= 1 - \frac{mc_V \left(T_4 - T_1\right)}{mc_V \left(T_3 - T_2\right)}$$
$$= 1 - \frac{T_4 - T_1}{T_3 - T_2}$$

In this case the ratios of compression and expansion are the same because:

$$V_1 = V_4 \quad \text{and} \quad V_2 = V_3$$

$$\text{Also, since } \frac{T_2}{T_1} = \left\{\frac{V_1}{V_2}\right\}^{\gamma-1} = r^{\gamma-1}$$

$$\text{and } \frac{T_3}{T_4} = \left\{\frac{V_4}{V_3}\right\}^{\gamma-1} = r^{\gamma-1}$$

$$\text{then } \frac{T_2}{T_1} = \frac{T_3}{T_4} \quad = r^{\gamma-1}$$

$$\text{hence, } T_3 = T_4 r^{\gamma-1} \quad \text{and} \quad T_2 = T_1 r^{\gamma-1}$$

$$\text{therefore, } T_3 - T_2 = r^{\gamma-1}\left(T_4 - T_1\right)$$

Substituting this value of $(T_3 - T_2)$ into the general expression for the ideal thermal efficiency, $(T_4 - T_1)$ cancels, leaving:

$$\text{Ideal thermal efficiency} = 1 - \frac{1}{r^{\gamma-1}}$$

If y is taken as 14 (for air) then this is also the Air Standard Efficiency.

$$\text{Also, since } r^{\gamma-1} = \frac{T_2}{T_1} = \frac{T_3}{T_4}$$

$$\text{then, ideal thermal efficiency} = 1 - \frac{T_1}{T_2}$$

$$= 1 - \frac{T_4}{T_3}$$

On examination of the previous expression it will be seen that the greater the value of r, the greater will be the efficiency, hence the trend for higher ratios of compression in modern petrol engines. The majority of petrol engines, however, take in a mixture of petrol vapour and air during the induction stroke and this is compressed during the compression stroke. Being an explosive mixture it will burst into flame without the assistance of an electric spark or other means if it reaches its auto ignition temperature. Therefore, if the ratio of compression is too high for the grade of petrol used, pre-ignition can take place.

Figure 8.2 shows the relationship between the ideal thermal efficiency and the ratio of compression in a constant volume cycle. From the graph we see that, although the efficiency increases with higher compression ratios, the rate of increase in efficiency becomes less as the compression ratio is increased, and there is no appreciable gain by increasing the compression ratio above about 16:1 (practical limitations to about 10:1).

Example 8.1. The compression ratio of an engine working on the constant volume cycle is 9.3:1. At the beginning of compression the temperature is 31°C and at the end of combustion the temperature is 1205°C. Taking compression and expansion to be adiabatic and the value of y as 1.4, calculate (i) the temperature at the end of compression, (ii) the temperature at the end of expansion and (iii) the ideal thermal efficiency.

Referring to Figure 8.2,

$V_1 = 9.3$ and $V_4 = 9.3$

$V_2 = 1$ and $V_3 = 1$

$T_1 = 304 \text{ K}$

$T_3 = 1478 \text{ K}$

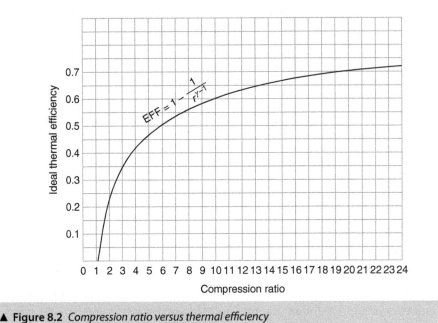

▲ **Figure 8.2** *Compression ratio versus thermal efficiency*

Compression period:

$$\frac{T_2}{T_1} = \left\{\frac{V_1}{V_2}\right\}^{\gamma-1}$$

$$\therefore\ T_2 = 304 \times 9.3^{0.4} = 741.8\ \text{K}$$

∴ Temperature at end of compression

$$= 468.8°\text{C} \quad \text{Ans.} \quad (\text{i})$$

Expansion period:

$$\frac{T_3}{T_4} = \left\{\frac{V_4}{V_3}\right\}^{\gamma-1}$$

from which T_4 can be calculated, but, as ratios of compression and expansion are the same, and follow the same law, then an easier method is:

$$\frac{T_2}{T_1} = \frac{T_3}{T_4}$$

$$T_4 = \frac{304 \times 1478}{741.8} = 605.6\ \text{K}$$

∴ Temperature at end of expansion

$$= 605.6 - 273 = 332.6°\text{C} \quad \text{Ans.} \quad (\text{ii})$$

The ideal thermal efficiency can now be calculated from any of the expressions given above, thus:

$$\text{Ideal thermal efficiency} = 1 - \frac{T_4 - T_1}{T_3 - T_2} \quad \text{or} \quad 1 - \frac{1}{r^{\gamma - 1}}$$

$$\text{or} \quad 1 - \frac{T_1}{T_2} \quad \text{or} \quad 1 - \frac{T_4}{T_3}$$

Taking the last expression,

$$\text{Ideal thermal efficiency} = 1 - \frac{605.6}{1478} = 1 - 0.4099$$

$$= 0.5901 \text{ or } 59.01\% \quad \text{Ans.} \quad \text{(iii)}$$

Diesel Cycle

The term *constant pressure cycle* refers to a cycle where the pressure remains constant during the two periods of heat energy supply and rejection. In the diesel cycle, however, heat is supplied at constant pressure, but rejection of heat takes place at constant volume. The diesel cycle is a *modified constant pressure cycle*, and a marine example of engines operating in a diesel cycle is the slow-speed crosshead main engine.

Referring to Figure 8.3, the cycle of operations commences with a volume of air V_1 at a pressure p_1 and temperature T_1. The air is compressed adiabatically to a volume V_2 and the pressure and temperature rise to p_2 and T_2. The piston is now at the top (inward end) of its stroke and heat is supplied at such a rate to maintain the pressure constant as the piston moves down the cylinder for a fraction of the power stroke. At the end of the heat supply period the volume is V_3, the temperature has been further increased to T_3, and the pressure represented by p_3 is the same as p_2. The air now expands adiabatically for the remainder of the power stroke until the final volume V_4 is the same as the initial volume V_1, the pressure and temperature falling during expansion to p_4 and T_4. Finally, the cycle is completed by the rejection of heat at constant volume to the initial conditions.

$$\text{Ideal thermal efficiency} = 1 - \frac{\text{Heat rejected}}{\text{Heat supplied}}$$

$$= 1 - \frac{mc_V(T_4 - T_1)}{mc_p(T_3 - T_2)}$$

$$= 1 - \frac{1}{\gamma}\left\{\frac{T_4 - T_1}{T_3 - T_2}\right\}$$

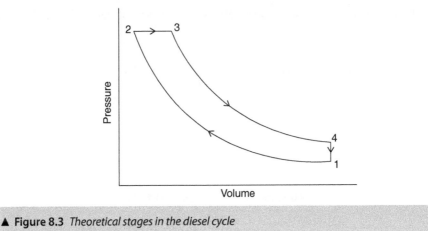

▲ **Figure 8.3** *Theoretical stages in the diesel cycle*

Example 8.2. The compression ratio in a diesel engine is 13:1 and the ratio of expansion is 6.5:1. At the beginning of compression the temperature is 32°C. Assuming adiabatic compression and expansion, calculate the temperatures at the three remaining cardinal points of the cycle, and the ideal thermal efficiency, taking the specific heats at constant pressure and constant volume as 1.005 and 0.718 kJ/kg K, respectively.

Referring to Figure 8.3:

$$V_1 = 13 \quad V_4 = 13 \quad V_2 = 1$$

$$\frac{V_4}{V_3} = \text{ratio of expansion} = 6.5$$

$$\therefore V_3 = \frac{V_4}{6.5} = \frac{13}{6.5} = 2$$

$$T_1 = 305 \text{ K}$$

$$\gamma = \frac{c_P}{c_V} = \frac{1.005}{0.718} = 1.4$$

First stage, adiabatic compression:

$$\frac{T_2}{T_1} = \left\{ \frac{V_1}{V_2} \right\}^{\gamma-1}$$

$$\therefore T_2 = 305 \times 13^{0.4} = 850.9 \text{ K}$$

∴ Temperature at end of compression

$$= 577.9°C \quad \text{Ans.} \quad (i)$$

Second stage, heating at constant pressure:

$$\frac{T_3}{T_2} = \frac{V_3}{V_2} \text{ (Charles' law)}$$

$$T_3 = 850.9 \times 2 = 1701.8 \text{ K}$$

∴ Temperature at end of combustion

$$= 1428.8°C \quad \text{Ans.} \quad (ii)$$

Third stage, adiabatic expansion:

$$\frac{T_4}{T_3} = \left\{\frac{V_3}{V_4}\right\}^{\gamma-1}$$

$$T_4 = 1701.8 \times \left\{\frac{2}{13}\right\}^{0.4}$$

$$= \frac{1701.8}{6.5^{0.4}} = 804.8 \text{ K}$$

∴ Temperature at end of expansion

$$= 531.8°C \quad \text{Ans.} (iii)$$

Ideal thermal efficiency $= 1 - \dfrac{\text{Heat rejected}}{\text{Heat supplied}}$

$$= 1 - \frac{1}{\gamma}\left\{\frac{T_4 - T_1}{T_3 - T_2}\right\}$$

$$= 1 - \frac{1}{1.4}\left\{\frac{804.8 - 305}{1701.8 - 850.9}\right\}$$

$$= 1 - \frac{1}{1.4} \times \frac{499.8}{850.9}$$

$$= 1 - 0.4196$$

$$= 0.5804 \text{ or } 58.045\% \quad \text{Ans.} \quad (iv)$$

In the ideal diesel cycle, where the compression and expansion are both adiabatic, the ideal thermal efficiency can be expressed in terms of the ratio of compression and a comparison can then be made with the efficiency of a constant volume cycle of the same ratio of compression.

Expressing all temperatures in terms of T_1, substituting and simplifying:

$$\frac{T_2}{T_1} = \left\{\frac{V_1}{V_2}\right\}^{\gamma-1}$$

$$\therefore T_2 = T_1 r^{\gamma-1}$$

$$\frac{T_3}{T_2} = \frac{V_3}{V_2}$$

Let this ratio of burning period volumes be represented by , then,

$$\frac{T_3}{T_2} = \rho$$

$$\therefore T_3 = T_2\rho = T_1 r^{\gamma-1}\rho$$

$$\frac{T_4}{T_3} = \left\{\frac{V_3}{V_4}\right\}^{\gamma-1}$$

Since $\dfrac{V_3}{V_2} = \rho$ and $\dfrac{V_4}{V_2} = r$, then $\dfrac{V_3}{V_4} = \dfrac{\rho}{r}$

$$\therefore \frac{T_4}{T_3} = \left\{\frac{\rho}{r}\right\}^{\gamma-1}$$

$$T_4 = T_3 \times \left\{\frac{\rho}{r}\right\}^{\gamma-1}$$

$$= T_1 r^{\gamma-1}\rho \times \left\{\frac{\rho}{r}\right\}^{\gamma-1}$$

$$= T_1 \rho^\gamma$$

Ideal thermal efficiency $= 1 - \dfrac{1}{\gamma}\left\{\dfrac{T_4 - T_1}{T_3 - T_2}\right\}$

$$= 1 - \frac{1}{\gamma}\left\{\frac{T_1\rho^\gamma - T_1}{T_1 r^{\gamma-1}\rho - T_1 r^{\gamma-1}}\right\}$$

$$= 1 - \frac{1}{\gamma} \times \frac{1}{r^{\gamma-1}}\left\{\frac{\rho^\gamma - 1}{\rho - 1}\right\}$$

If $\gamma = 1.4$, this is also the Air Standard Efficiency.

Comparing this expression with the ideal thermal efficiency of the constant volume cycle in terms of r, it will be seen that, for the same ratio of compression, the constant volume cycle has the higher thermal efficiency. This does not mean, however, that a petrol engine working on the constant volume cycle is more efficient than a diesel engine working on the modified constant pressure cycle, because, in the former, an explosive mixture is compressed and there is a limit to the ratio of compression, whereas air is only compressed in a diesel engine and the ratio of compression can be as high as required.

Dual Combustion Cycle

In most high-speed compression–ignition engines, combustion takes place partly at constant volume and partly at constant pressure, and therefore the cycle is referred to as *dual-combustion* (or mixed).

Figure 8.4 shows the ideal dual-combustion cycle. Commencing with a volume of air, V_1 at pressure $p1$ and temperature T_1 the air is compressed adiabatically to a volume V_2 and the pressure and temperature rise to p_2 and T_2, respectively. Heat energy is now supplied at constant volume, the pressure and temperature are increased to p_3 and T_3, respectively, while the volume remains unchanged so that V_3 is equal to V_2. The supply of heat energy is continued at such a rate as to maintain the pressure constant while the piston moves outward until the volume is V_4, the temperature is further increased to T_4 and the pressure p_4 remains the same as p_3. Now adiabatic expansion takes place until the volume V_5 is the same as the initial volume V_1, the pressure and temperature falling due to expansion to p_5 and T_5, respectively. Finally, heat is rejected at constant volume and the pressure and temperature fall to the initial conditions of p_1 and T_1.

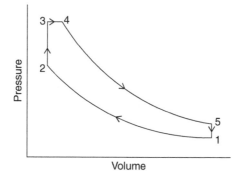

▲ **Figure 8.4** *The ideal dual-combustion cycle*

Ideal thermal efficiency $= 1 - \dfrac{\text{Heat rejected}}{\text{Heat supplied}}$

$$= 1 - \frac{mc_V(T_5 - T_1)}{mc_V(T_3 - T_2) + mc_p(T_4 - T_3)}$$

$$= 1 - \frac{(T_5 - T_1)}{(T_3 - T_2) + \gamma(T_4 - T_3)}$$

The above expression can be converted into terms of the ratio of compression in a similar manner as the previous cycles, thus:

If r = ratio of compression = V_1/V_2

Γ = ratio of specific heats = cp/cv

ρ = ratio of burning period, or 'cut-off' ratio = V_4/V_3

a = ratio of pressure increase at constant volume = p_3/p_2

Ideal thermal efficiency $= 1 - \dfrac{1}{r^{\gamma-1}} \left\{ \dfrac{a\rho^\gamma - 1}{(a - 1) + \gamma a(\rho - 1)} \right\}$

Note: (i) if $a = 1$, the above becomes a pure diesel cycle.

(ii) if $\rho = 1$, it becomes a pure constant–volume cycle.

Since compression–ignition oil engines depend upon the temperature of the air at the end of compression to ignite the fuel injected into the cylinder, the compression ratio must be fairly high, usually not less than about 14 to give the necessary temperature rise during compression.

The higher compression pressures developed in this type of engine limit the use of constant volume combustion, since the maximum pressure in the cycle is limited by the consideration of strength.

As the maximum pressure is limited, increasing the compression ratio reduces the amount of fuel burned at constant volume, so that more must be burned at constant pressure and thus the gain due to increased compression ratio is partly nullified.

Carnot Cycle

This is a purely theoretical cycle devised by the French scientist Sadi Carnot. Although it is not possible from practical considerations for an engine to work on this cycle, it has a higher theoretical thermal efficiency than any other working between the same temperature limits and, therefore, provides a useful standard for comparing the performance of other heat engines. Students will be able to determine that there is a direct correlation between the efficiency of a diesel engine and the level of CO_2 in the exhaust gas. Since the 1950s the efficiency of IC engines has been improving as measured by the specific fuel consumption and the level of CO_2 in the exhaust. The Carnot cycle is the theoretical maximum and according to some manufacturers modern engines are approaching that efficiency level. Therefore any further improvements in thermal efficiency of a marine power plant will need to come from waste heat recovery systems.

Referring to the P–V diagram, Figure 8.5, it is usual to explain this cycle by commencing at state point A. Gas has been previously compressed in the cylinder by the piston moving inward and, at A, the piston is at the 'top' of its stroke. The pressure and temperature are high, the value of the latter being represented by T_1. As the piston is pushed outward, doing work, heat is supplied to the gas from an external hot source at such a rate as to maintain its temperature constant, and during this period the gas therefore expands

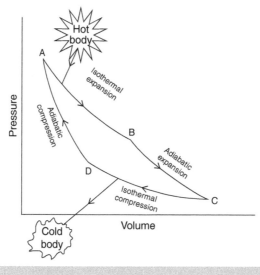

▲ **Figure 8.5** *The P–V diagram for the Carnot cycle*

isothermally until point B is reached. At this point the heat supply is cut off and no heat is given to or rejected from the gas as the piston moves on to the end of the stroke at C. Hence during this period, the gas expands adiabatically as it does work and therefore the temperature falls. The temperature of the gas at point C is represented by T_2. The piston now moves inward to compress the gas from C to D and during this period it is assumed that any generated heat due to compression can flow out of the gas into a cold 'sink'. That is, the gas rejects heat energy to a cold external source at such a rate to maintain the temperature constant at T_2. This is isothermal compression. At point D, the flow of heat out of the gas is stopped and from D to A the gas is compressed adiabatically while the piston completes its stroke, the temperature of the gas rising to the initial temperature T_1.

Hence, the four stages of the Carnot cycle are described briefly as follows:

A to B	Isothermal expansion of the gas during which the amount of heat supplied is equal to the work done. Letting r = ratio of isothermal expansion, heat supplied = $p_A V_A$ $\ln r = mRT_1 \ln r$.
B to C	Adiabatic expansion of the gas during which no heat is supplied or rejected.
C to D	Isothermal compression. During this period heat is rejected from the gas, the quantity of heat being the equivalent of the work done on the gas, and, since the ratio of isothermal compression must be the same as the ratio of isothermal expansion to form a closed cycle, then, heat rejected = $p_c V_c \ln r = mRT_2 \ln r$.
D to A	Adiabatic compression during which no heat is supplied or rejected.

Therefore:

$$\text{Ideal thermal efficiency} = \frac{\text{Heat supplied} - \text{Heat rejected}}{\text{Heat supplied}}$$

$$= 1 - \frac{\text{Heat rejected}}{\text{Heat supplied}}$$

$$= 1 - \frac{mRT_2 \ln r}{mRT_1 \ln r}$$

$$= 1 - \frac{T_2}{T_1}$$

$$= \frac{T_1 - T_2}{T_1}$$

This expression for the Carnot efficiency shows that, to obtain the highest efficiency, heat should be taken in at the highest possible temperature (T_1) and rejected at the lowest possible temperature (T_2). This conclusion is applicable in the design of any heat engine, but the actual achievement of this cycle is constrained by material science and the physical constructional constraints of current engine design.

Reversed Carnot Cycle

The Carnot cycle is theoretically reversible and if applied in reverse manner would act as a refrigerator by taking heat from a cold region and maintaining it at a low temperature, as shown in Figure 8.6.

Referring to Figure 8.6 and commencing at state point A, the four stages of the reversed Carnot cycle consist of:

(i)   Work done by the gas while it expands adiabatically from A to D and the temperature falls from T_1 to T_2. No heat is given to or taken from the gas during this process.

(ii) Further work done by the gas as it expands isothermally from D to C, a quantity of heat is taken in by the gas (from the cold body) equal to the work done, to maintain the temperature constant at T_2.

(iii) Adiabatic compression of the gas from C to B, no heat being given to or taken from the gas, therefore the temperature increases from T_2 to T_1.

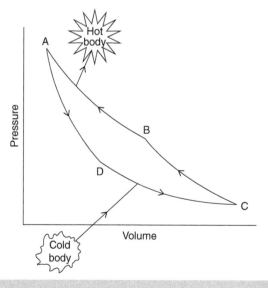

▲ **Figure 8.6** *The P–V diagram for the reversed Carnot cycle*

(iv) Isothermal compression from B to A during which heat is rejected from the gas (to the hot body) to maintain the temperature constant at T_1.

Thus an engine working on the reversed Carnot cycle would require to be driven and, as heat would be continually taken from a cold region and sent out to a hotter region, it would therefore act as a refrigerating machine. The measure of the 'efficiency' of refrigeration is known as the

$$\frac{\text{Quantity of heat extracted}}{\text{Heat equivalent of work done to extract the heat}}$$

$$= \frac{mRT_2 \ln r}{mRT_1 \ln r - mRT_2 \ln r}$$

$$= \frac{T_2}{T_1 - T_2}$$

Other Ideal Cycles

When considering vapour power cycles, for steam turbines, the Rankine cycle is used (see Chapter 12). With gas power cycles, for gas turbines, the Joule (Brayton) cycle is used (also see Chapter 12).

Operation on other cycles includes Stirling (constant volumes and isothermals), Ericsson (constant pressures and isothermals), Atkinson (modified constant volume; see Test Example 7) and theoretical variations (see Test Example 8).

Mean Effective Pressure

The ideal cycles can be analysed to calculate the indicated m.e.p. from the diagram (see also page 103). Consider, as example, the Diesel cycle and refer to Figure 8.3.

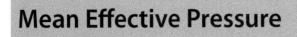

$$\text{m.e.p.} = \frac{p_2(V_3 - V_2) + \dfrac{p_3 V_3 - p_4 V_4}{\gamma - 1} - \dfrac{p_2 V_2 - p_1 V_1}{\gamma - 1}}{V_1 - V_2}$$

(The m.e.p. will be in kN/m^2 if the work done, i.e., the area of the diagram is in U, i.e., $kN/m^2 \times m^3$ and the 'length' of the diagram in m^3. Pressures can be in bars, and volumes represented by stroke lengths or ratios giving the m.e.p. in bars – the work done is not then, of course, in consistent units.)

Non-ideal Cycles

The ideal cycle is sometimes used as the reference model for the calculation of properties at cardinal points, evaluation of m.e.p., etc. using a diagram and polytropic processes (see Test Examples 9 and 10). With the focus on Marpol Annex VI and on emissions from diesel exhaust, other non-conventional cycles are being used more and more in engines to obtain better environmental acceptance.

Miller Cycle

This engine cycle cannot be used effectively unless an engine has full electronic control of both the fuel injection process and the ability to vary the valve operating timing. The reason for using this relatively inefficient combustion process is to reduce the peak temperature of the combustion process as this is responsible for over 90% of the thermal nitrous oxide (NOx) formation.

Therefore, manufacturers are using 'primary' combustion measures to dramatically reduce the peak temperatures in the combustion chamber without incurring fuel consumption penalties or, if possible, at improved fuel efficiency. To achieve this, a range of engine modifications have been used, including:

- further cooling of the charge air;
- improved re-entrant piston bowls;
- low swirl inlet ports;
- higher compression ratio CRs;
- higher fuel injection pressures and improved injector nozzle spray patterns;
- revised fuel injection timing; and
- a combination of revised 'Miller cycle' valve timing and high efficiency, high pressure turbo-charging, including in some cases two-stage turbo-charging.

The Miller cycle involves the early closure of the inlet valve, allowing the air entering the cylinder to expand and cool just before compression. This cooling action reduces the temperature peaks during combustion which is the major cause of thermal NOx production during combustion.

However, the shorter inlet valve opening period would normally mean a lower mass of combustion air entering the cylinder and hence reduced engine power and torque. To counter this effect, higher pressure turbo-charging ensures that an equal or – in the case of MAN Diesel's new technology package – greater – amount of air can enter the cylinder in the shorter time available. During trials using an intensive Miller cycle under full load conditions and turbo-charger pressure ratios of 6.5:7, MAN Diesel has recorded reductions in NOx of over 30%, reductions in fuel consumption as great as 8% and an increase in specific power output of 15%.

Test Examples 8

1. The compression ratio of a petrol engine working on the constant volume cycle is 8.5. The pressure and temperature at the beginning of compression are 1 bar and 40°C and the maximum pressure of the cycle is 31 bar. Taking compression for the air–petrol gas mixture to follow the adiabatic law $pV^{1.35} = C$, calculate (i) the pressure at the end of compression, (ii) temperature at end of compression and (iii) temperature at end of combustion.

2. In an ideal constant volume cycle the temperature at the beginning of compression is 50°C. The volumetric compression ratio is 5:1. If the heat supplied during the cycle is 930 kJ/kg of working fluid, calculate:

    (i) the maximum temperature attained in the cycle,

    (ii) work done during the cycle/kg of working fluid, and

    (iii) the ideal thermal efficiency of the cycle.

    Take $\gamma = 1.4$ and $c_v = 0.717$ kJ/kg K.

3. In an air-standard Otto cycle the pressure and temperature of the air at the start of compression are 1 bar and 330 K, respectively. The compression ratio is 8:1 and the energy added at constant volume is 1250 kJ/kg.

    (i) Calculate:

    (a) the maximum temperature in the cycle,

    (b) the maximum pressure in the cycle.

(i) Draw the cycle:

(a) on a pressure–volume diagram,

(b) on a temperature–entropy diagram.

For air: $c_p = 1005$ J/kg K; $c_v = 718$ J/kg K.

4. In an air-standard diesel cycle the pressure and temperature of the air at the start of compression are 1 bar and 330 K, respectively. The compression ratio is 16:1 and the energy added at constant pressure is 1250 kJ/kg.

Calculate:

(i) the maximum pressure in the cycle,

(ii) the maximum temperature in the cycle,

(iii) the cycle efficiency,

(iv) the mean effective pressure.

For air: $c_p = 1005$ J/kg K; $c_v = 718$ J/kg K

5. The compression ratio of an engine working on the dual-combustion cycle is 10.7. The pressure and temperature of the air at the beginning of compression is 1 bar and 32°C, respectively. The maximum pressure and temperature during the cycle is 41 bar and 1593°C, respectively. Assuming adiabatic compression and expansion, calculate the pressures and temperatures at the remaining cardinal points of the cycle and the ideal thermal efficiency. Take the values, $c_v = 0.718$, and $c_p = 1\,005$ kJ/kg K.

6. A heat engine is to be operated, using the Carnot cycle, with maximum and minimum temperatures of 1027°C and 27°C, respectively.

(i) Calculate the efficiency of the cycle,

(ii) The working fluid within the cycle is to be steam,

(a) state the difficulty associated with the practical operation of the cycle,

(b) state the modification required to the cycle to allow practical operation.

7. Gas initially at a pressure of 1 bar and temperature 60°C undergoes the following cycle:

(a) adiabatic compression through a compression ratio of 4.5:1;

(b) heating at constant volume through a pressure ratio of 1.35:1;

(c) adiabatic expansion to initial pressure;

(d) constant pressure cooling to initial volume.

If $c_p = 1000$ J/kg K and $c_v = 678$ J/kg K for the gas, determine the thermal efficiency of the cycle.

8. 1 kg of air at a pressure and temperature of 1 bar and 15°C, respectively, initially undergoes the following processes in a cycle: (1.) Isothermal compression to 2 bar. (2.) Polytropic compression from 2 bar to 4 bar. (3.) Isentropic expansion from 4 bar to initial condition.

   Sketch the pV diagram of the cycle and calculate for each process (i) the work transfer (ii) the heat transfer. $R = 287$ J/kg K and $y = 1.4$ for air.

9. The compression ratio of a diesel engine is 15:1. Fuel is admitted for one-tenth of the power stroke and combustion takes place at constant pressure. Exhaust commences when the piston has travelled nine-tenths of the stroke. At the beginning of compression the temperature of the air is 41°C. Assuming compression and expansion to follow the law $pVn = C$ where $n = 1.34$, calculate the temperatures at the (i) end of compression, (ii) end of combustion and (iii) beginning of exhaust.

10. A two-stroke, single cylinder engine, operating on 'diesel cycle', has a volumetric compression ratio of 12:1 and a stroke volume of 0.034 m³. The pressure and temperature at the start of compression are 1 bar and 80°C, respectively, while the maximum temperature at the end of heat reception is 1650°C.

    If the compression is according to the law $pV^{1.36} = $ constant and the expansion follows the law $pV^{1.4} = $ constant, calculate:

    (i) the indicated mean effective pressure (m.e.p.) for the cycle,

    (ii) the power developed at 200 cycles/min.

RECIPROCATING
AIR COMPRESSORS

Compressed air, at various different pressures, is used for many purposes on-board ships. These uses could be, for example, scavenging, supercharging or turbo-charging, starting diesel engines, and as the operating fluid for many automatic control systems or hand tools. Air compressors producing medium and high pressures are usually of the reciprocating type and may be single- or multi-stage. Rotary types are common for large quantities of air at low pressures.

Figure 9.1 shows diagrammatically a single-stage, single-acting reciprocating compressor, and its P–V diagram illustrating the cycle. *Note*: the inlet and outlet valves operate when there is a pressure differential across them. This makes for a very simple and lightweight arrangement.

Commencing at point A, the cycle of operations is as follows: A to B, compression period; with all valves closed the piston moves inward and the air which was previously drawn into the cylinder is compressed. Compression continues until the air pressure is sufficiently high to force the discharge valves open against their pre-set compression springs. Thus, at point B, the discharge valves open, and the compressed air is discharged at constant pressure for the remainder of the inward stroke, that is, from B to C. At point C the piston has completed its inward stroke and changes direction to move outward. Immediately after the piston begins to move back there is a drop in pressure of the compressed air left in the clearance space, the discharge valves close and from C to D this air expands. At point D the pressure has fallen to less than the atmospheric pressure and the lightly sprung suction valves are opened by the greater pressure of the atmospheric air. Air is drawn into the cylinder for the remainder of the outward stroke, that is, from D to A.

Example 9.1. The stroke of the piston of an air compressor is 250 mm (see Figure 9.2) and the clearance volume is equal to 6% of the stroke volume. The pressure of the air

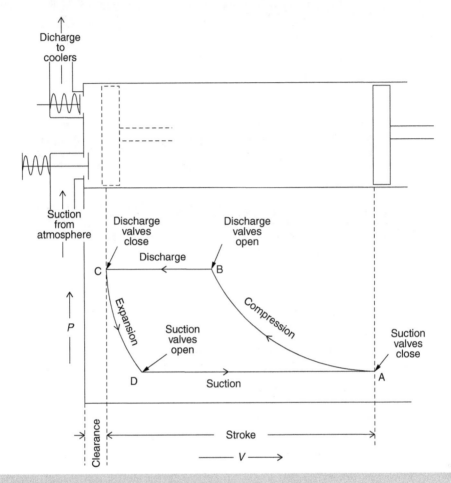

▲ **Figure 9.1** *Mechanical arrangement and P–V diagram for reciprocating compressor*

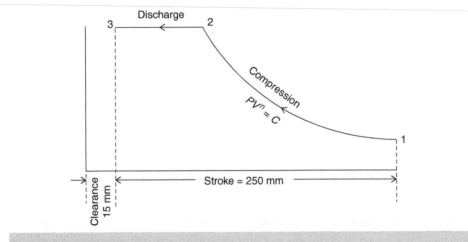

▲ **Figure 9.2** *P–V diagram for the air compressor in Example 9.1*

at the beginning of compression is 0.98 bar and it is discharged at 3.8 bar. Assuming compression to follow the law $pV^n = $ constant, where $n = 1.25$, calculate the distance moved by the piston from the beginning of its pressure stroke before the discharge valves open, and express this as a percentage of the stroke.

$$\text{Clearance length} = 6\% \text{ of } 250 = 15$$

$$p_1 = 0.98 \quad p_2 = 38 \quad V_1 = 250 + 15 = 265$$

$$p_1 V_1^{1.25} = p_2 V_2^{1.25}$$

$$0.98 \times 265^{1.25} = 3.8 \times V_2^{1.25}$$

$$V_2^{1.25} = \frac{0.98 \times 265^{1.25}}{3.8}$$

$$V_2 = 265 \times \sqrt[1.25]{\frac{0.98}{3.8}} = 89.6$$

Distance moved by piston from beginning of stroke to point where discharge valves open is represented by $V_1 - V_2$.

$$V_1 - V_2 = 265 - 89.6 = 1754 \text{ mm} \quad \text{Ans.} \quad (i)$$

Expressed as a percentage of the stroke of 250 mm

$$= \frac{175.4}{250} \times 100 = 70.16\% \quad \text{Ans.} \quad (ii)$$

Example 9.2. The diameter of an air compressor cylinder is 140 mm (see Figure 9.3), the stroke of the piston is 180 mm and the clearance volume is 77 cm³. The pressure and temperature of the air in the cylinder at the end of the suction stroke and beginning of compression is 0.97 bar and 13°C, respectively. The delivery pressure is

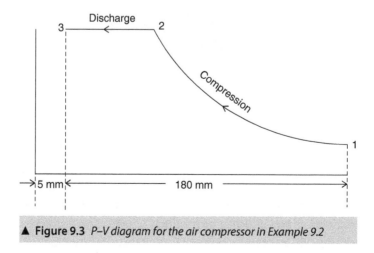

▲ **Figure 9.3** *P–V diagram for the air compressor in Example 9.2*

constant at 4.2 bar. Taking the law of compression as $pV^{1.3}$ = constant, calculate (i) for what length of stroke air is delivered, (ii) the volume of air delivered per stroke, in litres, and (iii) the temperature of the compressed air.

$$\text{Clearance length } [\text{mm}] = \frac{\text{Clearance volume } [\text{mm}^3]}{\text{Area of cylinder } [\text{mm}^2]}$$

$$= \frac{77 \times 10^3}{0.7854 \times 140^2} = 5 \text{ mm}$$

$$p_1 V_1^{1.3} = p_2 V_2^{1.3}$$

$$0.97 \times (180 + 5)^{1.3} = 4.2 \times V_2^{1.3}$$

$$V_2^{1.3} = \frac{0.97 \times 185^{1.3}}{4 \cdot 2}$$

$$V_2 = 185 \times^{1.3} \frac{\sqrt{0.97}}{4.2}$$

$$= 59.94 \text{ mm}$$

$$\text{Delivery period} = V_2 - V_3 = 59.94 - 5$$

$$= 54.94 \text{ mm of stroke} \quad \text{Ans.} \quad (\text{i})$$

$$\text{Volume delivered} = \text{Area} \times \text{Length}$$

$$= 0.7854 \times 140^2 \times 54.94$$

$$= 8.457 \times 10^5 \text{ mm}^3$$

$$10^6 \text{ mm}^3 = 1 \text{l}$$

$$\text{Volume delivered} = 0.8457 \text{l} \quad \text{Ans.} \quad (\text{ii})$$

$$\frac{p_1 V_1}{T_1} = \frac{p_2 V_2}{T_2}$$

$$\frac{0.97 \times 185}{286} = \frac{4.2 \times 59.94}{T_2}$$

$$T_2 = \frac{286 \times 4.2 \times 59.94}{0.97 \times 185} = 401.2 \text{ K}$$

$$\therefore \text{ Temperature at end of compression}$$

$$= 128.2°C \quad \text{Ans.} \, (\text{iii})$$

Effect of Clearance

Clearance is necessary between the piston face and the cylinder head and the suction and discharge valves, to avoid physical contact. The clearance should, however, be kept to a minimum because the volume of compressed air left in the clearance space, at the end of the inward stroke, must be expanded on the outward stroke to below atmospheric pressure, before the suction valves can open. If the suction valve does not open at the correct time, then the volume of air drawn into the cylinder during the suction stroke will be low and compressor performance will be poor. This effect is illustrated in Figure 9.4.

An ideal compressor could be built having no clearance, as in Figure 9.4(a). The suction valves would open immediately as the piston begins to move on its outward stroke and air would be drawn into the cylinder for the whole stroke. Practically this is not currently an option.

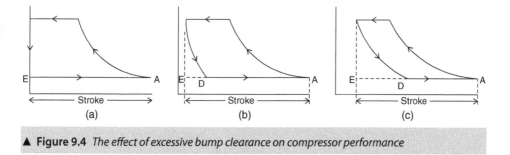

▲ **Figure 9.4** *The effect of excessive bump clearance on compressor performance*

Figure 9.4(b) shows the effect of a small clearance. The small volume of compressed air left in the clearance space is quickly expanded to sub-atmospheric pressure to allow the suction valves to open, and air is drawn into the cylinder from D to A which is a large proportion of the full stroke E to A. (*Note*: as explained on page 139, the valves on this type of air compressor are not operated with a mechanical mechanism but by pressure differential.)

Figure 9.4(c) shows the effect of excessive clearance. There is a comparatively large volume of air left in the clearance space and it requires a considerable movement of the piston on its outward stroke to expand this air to a pressure below atmospheric. The suction period D to A is an inefficient proportion of the stroke E to A.

The ratio between the volume of air drawn into the cylinder during the suction stroke and the full stroke volume swept out by the piston, is the *volumetric efficiency* of the compressor. Students will be able to appreciate that even a small increase in the 'clearance' between the top of the piston and the cylinder head will have an adverse effect on the efficiency of the air compressor. This could be caused by wear in any of the internal bearings related to the piston, connecting rods and crankshaft. It is common to measure the 'bump clearance' as part of regular maintenance, to ensure that any increase in measurement can be monitored and a reduction in performance anticipated.

$$\text{Volumetric efficiency} = \frac{\text{Volume of air drawn in per stroke}}{\text{Stroke volume}}$$

and this is represented by the ratio $\dfrac{DA}{EA}$.

Work Done Per Cycle

Neglecting clearance

As previously shown, the area of a pV diagram represents work done, and if the pressure is in kN/m² and the volume in cubic metres, then the area of the pV diagram and the work done per cycle is in kJ.

Referring to Figure 9.5 (neglecting clearance):

Net area = Net work done on the air per cycle

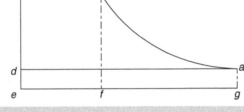

▲ **Figure 9.5** *Work done on the air per cycle*

Area $abcd$ = Area $bcef$ + Area $abfg$ − Area ade

Let p_2 = Delivery pressure, V_2 = Volume bc

p_1 = Suction pressure, V_1 = Volume da

Area under curve = $\dfrac{p_2 V_2 - p_1 V_1}{n-1}$, therefore:

$$\begin{aligned}
\text{Work/cycle} &= p_2 V_2 + \frac{p_2 V_2 - p_1 V_1}{n-1} - p_1 V_1 \\
&= \frac{p_2 V_2 (n-1) + p_2 V_2 - p_1 V_1 - p_1 V_1 (n-1)}{n-1} \\
&= \frac{n p_2 V_2 - p_2 V_2 + p_2 V_2 - p_1 V_1 - n p_1 V_1 + p_1 V_1}{n-1} \\
&= \frac{n}{n-1}\left(p_2 V_2 - p_1 V_1\right)
\end{aligned}$$

$$p_1 V_1 = mRT_1,$$
$$p_2 V_2 = mRT_2$$

$$\frac{T_2}{T_1} = \left\{ \frac{p_2}{p_1} \right\}^{\frac{n-1}{n}} = \left\{ \frac{V_1}{V_2} \right\}^{n-1}$$

By substitution, the work per cycle can be expressed in other terms to suit available data, such as:

$$
\begin{aligned}
\text{Work/cycle} &= \frac{n}{n-1}\left(p_2 V_2 - p_1 V_1 \right) \\[2mm]
&= \frac{n}{n-1} mR\left(T_2 - T_1 \right) \\[2mm]
&= \frac{n}{n-1} mRT_1 \left\{ \frac{T_2}{T_1} - 1 \right\} \quad \text{or} \\[2mm]
&= \frac{n}{n-1} p_1 V_1 \left\{ \frac{T_2}{T_1} - 1 \right\} \\[2mm]
&= \frac{n}{n-1} mRT_1 \left[\left\{ \frac{p_2}{p_1} \right\}^{\frac{n-1}{n}} - 1 \right] \quad \text{or} \\[2mm]
&= \frac{n}{n-1} p_1 V_1 \left[\left\{ \frac{p_2}{p_1} \right\}^{\frac{n-1}{n}} - 1 \right] \\[2mm]
&= \frac{n}{n-1} mRT_1 \left[\left\{ \frac{V_1}{V_2} \right\}^{n-1} - 1 \right] \quad \text{or} \\[2mm]
&= \frac{n}{n-1} p_1 V_1 \left[\left\{ \frac{V_1}{V_2} \right\}^{n-1} - 1 \right]
\end{aligned}
$$

Example 9.3. The cylinder of a single-acting compressor is 225 mm in diameter and the stroke of the piston is 300 mm. It takes in air at 0.96 bar and delivers it at 4.8 bar and makes four delivery strokes per second. Assuming that compression follows the law pVn = constant and neglecting clearance, calculate the theoretical power required to drive the compressor when the value of the index of the law of compression is (i) 1.2, (ii) 1.35 (as shown in Figure 9.6).

$$V_1 = 0.7854 \times 0.2252 \times 0.3 = 0.01193 \text{ m}^3$$

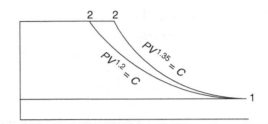

▲ **Figure 9.6** *Compression index at (i) 1.2 and (ii) 1.35*

When $n = 1.2$:

$$p_1V_1^{1.2} = p_2V_2^{1.2}$$
$$0.9 \times 0.01193^{1.2} = 4.8 \times V_2^{1.2}$$
$$V_2^{1.2} = \frac{0.96 \times 0.01193^{1.2}}{4.8} = \frac{0.01193^{1.2}}{5}$$
$$V_2 = \frac{0.01193}{\sqrt[1.2]{5}} = 0.00312 \text{ m}^3$$

$$\text{Work done per cycle} = \frac{n}{n-1}\left(p_2V_2 - p_1V_1\right)$$
$$= \frac{1.2}{0.2}\left(4.8 \times 10^2 \times 0.00312 - 0.96 \times 10^2 \times 0.001193\right)$$
$$= 2.114 \text{ kJ}$$

$$\text{Power}\left[\text{kW}\right] = \text{Work done per second}\left[\text{kJ/s}\right]$$
$$= \text{kJ/cycle} \times \text{cycle/s}$$
$$= 2.114 \times 4 = 8.456 \text{ kW} \quad \text{Ans.}\left(\text{i}\right)$$

Note that although any convenient units can be used for pressure and volume when they appear on both sides of an equation, such as pressure in bars, and volume represented by length on a pV diagram (as in the two previous examples), it is again emphasised that the pressure must be in kN/m² (1 bar = 102 kN/m²) and the volume in m³ when kJ of work is required by their product.

When $\pi = 1.35$

$$p_1V_1^{1.35} = p_2V_2^{1.35}$$
$$0.96 \times 0.01193^{1.35} = 4.8 \times V_2^{1.35}$$
$$V_2 = \frac{0.01193}{\sqrt[1.35]{5}} = 0.003621 \text{ m}^3$$

$$\text{Work done per cycle} = \frac{n}{n-1}\left(p_2 V_2 - p_1 V_1\right)$$

$$= \frac{1.35}{0.35}\left(4.8 \times 10^2 \times 0.003621 - 0.96 \times 10^2 \times 0.01193\right)$$

$$= 2.287 \text{ kJ}$$

$$\text{Power} = 2.287 \times 4 = 9.148 \text{ kW} \quad \text{Ans. (ii)}$$

Note that the *mass* of air delivered per stroke is the same in each case. The greater volume indicated by the value of V_2 when n is 1.35 is due purely to the air being at a higher temperature at the end of compression than it is when $n = 1.2$.

It can also be seen from the above that the nearer isothermal compression can be approached, the less work will be required to compress and deliver a given mass of air.

Including clearance

Referring to Figure 9.7 at the end of the compression and delivery stroke, the clearance space of volume V_3 is full of air at the delivery pressure of p_2 and temperature T_2. As the piston moves outward, the air does work on the piston as it expands to volume V_4 and the pressure falls to p_4. Theoretically when p_4 is equal to p_1 the suction valves lift to begin the suction period.

The indicated work done by the piston on the air is therefore represented by the net area 1234.

$$\text{Work/cycle} = \frac{n}{n-1}\left(p_2 V_2 - p_1 V_1\right) - \frac{n}{n-1}\left(p_3 V_3 - p_4 V_4\right)$$

$$= \frac{n}{n-1}p_1 V_1\left[\left\{\frac{p_2}{p_1}\right\}^{\frac{n-1}{n}} - 1\right] - \frac{n}{n-1}p_4 V_4\left[\left\{\frac{p_3}{p_4}\right\}^{\frac{n-1}{n}} - 1\right]$$

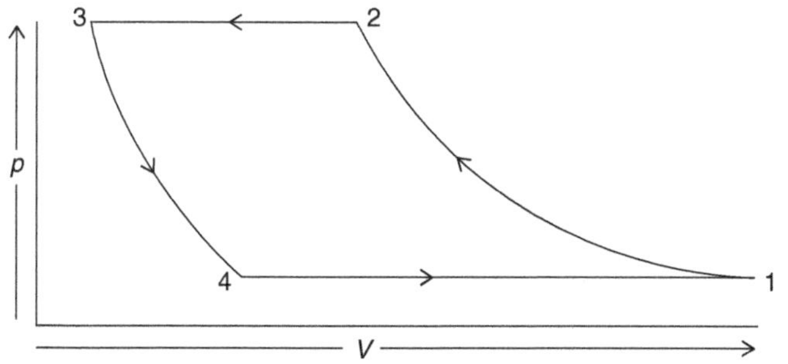

▲ **Figure 9.7** *Work done per cycle*

Taking the index of expansion to be equal to the index of compression and assuming $p_4 = p_1, p_3 = p_2, T_4 = T_1, T_3 = T_2$ then:

$$\text{Work/cycle} = \frac{n}{n-1}p_1V_1\left[\left\{\frac{p_2}{p_1}\right\}^{\frac{n-1}{n}} - 1\right] - \frac{n}{n-1}p_1V_4\left[\left\{\frac{p_2}{p_1}\right\}^{\frac{n-1}{n}} - 1\right]$$

$$= \frac{n}{n-1}p_1\left(V_1 - V_4\right)\left[\left\{\frac{p_2}{p_1}\right\}^{\frac{n-1}{n}} - 1\right]$$

$V_1 - V_4$ is the volume of air taken in per cycle at the pressure p_1 and temperature T_1, letting m = mass of this air:

$$\text{Work/cycle} = \frac{n}{n-1}mRT_1\left[\left\{\frac{p_2}{p_1}\right\}^{\frac{n-1}{n}} - 1\right]$$

Comparing this with the work/cycle when there is no clearance, we see that for a given mass the work is the same in each case, thus, the work done per unit mass of air delivered is not affected by the size of the clearance space.

Multi-stage Compression

By compressing the air in more than one stage and intercooling between stages, the practical compression curve can approach the isothermal 'ideal' curve more closely, hence reducing the work required per kg of air compressed.

Figure 9.8 shows the p–V diagrams neglecting clearance for two-stage and three-stage compression, respectively, the shaded areas representing the work saved in each case compared with single-stage compression.

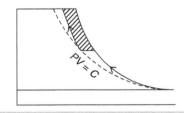
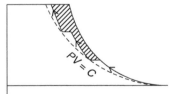

▲ **Figure 9.8** *Multi-stage compression*

To obtain maximum efficiency from a multi-stage compressor, that is, to do the least work to compress and deliver a given mass of air,

(i) the air should be intercooled to as near the initial temperature as possible; and

(ii) the pressure ratio in each stage should be the same.

Multi-stage compressors may consist simply of separate compressor cylinders that may be arranged in tandem. Figure 9.9 shows a diagrammatic arrangement of a three-stage tandem air compressor. Some arrangements, however, use 'conical' shaped pistons.

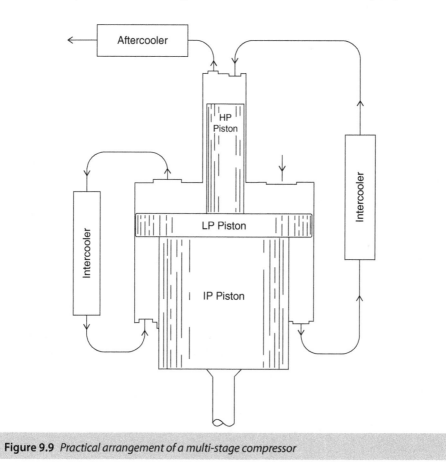

▲ **Figure 9.9** *Practical arrangement of a multi-stage compressor*

It is important to note that, when calculating the volume of atmospheric air drawn into the low-pressure cylinder of a tandem compressor, the *effective* area of suction is the difference between the area of the low pressure and the area of the high pressure. Thus, if D = diameter of low pressure piston, and d = diameter of high pressure (taking π/4 as 0.7854), then:

$$\text{Effective area} = 0.7854\left(D^2 - d^2\right)$$

Example 9.4 (see Figure 9.9). In a single-acting, three-stage tandem air compressor, the piston diameters are 70, 335 and 375 mm diameter, respectively, the stroke is 380 mm and it is driven directly from a motor running at 250 rev/min. The suction pressure is atmospheric (1.013 bar) and the discharge is 45 bar gauge pressure. Assuming that the air delivered to the reservoirs is cooled down to the initial suction temperature and taking the volumetric efficiency as 90%, calculate the volume of compressed air delivered to the reservoirs per minute (again taking π/4 as 0.7854).

$$\text{Effective area of L P} = 0.7854(0.375^2 - 0.07^2)$$
$$= 0.7854 \times 0.445 \times 0.305 \text{ m}^2$$
$$\text{Stroke volume} = \text{Area} \times \text{Stroke}$$
$$\therefore \text{ A volume } [\text{m}^3] \text{ of air drawn in per stroke}$$
$$= 0.7854 \times 0.445 \times 0.305 \times 0.38 \times 0.9$$
$$= 0.03646 \text{ m}^3$$
$$\text{Air drawn in per minute} = 0.03646 \times 250 = 9.115 \text{ m}^3$$

This is the volume of air taken into the compressor per minute at 1.013 bar. The volume delivered per minute at 46.013 bar is at the same temperature; therefore, since pressure × volume = constant, volume varies inversely as the absolute pressure, then:

$$\text{Volume delivered} = 9.115 \times \frac{1.013}{46.013}$$
$$= 0.2006 \text{ m}^3/\text{min} \quad \text{Ans.}$$

Example 9.5. The cylinders of a single-acting, two-stage tandem air compressor are 55 and 215 mm diameter, respectively, and the stroke is 230 mm. It is connected to three air storage receivers of equal size, and of internal dimensions 300 mm diameter and 1.5 m long overall with hemispherical ends. Taking atmospheric pressure as 1 bar and assuming a volumetric efficiency of 0.88, calculate the time required to pump up the bottles to a pressure of 31 bar gauge from empty, when running at 125 rev/min.

$$\text{Length of cylindrical part of bottles} = \text{Overall length} - \text{Diameter}$$
$$= 1.5 - 0.3 = 1.2 \text{ m}$$
$$\text{Volume of three bottles} = 3\left(\frac{\pi}{6} \times 0.3^3 + \frac{\pi}{4} \times 0.3^2 + 1.2\right)$$
$$= 3\pi \times 0.3^2 (0.05 + 0.3)$$
$$= 0.297 \text{ m}^3$$

To produce 0.297 m³ of air at an absolute pressure of 31 + 1 = 32 bar, the volume of atmospheric pressure air (termed 'free air') required, at the same temperature, is inversely proportional to the pressure:

$$\text{Volume of atmospheric} = 0.297 \times \frac{32}{1} = 9.504 \text{ m}^3$$

However, 'empty' air receivers contain their own volume of air at atmospheric pressure, therefore, volume of air to be taken into compressor from atmosphere

$$= 9.504 - 0 - 207 = 9.207 \text{ m}^3$$

Volume of air taken into compressor per minute

$$= 0.7854 \left(0.215^2 - 0.055^2\right) \times 0.23 \times 0.88 \times 125$$
$$= 0.8584 \text{ m}^3/\text{min}$$
$$\therefore \text{Time required} = \frac{9.207}{0.8584} = 10.73 \text{ min} \quad \text{Ans.}$$

FREE AIR DELIVERY. It is convenient to use free air delivery as a practical comparison between air compressors. Free air delivery is the rate of volume flow measured at inlet pressure and temperature.

At inlet (i.e., before entering the compressor) $pV = mRT$

If V is the free air delivery in m³/s.

If m is the mass flow in kg/s.

$\therefore pV = mRT$

or $V = \dfrac{mRT}{p}$

or $m = \dfrac{pV}{RT}$ either may be required when solving problems

E.g., Power input/cycle $= \dfrac{n}{n-1} mR\left(T_2 - T_1\right)$ [kW]

ISOTHERMAL EFFICIENCY. It has been stated that the nearer the isothermal compression can be approached the less will be the work required to compress and deliver the air. Hence another compressor comparison is *isothermal efficiency*.

$$\text{Isothermal efficiency} = \frac{\text{Power input with isothermal compression}}{\text{Actual power input}} \times 100\%$$

INDICATED MEAN EFFECTIVE PRESSURE. Taking the work per cycle as at Figure 9.5:

$$\text{m.e.p. (kN/m}^2\text{)} = \frac{\text{Area of diagram (kJ)}}{\text{Length of diagram (m}^3\text{)}}$$

MINIMUM WORK. For reversible polytropic compression the total work required is a minimum when the work is equally divided between the stages. For a two-stage machine with perfect intercooling, by differentiating the total work expression with respect to the intermediate pressure p_2, it can be shown that:

$$p_2 = \sqrt{p_1 p_3}$$

Example 9.6. A two-stage, single-acting air compressor takes air at 1 bar, 15°C and delivers 0.075 kg/s at 9 bar. The compression and expansion in both stages is according to the law $pV^{1.3}$ = constant. The compressor is designed for minimum work with perfect intercooling.

Calculate:

(i) the power input to the compressor,

(ii) the heat rejected in the intercooler.

$$\text{For air: } R = 287 \text{ J/kg K, } c_p = 1005 \text{ J/kg K}$$
$$p_2 = \sqrt{p_1 p_3}$$
$$= \sqrt{9}$$
$$= 3$$
$$\frac{T_2}{T_1} = \left\{ \frac{p_2}{p_1} \right\}^{\frac{n-1}{n}}$$
$$T_2 = 288 \times 3^{\frac{0.3}{1.3}}$$
$$= 370.9 \text{ K}$$

$$\text{Work done per stage} = \frac{n}{n-1} mRT \left(T_2 - T_1 \right)$$
$$= \frac{1.3}{0.3} \times 0.075 \times 0.287 (370.9 - 288)$$
$$= 7.732 \text{ kW}$$

$$\text{Power input to compressor} = 2 \times 7.732$$
$$= 15.464 \text{ kW} \quad \text{Ans. (i)}$$

$$\text{Heat rejected in intercooler} = mC_p (370 - 288)$$
$$= 0.075 \times 1.005 \times 82.9$$
$$6.249 \text{ kW} \quad \text{Ans. (ii)}$$

Test Examples 9

1. In a single-stage air compressor the diameter of the cylinder is 250 mm, the stroke of the piston is 350 mm and the clearance volume is 900 cm³. Air is drawn in at a pressure of 0.986 bar and delivered at 4.1 bar. Taking the law of compression to be $pV^{1.25}$ = constant, calculate the distance travelled by the piston from the beginning of its compression stroke when the delivery valves open.

2. A four-cylinder, single-acting compressor of 152 mm bore and 105 mm stroke runs at 12 rev/s and compresses air from 1 bar to 8 bar. The clearance volume is 8% of the swept volume and both compression and expansion are according to the law $pV^{1.3}$ = constant.

    (i) Sketch the cycle on a pressure volume diagram.

    (ii) Calculate:

        (a) the volumetric efficiency,

        (b) the free air delivery.

3. A single-stage air compressor of 180 mm stroke and 140 mm bore compresses air from 1 bar, 15°C to 8.5 bar. The mass of air in the cylinder at the beginning of the compression stroke is 0.0035 kg. The air in the clearance volume expands according to the law $pV^{1.32}$ = constant.

    (i) Sketch the cycle on a pV diagram,

    (ii) Calculate:

        (a) the clearance volume,

        (b) the volumetric efficiency.

    For air $R = 287$ J/kg K.

4. A three-cylinder, single-stage compressor with negligible clearance volume and 100 mm bore, 120 mm stroke produces a free air delivery of 1.2 m³/min. It compresses air from 1 bar to 6.6 bar and the compression process may be assumed to be polytropic with an index of 1.3. The mechanical efficiency is 90%.

    (i) Sketch the process on a pV diagram.

    (ii) Calculate:

        (a) the operating speed,

        (b) the power input required.

5. A motor-driven, three-stage, single-acting tandem air compressor runs at 170 rev/min. The high pressure and low pressure cylinder diameters are 75 mm and 350 mm diameter, respectively, and the stroke is 300 mm. Find the time to pump up air reservoirs of 22 m³ total capacity from 19 bar gauge to 30 bar gauge, taking the volumetric efficiency as 0.92 and atmospheric pressure as 1.01 bar.

6. In a single-acting air compressor, the clearance volume is 364 cm³, diameter of cylinder 200 mm, stroke 230 mm and it runs at 2 rev/s. It receives air at 1 bar and delivers it at 5 bar, the index of compression and expansion being 1.28. Calculate the indicated power, the mean indicated pressure and the volumetric efficiency.

7. The pressure and temperature of the air at the beginning of compression in a single-stage, double-acting air compressor are 0.98 bar and 24°C, respectively, the pressure ratio is 4.55 and the index of compression and expansion is 1.25. The stroke is 1.2 × cylinder diameter, clearance equal to 5% of the stroke and the compressor runs at 8 rev/s. If it takes air from the atmosphere at 1.013 bar, 16°C, at the rate of 5 m³/min, calculate (i) the compressor power, (ii) volumetric efficiency and (iii) the dimensions of the cylinder.

8. A single-acting, two-stage air compressor delivers air at 16 bar. Inlet conditions are 1 bar, 33°C and the free air delivery is 17 m³/min.

   If the compressor is designed for minimum work and complete intercooling, determine if the compression law is $pV^{1.3} = constant$:

   (i) the power required to drive the first stage,

   (ii) the heat rejected in the intercooler per minute. *Note*: $c_p = 1005$ J/kg K and $R = 287$ J/kg K.

9. The free air delivery of a single-stage, double-acting, reciprocating air compressor is 0.6083 m³/s measured at 1.013 bar and 15°C. Compressor suction conditions are 0.97 bar and 27°C, delivery pressure is 4.85 bar. Clearance volume is 6% of stroke volume, stroke and bore are equal.

   If the index of compression and expansion is 1.32 and the machine runs at 5 rev/s determine:

   (i) the required cylinder dimensions,

   (ii) the isothermal efficiency of the compressor. $R = 287$ J/kg K.

10. A multi-stage air compressor is fitted with perfect intercoolers and is designed for minimum work. It takes in air at 1 bar, 35°C and delivers it at 100 bar. If the maximum temperature at any point during compression is not to exceed 95°C and the compression and expansion in each stage obeys the law $pV^{1.3} = constant$, determine:

   (i) the minimum number of stages required,

   (ii) the indicated power input required for a mass flow rate of 0.1 kg/s.

Take $R = 0.287$ kJ/kg K for air.

10

STEAM

The 'golden' age of steam has now passed for ships as it has for other forms of transport, and again the replacement is the internal combustion (IC) engine, almost 100% in the form of diesel engines. However, in the quest for greater efficiency from the overall power plant it has been found that steam still has a role to play.

This is the case when the direct burning of hydrocarbons is used to compare one system with the next. To extract the energy from the fuel the steam system must complete more step changes to finally reach the output than does the IC engine. However, systems that have been developed and are continually being refined use the IC engine as the primary energy source and utilise Waste Heat Recovery Systems (WHRS) to produce steam and extract more of the energy that would normally be 'wasted' (see Figure 10.1). Steam continues to be used at sea as auxiliary plant where the steam produced is used for heating purposes. Therefore, for the marine engineer, knowledge about the working principles and constructional details of a steam plant is still very important.

The first consideration is the normal working conditions within a steam plant; the consumers use the steam at the same rate at which it is generated in the boilers, and therefore the steam is generated at constant pressure.

The boiling point of water is the temperature at which water changes into steam. This natural characteristic depends strictly upon the pressure exerted on it. A few examples:

PRESSURE [bar]	0.04	1.013	10	20	30
BOILING POINT [°C]	29	100	179.9	212.4	233.8

This means that the temperature of the steam being produced at specific pressure inside a boiler is the same as the natural boiling temperature at that pressure. Therefore, as water at 30 bar pressure does not begin to boil until its temperature reaches 233.8°C, for steam to be generated with a system pressure of 30 bar the temperature must be at 233.8°C.

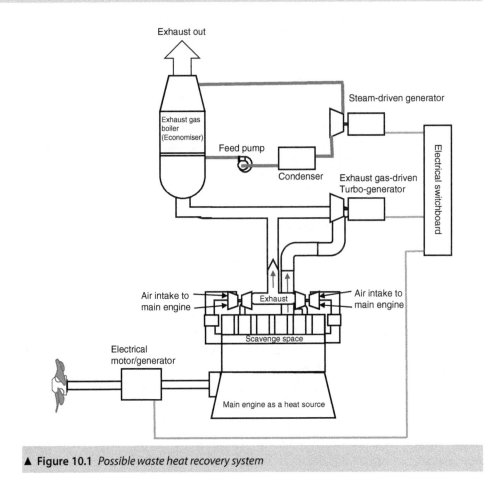

▲ Figure 10.1 *Possible waste heat recovery system*

DEFINITIONS. Steam which is in physical contact with the boiling water from which it has been generated is termed *saturated steam*; its temperature is the same as the boiling water and this is referred to as the *saturation temperature* (t_{sat} or t_s). The boiling water is referred to as *saturated liquid* because it is at the saturation temperature corresponding to that particular pressure (p_{sat}) under which evaporation takes place.

If the vapour produced is pure steam, it is called *dry saturated steam*. If the steam contains water (usually very fine particles held in suspension in the form of a mist), it is called *wet saturated steam* which is more often briefly referred to as *wet steam*. The quality of wet steam is expressed by its *dryness fraction* (JC) which is the ratio of the mass of pure steam in a given mass of the steam-plus-water mixture.

In order to increase the temperature of steam above its saturation temperature, the steam must be taken away from direct contact with the water from which it was generated and heated externally, usually as it passes through superheater elements heated by high temperature flue gases. If the steam is still wet as it enters the superheaters, the particles of water must first be evaporated to produce dry steam before the temperature is increased. Steam whose temperature is higher than its saturation temperature (which depends upon its pressure) is termed *superheated steam*, and the difference in temperature between the superheated steam and its saturation temperature is referred to as the *degree of superheat*.

FORMATION OF STEAM. Consider generating a unit mass of steam in a boiler, at constant pressure p, from the initial stage of pumping in a unit mass of the feed water.

Firstly, the water is forced into the boiler at pressure p [kN/m²] and, representing the volume of unit mass [1 kg] of the water as v_1 [m³] then the flow energy given to the water to enable it to enter the boiler is pv_1 [kJ]. (Students will note that feed water will NOT enter the boiler unless it is pumped in at a pressure that is greater than the pressure of the water/steam in the boiler. This means that the performance of the feed water pump is vital to the safe operation of the boiler.)

Heat energy is then transferred to the water and its temperature is raised from feed temperature to evaporation temperature (boiling point). Most of the heat energy received by the water is to increase its temperature and therefore increase its stored up energy, called *internal energy* and represented by u, but some is used to do the work of increasing its volume from v_1 to v_f as it expands against the pressure p, the work done being $p(v_f - v_1)$. If u_1 represents the internal energy initially in the feed water, u_f the internal energy in the water at evaporation temperature, and h_f the heat energy supplied, then,

---

Heat energy transferred = Increase in internal energy + External work done

---

$$h_f = \left(u_f - u_1\right) + p\left(v_f - v_1\right)$$

On reaching the boiling point, the water evaporates into steam as it absorbs more heat energy, thus changing its physical state from a liquid into a vapour. During the evaporating process the pressure and temperature remain constant, and the steam produced is saturated steam at the same pressure and temperature as the boiling point of water. During the change, considerable expansion takes place, the volume of the steam being many times greater than the volume occupied by the water from which it was generated. Some of the heat energy transferred during this stage is to evaporate the water into steam, freeing the molecules and increasing its internal energy from u_f to u_g. The remainder of the heat energy is used in expanding the volume from v_f to v_g against the pressure p, the work done being $p(v_g - v_f)$. Hence, if h_{fg} represents the heat energy transferred, then,

$$h_{fg} = \left(u_g - u_f \right) + p \left(v_g - v_f \right)$$

ENTHALPY. For a constant pressure process, the heat energy transferred, which is the sum of the internal energy and the work done, is termed *enthalpy*. Thus, enthalpy is an energy function of a constant pressure process and, in quantities per unit mass, is defined by the equation:

$$h = u + pv,$$

where h = specific enthalpy [kJ/kg]

u = specific internal energy [U/kg]

v = specific volume [m³/kg]

p = absolute pressure [kN/m²]

For a mass of m kg, the appropriate symbols are written in capitals:

$$H = U + pV$$

Suffixes distinguish between the properties of liquid and vapour; some of these have already been introduced in the previous chapter:

f refers to saturated liquid;

g refers to saturated vapour;

fg refers to evaporation from liquid to vapour.

Thus:

$h_f =$	specific enthalpy of saturated liquid, that is, when the liquid is at its saturation temperature
$h_g =$	specific enthalpy of dry saturated vapour
$h_{fg} = h_g - h_f =$	specific enthalpy of evaporation, that is, the change of specific enthalpy to evaporate unit mass of saturated liquid to dry saturated vapour
$h =$	specific enthalpy of either liquid or vapour at any other state

Since there is no absolute value for internal energy, there is none for enthalpy, but changes of the property can be measured and the convenient datum of water at 0°C is chosen from which to measure enthalpy and other properties of water and steam. These values are all set out in 'steam tables'.

Steam Tables

Much experimental work has been done on the properties of steam and the results are published in various forms. Those used in this book are from *Thermodynamic and Transport Properties of Fluids: SI Units*, 5th edition, arranged by G.F.C. Rogers and Y.R. Mayhew, Blackwell Publishing, Oxford.

Sets of tables are also now available online – please search for 'steam tables' online. Using this book as a guide to the properties of water/steam, students will be able to then identify the values in the online versions of steam tables.

Students will need to refer to the book or a similar set of tables for information while studying the following examples. Note that in these tables, p represents the pressure in bars. 1 bar = 10^5 N/m^2 = 10^2 kPa.

The properties of saturated water and steam are tabulated on the first few pages with the first table being for the convenience of reference for temperatures up to 100°C. The remainder are set out with reference to pressure. The properties of superheated steam at various pressures and temperatures are tabulated next.

WATER. The properties of liquids depend almost entirely on the temperature, and therefore when looking up the properties of water, we look at its particular temperature and disregard its pressure.

Example 10.1. Read from the tables the specific enthalpy of water at

(a) pressure of 1 bar and temperature 60°C

(b) ... 5 bar ... 88°C

(c) ... 10 bar ... 130°C

(d) ... 12 bar ... 188°C.

(a) 60°C: see page 2, h = 251.1 kJ/kg

(b) 88°C: see page 3, h = 369 kJ/kg

(c) 130°C: see page 4, h =546 kJ/kg

(d) 188°C: see page 4, h_f =798 kJ/kg

Note that for case (d) we can find that the temperature of 188°C is the saturation temperature corresponding to its pressure of 12 bar, and therefore we use the suffix f to signify that the water is at its saturation temperature.

STEAM. Example 10.2. Find the temperature and specific enthalpy of saturated water and dry saturated steam at pressures of (a) 1.01325 bar, (b) 10 bar, (c) 20 bar, (d) 42 bar.

(a) 1.01325 bar, see page 2, $t_s = 100°C$, $h_f = 419.1$, $h_g = 2675.8$

(b) 10 bar, see page 4, $t_s = 179.9°C$, $h_f = 763$, $h_g = 2778$

(c) 20 bar, see page 4, $t_s = 212.4°C$, $h_f = 909$, $h_g = 2779$

(d) 42 bar, see page 5, $t_{s'} = 253.2°C$, $h_f = 1102$, $h_g = 2800$

WET STEAM. When steam is 'wet' it means that it is a mixture of dry saturated steam and saturated water. Thus, if the dryness fraction is represented by x, then 1 kg of wet steam is composed of x kg of dry saturated steam, and the remainder, which is $(1 - x)$ kg, is water. Hence only x kg out of each 1 kg has been evaporated and the change of enthalpy to do this is $x \times$ specific enthalpy of evaporation, written xh_{fg}.

Therefore, specific enthalpy of wet steam of dryness fraction x is:

$$h = h_f + xh_{fg}$$

Example 10.3. Find (i) the enthalpy and (ii) volume of 1 kg of wet saturated steam at a pressure of 0.2 bar and dryness fraction 0.85, and (iii) find also the additional heat energy required to completely dry the steam.

0.2 bar, page 3,

$$h = h_{f + }h_{fg}$$
$$= 251 + 0.85 \times 2358$$
$$= 251 + 2004.3$$
$$= 2255.3 \text{ kJ/kg} \quad \text{Ans.(i)}$$

v_g is the volume occupied by 1 kg of dry saturated steam, and therefore the volume occupied by x kg of dry saturated steam is xv_g. The value of v_g is given on page 3 as 7.648 m³ for a pressure of 0.2 bar.

x_f is the volume occupied by 1 kg of saturated water, and therefore the volume occupied by $(1 - x)$ kg of saturated water is $(1 - x)v_f$. The value of v_f is not listed on page 3, but a near figure can be found on page 10, which gives $v_f \times 10^2 = 0.1017$ for a pressure of 01992 bar.

Hence, the volume of 1 kg of wet steam is

$$v = xv_g + (1 - x)v_f$$
$$= 0.85 \times 7.648 + 0.15 \times 0.001017$$
$$= 6.5008 + 0.00015255$$

We see that the volume of the water is comparatively very small and therefore, for most practical cases, can be neglected. Hence, the specific volume of wet steam is usually taken as:

$$V = xV_g$$

and for this example it is:

$$V = 0.85 \times 7.648$$
$$= 6.5008 \text{ m}^3/\text{kg} \quad \text{Ans. (ii)}$$

To dry the steam, (1 − 0.85) kg of water is to be evaporated,

$$\text{Increase of enthalpy} = 0.15 \times h_{fg}$$
$$= 0.15 \times 2358$$
$$= 353.7 \text{ kJ} \quad \text{Ans. (iii)}$$

Alternatively, it is the difference between the enthalpy of dry saturated steam and wet steam:

$$= h_g - h$$
$$= 2609 - 2255.3$$
$$= 353.7 \text{ kJ}$$

Example 10.4. If the specific enthalpy of wet saturated steam at a pressure of 11 bar is 2681 kJ/kg, find its dryness fraction.

The specific enthalpy of wet steam of dryness x is:

$$h = h_f + xh_{fg}$$
$$xh_{fg} = h - h_f$$
$$\text{hence, } x = \frac{xh_{fg}}{h_{fg}} = \frac{h - h_f}{h_{fg}}$$
$$= \frac{2681 - 781}{2000} \quad (\text{page 4})$$
$$= \frac{1900}{2000} = 0.95 \quad \text{Ans.}$$

Example 10.5. Find the heat transfer required to convert 5 kg of water at a pressure of 20 bar and temperature 21°C into steam of dryness fraction 0.9, at the same pressure.

Water 21°C, page 2, $h = 88$ kJ/kg

Steam 20 bar, page 4,

$h = hf + xhf_g$

$= 909 + 0.9 \times 1890 = 2610$ kJ/kg

Change of enthalpy, water to steam

$= 2610 - 88 = 2522$ kJ/kg

Heat transfer = total change of enthalpy (for 5 kg) $= 5 \times 2522 = 12610$ kJ   Ans.

SUPERHEATED STEAM. As previously stated, steam is said to be *superheated* when its temperature is higher than the saturation temperature corresponding to its pressure. In practice, steam is superheated at constant pressure, the saturated steam being taken from the boiler steam space and passed through the superheater tubes where it receives additional heat energy to dry the steam and raise its temperature. The properties of superheated steam are given on pages 6–8 of the tables.

Example 10.6. Find the specific volume and specific enthalpy of superheated steam (i) at a pressure of 10 bar and temperature 250°C, (ii) at a pressure of 100 bar and temperature 400°C.

10 bar 250°C, page 7,

$\left.\begin{array}{l} v = 0.2328 \text{ m}^3/\text{kg} \\ h = 2944 \text{ kJ/kg} \end{array}\right\}$ Ans. (i)

100 bar 400°C, page 8

$\left.\begin{array}{l} v = 0.02639 \text{ m}^3/\text{kg} \\ h = 3097 \text{kJ/kg} \end{array}\right\}$ Ans. (ii)

Note that at the higher pressures of 80 bar and upwards (page 8) the values of the specific volume are given as $v \times 10^2$, that is, the listed values are $10^2 \times$ actual value.

Example 10.7. Find the specific volume and specific enthalpy of steam at a pressure of 7 bar having 85°C of superheat, and determine the mean specific heat of the superheated steam over this range.

7 bar, page 7, $f_{sat} = 165$°C, $h_g = 2764$ kJ/kg

Temperature of superheated steam

$= 165 + 85 = 250$°C

$v = 0.3364$ m³/kg   Ans. (i)

$h = 2955$ kJ/kg   Ans. (ii)

Heat energy $[kJ]$ added to superheat the steam

$$= \text{Mass [kg]} \times c_p\,[kJ/kg\ K] \times \theta\ [K]$$
$$= 2955 - 2764 = 1 \times c_p\,85$$
$$c_p = \frac{191}{85} = 2.247\ kJ/kg\ K \quad \text{Ans. (iii)}$$

INTERPOLATION. If properties are required at an intermediate pressure or temperature to those given in the tables, an estimate can be made by taking values from the nearest listed pressure or temperature above and below that required and assume a linear variation between the two. The following demonstrate the usual methods.

Example 10.8. Find the enthalpy per kg of steam at a pressure of 8 bar and temperature 270°C.

Properties at 8 bar are given on page 7, but 270°C lies between the listed temperatures of 250°C and 300°C.

At 8 bar 300°C, $h = 3057\ kJ/kg$
At 8 bar 250°C, $h = 2951\ kJ/kg$
Difference, increase 50°C, $h = 106\ kJ/kg$ increase
Given temperature of 270°C is 20° higher than 250°C,

difference for 20°C, $= \frac{20}{50} \times 106 = 42.4\ kJ/kg$

$\therefore$ At 8 bar, 270°C, $h = 2951 + 42.4$
$$= 2993.4\ kJ/kg \quad \text{Ans.}$$

Example 10.9. Find the specific enthalpy of steam at a pressure of 12 bar and temperature 350°C.

Values of pressures of 10 bar and 15 bar are given on page 7, temperature of 350°C is listed.

At 10 bar 350°C, $h = 3158\ kJ/kg$
At 15 bar 350°C, $h = 3148\ kJ/kg$
Difference, increase 5 bar, $h = 10\ kJ/kg$ decrease
Given pressure of 12 bar is 2 bar greater than 10 bar

Difference for 2 bar increase $= \frac{2}{5} \times 10 = 4\ kJ/kg$ less

$\therefore$ at 12 bar 350°C, $h = 3158 - 4$
$$= 3154\ kJ/kg \quad \text{Ans.}$$

Example 10.10. Find the specific enthalpy of steam at a pressure of 25 bar and temperature 380°C.

Nearest listed pressures and temperatures are 20–30 bar and 350–400°C, on page 7, this requires double interpolation.

$$20 \text{ bar } 400°C, h = 3158 \text{ kJ/kg}$$
$$30 \text{ bar } 400°C, h = 3231 \text{ kJ/kg}$$
For 10 bar increase $h = 17$ kJ/kg decrease
$$\text{For 5 bar increase } h = \frac{5}{10} \times 17 = 8.5 \text{ decrease}$$
$$\therefore \text{At 25 bar } 400°C, h = 3248 - 8.5$$
$$= 3239.5 \text{ kJ/kg} \quad \text{(i)}$$

$$20 \text{ bar } 350°C, h = 3138 \text{ kJ/kg}$$
$$30 \text{ bar } 400°C, h = 3117 \text{ kJ/kg}$$
For 10 bar increase $h = 21$ kJ/kg decrease
$$\text{For 5 bar increase } h = \frac{5}{10} \times 21 = 10.5 \text{ decrease}$$
$$\therefore \text{At 25 bar } 350°C, h = 3138 - 10.5$$
$$= 3127.5 \text{ kJ/kg} \quad \text{(ii)}$$

From (i) 25 bar 400°C, $h = 3239.5$ kJ/kg
From (ii) 25 bar 350°C, $h = 3127.5$ kJ/kg
For increase of 50°C, $h = 112$ kJ/kg increase
$$\text{For increase of 30°C, } h = \frac{30}{50} \times 112 = 67.2 \text{ increase}$$
$$\therefore \text{At 25 bar } 380°C, h = 3127.5 + 67.2$$
$$= 3194.7 \text{ kJ/kg} \quad \text{Ans.}$$

Mixing Steam and Water

In Chapter 2, examples were given involving the mixing of ice and water, and metals in liquids at different temperature, and now further examples will be given to include steam.

The same principles apply, namely, that unless otherwise stated it is assumed there is no transfer of heat energy from or to an outside source during the mixing process.

That is, the quantity of heat energy absorbed by the colder substance is equal to the quantity of heat energy lost by the hotter substance. In other words, when steam and water are mixed together, the total enthalpy of the water and steam before mixing is equal to the total enthalpy of the resultant mixture.

Example 10.11. 3 kg of wet steam at 14 bar and dryness fraction 0.95 are blown into 100 kg of water at 22°C. Find the resultant temperature of the water.

Water 22°C, page 2 of tables, $h = 92.2$ kJ/kg

Steam 14 bar, page 4,

$h = h_f + xhf_g$

$= 830 + 0.95 \times 1960 = 2692$ kJ/kg

Before mixing:

Enthalpy of 100 kg water $= 100 \times 92.2 = 9220$ kJ

Enthalpy of 3 kg steam $= 3 \times 2692 = 8076$

Total enthalpy $= 17296$ kJ

After mixing there are $100 + 3 = 103$ kg of water of total enthalpy 17296 U, and therefore specific enthalpy of resultant water is:

$$h = \frac{17296}{103} = 167.9 \text{ kJ/kg}$$

From page 2 of tables, 167.5 kJ/kg (which is sufficiently close) corresponds to a water temperature of 40°C.

Hence, resultant temperature $= 40°C$ Ans.

An alternative method is to work on the principle that heat energy absorbed by the cold water is equal to the heat energy lost by the steam.

When heat is taken from the steam, it first condenses at its saturation temperature of 195°C (page 4 of tables, at 14 bar) into water at the same temperature, the heat energy lost being the change of enthalpy of condensation which is mass $\times xh_{fg}$. This water condensate falls in temperature from 195°C to the final temperature of the mixture, the further heat energy lost being the product of the mass, specific heat and fall in temperature.

The heat energy gained by the cold water is the product of its mass, specific heat and rise in temperature.

Thus, let θ = final temperature of the water mixture, and assuming the mean specific heat of water to be 4.2 kJ/kg K, then:

$$\text{Heat energy gained by water} = \text{Heat energy lost by steam}$$
$$100 \times 4.2\,(\theta - 22) = 3\,\{0.95 \times 1960 + 4.2\,(195 - \theta)\}$$
$$420\,\theta - 9240 = 3\,(1862 + 819 - 4.2\,\theta)$$
$$420\,\theta - 9240 = 8043 - 12.6\,\theta$$
$$432.6\,\theta = 17283$$
$$\theta = 39.95°C$$

Example 10.12. Steam is tapped from an intermediate stage of a steam turbine at a pressure of 2.5 bar, its dryness fraction being 0.95, and passed to a contact feed heater. The hot water at 48°C is pumped into the heater where it mixes with the heating steam. Estimate the temperature of the feed water leaving the heater when the amount of steam tapped off is 9% of the steam supplied to the turbine.

Referring to Figure 10.2, let us assume the mass of steam supplied to the engine to be 1 kg, then 0.09 kg is tapped off to the heater. This leaves (1 − 0.09) = 0.91 kg of steam to continue through the engine, into the condenser, as water pumped into the feed heater. Being a *contact* heater, the 0.09 kg of heating steam makes contact and mixes with the 0.91 kg of condensate, making 1 kg of feed water to be pumped back into the boilers.

Tables page 4, steam 2.5 bar, $h_f = 535,$ $h_{fg} = 2182$

Tables page 2, water 48°C, $h = 200.9$

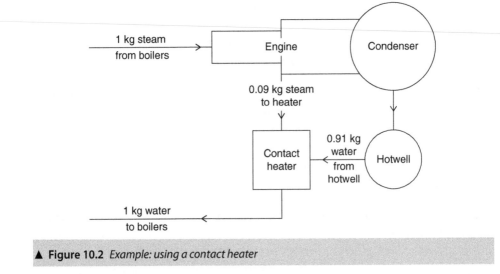

▲ **Figure 10.2** *Example: using a contact heater*

Enthalpy of steam entering heater:

$$= 0.09\left(h_f + xh_{fg}\right)$$
$$= 0.09\left(535 + 0.95 \times 2182\right) = 234.7 \text{ kJ}$$

Enthalpy of water entering heater:

$$= 0.91 \times 200.9 = 182.8 \text{ kJ}$$

Total enthalpy of steam and water entering heater:

$$= 234.7 + 182.8 = 417.5 \text{ kJ}$$

Total enthalpy of water leaving heater is that of 1 kg of feed water, and therefore the *specific* enthalpy of the feed water is 417.5 kJ/kg which we now look for in the tables and find the near figure of 417 kJ/kg on page 3 for a temperature of 99.6°C.

$$\therefore \text{Temperature of feed water} = 99.6°C \quad \text{Ans.}$$

Alternatively, for the latter part, we could assume a mean specific heat of 4.2 kJ/kg K for the feed water and take the enthalpy of water at 0°C as the heat energy to be transferred to it to raise its temperature from 0 to θ°C (mass × spec. heat × temp. rise), then:

$$1 \times 4.2 \times 9 = 417.5$$
$$\theta = \frac{417.5}{4.2} = 99.4°C$$

Example 10.13. Steam is bled from the main steam pipeline at a pressure of 17 bar and dryness fraction 0.98 to a surface feed heater, and the remainder passes through the engine. The condensate at 42°C from the engine condenser, and the drain from the feed heater, passes to the hotwell. Calculate the percentage of the main steam bled off to the heater if the temperature of the feed water to the boiler is 104.8°C (Figure 10.3).

The arrangement is illustrated in Figure 10.3. Being a *surface* heater, there is no mixing of the heating steam and feed water. In this type, the water passes through nests of tubes which are heated on the outside by the steam.

Let

H_A = enthalpy of steam entering heater

H_B = enthalpy of drain water leaving heater

H_C = enthalpy of water entering heater from hotwell

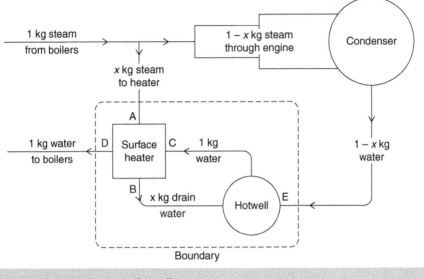

▲ Figure 10.3 *Example: using a hotwell*

H_D = enthalpy of water leaving heater (feed to boilers)

H_E = enthalpy of water entering hotwell from condenser

Enthalpy of drain water and condensate entering hotwell

$$= \text{Enthalpy of water leaving hotwell}$$
$$H_B + H_E = H_C \quad \text{(i)}$$

Enthalpy of heating steam and hotwell water entering heater

$$= \text{Enthalpy of drain and feed water leaving heater}$$
$$H_A + H_C = H_E + H_D \quad \text{(ii)}$$

Substituting value of H_C from (i) into (ii):

$$H_A + H_B + H_E = H_B + H_D$$
$$\therefore H_A + H_E = H_D$$

From this we can see that, since there is no heat energy transfer to or from an external source, the heater and hotwell with their connections can be considered as one common system, as shown by the dotted boundary line on Figure 10.3.

Let the mass of steam supplied from the boilers be 1 kg, then 1 kg of water is fed back to the boilers. Let's also assume that x kg of the supply steam is the amount bled off to the heater; then (1 − x) kg is the mass of steam passed on to the engine.

Steam 17 bar, tables page 4, $h_f = 872$, $h_{fg} = 1923$

Water 42°C, tables page 2, $h = 175.8$

Water 104.8°C tables page 4, $h = 439$

Enthalpy of x kg of heating steam entering system:

$$H_A = x(872 + 0.98 \times 1923) = 2757x \text{ kJ}$$

Enthalpy of $(1-x)$ kg of condensate entering system:

$$H_E = (1-x) \times 175.8 = 175.8 - 175.8x \text{ kJ}$$

Total enthalpy entering system:

$$H_A + H_E = 2757x + 175.8 - 175 - 8x$$
$$= 2581.2x + 175.8 \text{ kJ} \quad \text{(i)}$$

Enthalpy of 1 kg of feed water leaving system:

$$H_D = 439 \text{ kJ} \quad \text{(ii)}$$

Total entry $H_A + H_E = $ Total exit H_D

$$2581.2x + 175.8 = 439$$
$$2581.2x = 263.2$$
$$x = 0.102$$

As a percentage of the main steam supply:

$$\text{Bled steam} = 10.2\% \quad \text{Ans.}$$

Students will be able to note: Feed heating increases cycle efficiency.

Throttling of Steam

Throttling is the process of passing the steam through a restriction such as an orifice or a partially opened valve. This causes a phenomenon known as the 'wire-drawing' effect and the result is a reduction in pressure. There is not sufficient time for any appreciable heat transfer to take place between the steam and its surrounds and, as there are no moving parts, expansion takes place freely and no work is done by the steam. The difference in the velocity of the steam before and after is sufficiently small to be negligible. Consequently, if there is no transfer or conversion of heat energy, then there is no change in the enthalpy of the steam, and therefore enthalpy after throttling is equal to enthalpy before throttling.

One important application of throttling takes place in the steam reducing valve. This is a specially designed valve connected to the high pressure steam range, the restricted valve opening is spring controlled and set to maintain a near steady reduced pressure at the outlet, the reduced pressure steam being more suitable for use in driving pumps and other auxiliaries.

Example 10.14. Steam is passed through a reducing valve and reduced in pressure from 20 bar to 7 bar. Find the effect of throttling on the temperature and quality of the steam if the high pressure steam was (i) wet, having a dryness fraction of 0.9, (ii) dry saturated and (iii) superheated to a temperature of 300°C.

Tables pages 4 and 7

7 bar,	$t_s = 165°C$,	$h_f = 697$,	$h_{fg} = 2067$,	$h_g = 2764$
20 bar,	$t_s = 212.4°C$,	$h_f = 909$,	$h_{fg} = 1890$,	$h_g = 2799$,
				$H_{sup} = 3025$

(i) 20 bar, dryness 0.9:

$$h = h_f + xh_{fg}$$
$$= 909 + 0.9 \times 1890 = 2610$$

2610 is less than h_g for 7 bar, therefore reduced steam is wet

$\therefore$ 7 bar, dryness x, $h = h_f + xh_{fg}$

Enthalpy before = enthalpy after
$$2610 = 697 + x \times 2067$$
$$2067x = 1913$$
$$x = 0.9253$$

Effect of throttling:

Temperature is reduced from 212.4°C to 180°C $\Big\}$ Ans. (i)
Steam has 15° of superheat

(ii) 20 bar dry sat. $h_g = 2799$

7 bar dry sat. $h_g = 2764$

Enthalpy before = Enthalpy after

hence, enthalpy at 7 bar is 2799, which is 35 kJ/kg higher than its h_g and is therefore superheated.

From page 7 of tables, for 7 bar, $h = 2799$, lies between $h_g = 2764$ at 165°C and $h = 2846$ at 200°C.

By interpolation:

$$7 \text{ bar,} \quad h = 2846 \quad \text{for 200°C}$$
$$h = 2764 \quad \text{for 165°C}$$
$$\text{Difference in} \quad h = 82 \text{ kJ} \quad \text{for 35°C}$$

$$\text{Difference for 35 kJ} = \frac{35}{82} \times 35 = 14.94° \text{ say } 15°$$

Therefore, reduced pressure steam has 15°C of superheat, its temperature being $165 + 15 = 180°C$.

Effect of throttling:

$$\left.\begin{array}{l} \text{Temperature is reduced from 212.4°C to 165°C} \\ \text{Dryness is increased from 0.9 to 0.9253} \end{array}\right\} \text{Ans. (ii)}$$

(iii) 20 bar, temperature 300°C, $h = 3025$.

(steam has $300 - 212.4 = 87.6°C$ of superheat)

$h = 3025$ is more than h_g for 7 bar, and reduced pressure steam is therefore superheated.

$$\text{Enthalpy before} = \text{Enthalpy after}$$

From page 7 of tables, for 7 bar, $h = 3025$ lies between $h = 2955$ at 250°C and $h = 3060$ at 300°C.

$$h = 3060 \text{ for 300°C}$$
$$h = 2955 \text{ for 250°C}$$
$$\text{Difference in } h = 105 \text{ kJ for 50°C}$$

Therefore, difference in temperature for $(3025 - 2955) = 70$ kJ is

$$\frac{70}{105} \times 50 = 33.3°C$$
$$\therefore \text{Temperature} = 250 + 33.3 = 283.3°C$$
$$\text{Degree of superheat} = 283.3 - 165c$$
$$= 118.3°C$$

Effect of throttling:

$$\left.\begin{array}{l} \text{Temperature is reduced from } 300°C \text{ to } 283.3°C \\ \text{Degree of superheat is increased from } 87.6 \text{ to } 118.3 \end{array}\right\} \text{Ans. (iii)}$$

Note that in every case the effect of throttling is to either:

(i) reduce the pressure,

(ii) reduce the temperature,

(iii) increase the quality, that is, either to produce drier steam or to increase the degree of superheat.

With regard to finding the degree of superheat at the lower pressure, if the mean specific heat of superheated steam over this range were given, the degree of superheat could be obtained from:

Heat energy to raise temperature = Mass × Spec. heat × Temp. rise

Therefore, for 1 kg:

Increase of enthalpy above h = Mass × Spec. heat × Temp. rise

or $h - h_g = c_p(t - t_{sat})$

where t = temperature of the steam.

Taking the last case (iii) as an example, if the mean specific heat was given as 2.21 kJ/kg K:

$$h - h_g = c_p \times \text{Degree of superheat}$$

$$\text{Degree of superheat} = \frac{3025 - 2764}{2.21} = 118.1°C$$

Throttling Calorimeter

The throttling calorimeter, an instrument for measuring the dryness fraction of steam, is illustrated diagrammatically in Figure 10.4. The principle of operation depends upon the fact that if steam is reduced in pressure by a throttling process, the enthalpy after throttling is equal to the enthalpy before throttling, as previously explained.

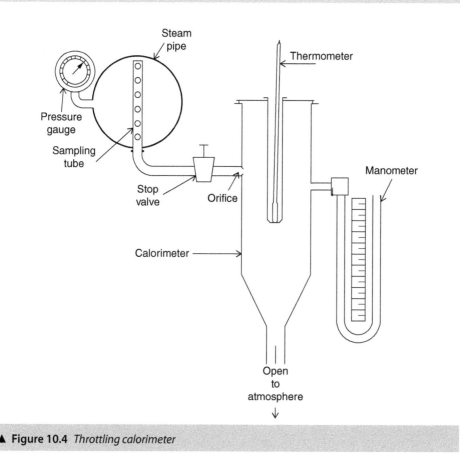

▲ Figure 10.4 *Throttling calorimeter*

Example 10.15. Steam at a pressure of 14 bar from the main steam pipe is passed through a throttling calorimeter. The pressure in the calorimeter is 1.2 bar and the temperature 119°C. Taking the specific heat of the low pressure superheated steam to be 20 kJ/kg K, calculate the dryness fraction of the main steam.

From tables page 4:

$$1.2 \text{ bar}, \quad t_s = 1048, \quad h = 2683$$
$$14 \text{ bar}, \quad h_f = 830, \quad h_{fg} = 1960$$

Enthalpy before throttling = Enthalpy after

$$830 + x \times 1960 = 2683 + 2(119 - 1048)$$
$$830 + 1960 = 2683x + 28.4$$
$$1960x = 1881.4$$
$$x = 0 - 9599 \text{ say } 0.96 \quad \text{Ans.}$$

It will be noted that the throttling calorimeter has its limitations. The dryness fraction can only be determined when the temperature of the throttled steam in the calorimeter is higher than its saturation temperature corresponding to its pressure, that is, when it is superheated. If the throttled steam was not superheated, the thermometer would record the saturation temperature whether it was dry or wet and its condition would not be known.

Separating Calorimeter

When very wet steam samples are to be tested an *approximate* value for the dryness fraction may be determined by passing the steam through a *separating calorimeter* (Figure 10.5).

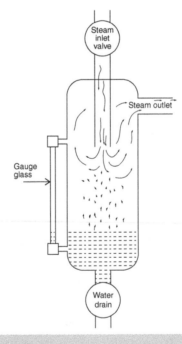

▲ **Figure 10.5** *A separating calorimeter*

The separated water collects in the bottom of the separator which is later drained off and measured. Let this be m_1. The remaining steam passes out through the outlet and is led to a small condenser where it is condensed into water, collected and measured. Let this be m_2.

From the sample of wet steam taken:

m_1 = mass of water separated from the sample
m_2 = mass of steam (assumed dry) remaining

$$\text{Dryness fraction} = \frac{\text{Mass of pure steam}}{\text{Total mass of wet steam (Steam + Water mixture)}}$$

$$= \frac{m_2}{m_2 + m_1}$$

However, the dryness fraction determined in this manner is only approximate because, although the steam leaving the separator is assumed to be dry, it will actually be slightly wet since perfect separation is not achieved.

Combined Separating and Throttling Calorimeter

In this arrangement, the outlet from the separator is connected directly to the inlet of the throttling calorimeter, and the condenser is connected to the exit of the throttling calorimeter.

$$X = x_1 \times x_2$$

$$\text{Dryness fraction} = \left\{ \begin{array}{l} \text{Dryness} \\ \text{fraction as} \\ \text{measured} \\ \text{by separating} \\ \text{calorimeter} \end{array} \right\} \times \left\{ \begin{array}{l} \text{Dryness} \\ \text{fraction as} \\ \text{measured} \\ \text{by throttling} \\ \text{calorimeter} \end{array} \right\}$$

Example 10.16. Steam at a pressure of 8 bar was tested by passing a sample through a combined separating and throttling calorimeter. The mass of water collected in the separator was 0.25 kg and the mass of condensate collected from the condenser after throttling was 2.5 kg. The pressure of the steam in the throttling calorimeter was 1.1 bar and its temperature 106.8°C. Taking the specific heat of the superheated steam after throttling to be 2.0 kJ/kg K, find the dryness fraction of the steam sample.

From tables page 4:

$$8 \text{ bar}, h_f = 721, h_{fg} = 2048$$
$$1.1 \text{ bar}, t_s = 102.3°C, h_g = 2680$$

Dryness fraction by separating calorimeter:

$$x_1 = \frac{m_2}{m_2 + m_1} = \frac{2.5}{2.5 + 0.25} = 0.909$$

Dryness fraction by throttling calorimeter:

Enthalpy before throttling = Enthalpy after
$$721 + x_2 \times 2048 = 2680 + 2\left(106.8 - 102.3\right)$$
$$x_2 \times 2048 = 1968$$
$$x_2 = 0.961$$

Dryness fraction of steam sample:

$$x = x_{1 \times x_2} = 0.909 \times 0.961 = 0.8736 \quad \text{Ans.}$$

Air in Condensers

It was stated in Chapter 5 under Dalton's law of partial pressures that the pressure exerted in a vessel occupied by a mixture of gases, or a mixture of gases and vapours, is equal to the sum of the pressures that each would exert if it alone occupied the whole volume of the vessel. The pressure exerted by each gas is termed a *partial pressure*.

Total pressure of mixture = Partial pressure due to air + Partial pressure due to steam

---

Applying this to condensers which contain a mixture of leakage air and steam:

This enables an estimate to be made on the amount of air leakage into condensers.

Example 10.17. The pressure in a condenser is 0.12 bar and the temperature is 43.8°C. (i) If the internal volume of the condenser is 8.5 m³, estimate the mass of air in the condenser, taking R for air = 0.287 kJ/kg K. (ii) If the dryness fraction of the steam is 0.9, calculate the mass of steam present.

Since there is always some water condensate in the condenser and the steam is in contact with this water, then the steam is at its saturation temperature and the pressure of the steam in the condenser depends upon this temperature. From the tables, page 3, we see that for a saturation temperature of 43.8°C the pressure of the steam is 0.09 bar:

$$\text{Partial pressure due to air} = \text{Total pressure} - \text{Steam pressure}$$
$$= 0.12 - 0.09$$
$$= 0.03 \text{ bar} = 3 \text{ kN/m}^2$$

From $pV = mRT$, mass of air present:

$$m = \frac{3 \times 8.5}{0.287 \times (43.8 + 273)}$$
$$= 0.2804 \text{ kg} \quad \text{Ans. (i)}$$

Also from tables page 3, for saturation temperature of 43.8°C, specific volume of dry saturated steam $v_g = 16.2 \text{ m}^3/\text{kg}$. For wet steam:

$$v = xv_g = 0.9 \times 16.2 = 14.58 \text{ m}^3/\text{kg}$$

Therefore, mass of steam in the condenser volume of 8.5 m³

$$= \frac{8.5}{14.58} = 0.583 \text{ kg} \quad \text{Ans. (ii)}$$

ABSOLUTE HUMIDITY. This is the ratio of the mass of water vapour to the mass of dry air (sometimes called specific humidity or moisture content).

$$\omega = \frac{m_{WV}}{m_A} = \frac{\rho WV}{P_A}$$
$$\omega\% = \frac{V_A}{V_{WV}} \times 100$$

(Relative humidity is sometimes used – ratio of partial pressures of vapour, i.e., actual to saturated.)

Test Examples 10

1. Steam at a pressure of 9 bar is generated in an exhaust gas boiler from feed water at 80°C. If the dryness fraction of the steam is 0.96, determine the heat transfer per kg of steam produced.

2. A turbo-generator is supplied with superheated steam at a pressure of 30 bar and temperature 350°C. The pressure of the exhaust steam from the turbine is 0.06 bar with a dryness fraction of 0.88.

   (i) Calculate the enthalpy drop per kilogram of steam through the turbine.

   (ii) If the turbine uses 0.5 kg of steam per second, calculate the power equivalent of the total enthalpy drop.

3. Steam enters the superheaters of a boiler at a pressure of 20 bar and dryness 0.98, and leaves at the same pressure at a temperature of 350°C. Find (i) the heat energy supplied per kilogram of steam in the superheaters, and (ii) the percentage increase in volume due to drying and superheating.

4. 1 kg of wet steam at a pressure of 8 bar and dryness 0.94 is expanded until the pressure is 4 bar. If expansion follows the law pV^n = a constant, where $n = 1.2$, find the dryness fraction of the steam at the lower pressure.

5. Dry saturated steam at a pressure of 2.4 bar is tapped off the inlet branch of a low pressure turbine to supply heating steam in a contact feed heater. The temperature of the feed water inlet to the heater is 42°C and the outlet is 99.6°C. Find the percentage mass of steam tapped off.

6. Wet saturated steam at 16 bar and dryness 0.98 enters a reducing valve and is throttled to a pressure of 8 bar. Find the dryness fraction of the reduced pressure steam.

7. Steam of mass 0.2 kg trapped within a cylinder is allowed to expand at constant pressure from an initial temperature 165°C and dryness fraction of 0.75 until its volume is doubled.

   Calculate:

   (i) the final temperature;

   (ii) the work done;

   (iii) the heat energy transferred.

8. A combined separating and throttling calorimeter was connected to a main steam pipe carrying steam at 15 bar and the following data recorded:

   Mass of water collected in separator = 0.55 kg;

   Mass of condensate after throttling = 10 kg;

   Press, of steam in throttling calorimeter = 1.1 bar;

   Temp. = 111°C.

   Taking the specific heat of the throttled superheated steam to be 2.0 kJ/kg K, find the dryness fraction of the main steam.

9. During the process of raising steam in a boiler, when the pressure was 1.9 bar gauge the temperature inside the boiler was 130°C, and when the pressure was 6.25 bar gauge the temperature was 165°C. If the volume of the steam space is constant at 4.25 m³, calculate the masses of steam and air present in each case. Take R for air = 0.287 kJ/kg K, atmospheric pressure = 1 bar and assume the steam is dry in each case.

10. A sample of air is saturated with water vapour (i.e., the water vapour in the air is saturated vapour) at 20°C, the total pressure being 1 bar. Calculate:

   (i) the partial pressures of the oxygen, nitrogen and water vapour;

   (ii) the absolute humidity.

   Air is 21% O_2 and 79% N_2 by volume and for air R = 287 J/kg K.

Note: Absolute humidity is the mass of water vapour per unit mass of dry air.

11

ENTROPY

Entropy (S.I. Unit 's') is the name given to a measure of a specific condition or state of 'order' or 'disorder' of the contents within a system. This order is sometimes also referred to as the 'randomness' of the contents.

When the theoretical understanding of 'heat energy' was being developed, it was observed that heat always naturally travelled from a hot body to a colder one and never the other way around, unless some sort of intervention was completed to change the natural process. The scientist Rudolf Clausius wanted a name for the leaking away of the heat energy to its surroundings. The name chosen and assigned to this natural phenomenon was 'entropy'.

Mechanical energy or work is the product of two quantities. This can be represented as an area of a graph made of rectangular axes, as shown in Figure 11.1(a). The graph is a diagram in which the ordinates of the axes represent pressure in kN/m_2 and volume in m_3. Therefore, the area enclosed represents mechanical work, due to the units on the horizontal axis being kN/m_2 and the units on the vertical axis being m_3, therefore the area is measured in $kN/m_2 \times m_3 = kNm = kJ$. The graph shows that the availability to do work depends upon the magnitude of the pressure.

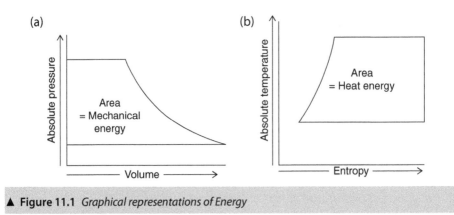

▲ **Figure 11.1** *Graphical representations of Energy*

In a similar manner, heat energy can be represented as an area on a graph. The availability to transfer heat energy depends upon the temperature and therefore the vertical ordinates of a heat energy diagram represent absolute temperature. The horizontal axis is termed *entropy*. The symbol for entropy is S and the change of entropy may be written as delta S (ΔS) with the units being kilojoules per kelvin [kJ/K]. Specific entropy, which is entropy per unit mass, is represented by the symbol s and has the units of kilojoules per kilogram kelvin [kJ/kg K].

Therefore, a diagram whose area represents heat energy, where the vertical ordinates represent absolute temperature *(T)* and the horizontal axis represents entropy *(S)*, is referred to as a temperature–entropy *(T–S)* diagram, as shown in Figure 11.1(b).

Entropy of Water and Steam

Subscripts to distinguish the specific entropy of saturated liquids and vapours are used as for other properties, that were explained in the previous chapter, thus:

s_f = Specific entropy of saturated liquid

s_g = Specific entropy of saturated vapour

$s_{fg} = s_g - s_f$ = Specific entropy of evaporation from liquid to vapour

s = Specific entropy of either liquid or vapour at any other state

In the same way that the values for internal energy and enthalpy are listed, the values of specific entropy measured from the datum of water at 0°C are also set out in the steam tables *(Thermodynamic and Transport Properties of Fluids: SI Units*, 5th edition, arranged by G. F. C. Rogers and Y. R. Mayhew) available through the online book sellers.

WATER. The thermal properties of liquids depend almost entirely on their temperature, therefore when looking up the properties of water we look for the properties listed against its particular temperature, disregarding its pressure unless it is saturated.

Example 11.1. Read from the tables the specific entropy of water at:

(a) Pressure of 1 bar and temperature 50°C

(b) Pressure of 4 bar and temperature 86°C

(c) Pressure of 8 bar and temperature 165°C

(d) Pressure of 14 bar and temperature 195°C

(a) Tables page 2, water 50°C, $s = 0.704$ kJ/kg K

(b) Tables page 3, water 86°C, $s = 1.145$ kJ/kg K

(c) Tables page 4, water 165°C, $s = 1.992$ kJ/kg K

(d) Tables page 4, water 195°C, $s_f = 2.284$ kJ/kg K

Note that when the s values for water at 50°C, 86°C and 165°C are checked the value of the pressure in the table is not the same as the example question. For case (d), however, it can be seen that the temperature of 195°C is the saturation temperature corresponding to the pressure of 14 bar, shown in the first column in the table, therefore we use the suffix f to signify that the water is at its saturation temperature.

STEAM. Example 11.2. Find the specific entropy of saturated water and dry saturated steam at pressures (a) 0.1 bar, (b) 5 bar, (c) 20 bar, (d) 50 bar.

(a) Tables page 2, 01 bar, $s_f = 0.649$, $s_g = 8.149$

(b) Tables page 4, 5 bar, $s_f = 1.860$, $s_g = 6.822$

(c) Tables page 4, 20 bar, $s_f = 2.447$, $s_g = 6.340$

(d) Tables page 5, 50 bar, $s_f = 2.921$, $s_g = 5.973$

Example 11.3. Find the specific entropy of wet steam (a) pressure 0.2 bar, dryness 0.8, (b) pressure 6 bar, dryness 0.9, (c) pressure 44 bar, dryness 0.95.

Specific entropy of wet steam of dryness fraction x is obtained in a similar manner as for specific enthalpy, previously explained (Example 10.2):

$$S = s_f + xs_{fg}$$

(a) Tables page 3, $s = 0.832 + 0.8 \times 7.075 = 6.492$

(b) Tables page 4, $s = 1.931 + 0.9 \times 4.830 = 6.278$

(c) Tables page 5, $s = 2.849 + 0.95 \times 3.180 = 5.870$

Example 11.4. Using the following formula, (i) calculate the entropy per kg of superheated steam at a pressure of 20 bar and temperature 300°C, taking the mean specific heat of water as 4.25 kJ/kg K and the mean specific heat of superheated steam as 2.58 kJ/kg K.

(ii) Compare the result with the value of entropy given in the tables.

$$S = C_w \ln\left(\frac{T_{sat}}{273}\right) + \frac{h_{fg}}{T_{sat}} + C_{sup} \ln\left(\frac{T}{T_{sat}}\right)$$

C_w = Mean specific heat of water

C_{sup} = Mean specific heat of superheated steam

T_{sat} = Saturation temperature absolute

T = Superheated steam temperature absolute

Superheated steam temp. = 573 K

Tables page 4, 20 bar, h_{fg} = 1890

Saturation temp. = 485.4 K

$$S = C_w \ln\left(\frac{T_{sat}}{273}\right) + \frac{h_{fg}}{T_{sat}} + C_{sup} \ln\left(\frac{T}{T_{sat}}\right)$$

$$= 4.25 \ln\left(\frac{485.4}{273}\right) + \frac{1890}{485.4} + 2.58 \ln\left(\frac{573}{485.4}\right)$$

$$= 4.25 \ln 1.778 + 3.894 + 2.58 \ln 1.18$$

$$= 2.446 + 3.894 + 0.427$$

$$= 6.767 \text{ kJ/kg K} \quad \text{Ans. (i)}$$

From tables, page 7, press. 20 bar, temp. 300°C:

$$s = 6.768 \text{ kJ/kg K} \quad \text{Ans. (ii)}$$

A TEMPERATURE–ENTROPY (T–S) DIAGRAM is a clear method of illustrating the process of generating steam.

Referring to Figure 11.2, consider 1 kg of water commencing at a temperature of 0°C = 273 K. As heat energy is transferred to the water its temperature rises and its entropy is increased, the heating of the water up to its boiling point is represented by the water curve AB. At any point on this water curve, the ordinate is the absolute temperature of the water, and the abscissa is its specific entropy (i.e., for 1 kg here) above that for water at 0°C. The heat energy transferred to the water to raise its temperature is the area of the diagram under the curve AB.

On reaching the saturation temperature corresponding to its pressure, point B, evaporation commences and this continues at constant temperature as the entropy is increased, therefore the evaporation process appears as a horizontal straight line. At point C evaporation is complete and dry saturated steam is produced. The latent heat energy of evaporation is represented by the area of the diagram under the evaporation line BC. Any intermediate point, such as q, along the evaporation line represents incomplete evaporation, that is, the steam is wet, the dryness fraction (X) is represented

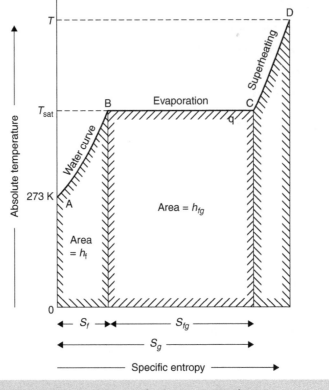

▲ **Figure 11.2** *Temperature–entropy diagram for the generation of steam*

by the ratio Bq/BC and the heat energy of evaporation is the area under that part of the evaporation line, which is xh_{fg}.

When further heat energy is transferred to dry saturated steam at constant pressure, it becomes superheated and the temperature rises as its entropy increases. This is shown by the superheat curve CD, the area of the diagram under this curve is the heat energy added to superheat the steam.

Temperature–Entropy Chart for Steam

A *T–S* chart is a complete diagram drawn up for a range of temperatures normally required by industrial systems. A simplified chart, for steam, is illustrated in Figure 11.3. The temperature scale here is in °C, but this could also be in °K. Entropy on this chart is also showing specific entropy. The features of this graph are described on the next page.

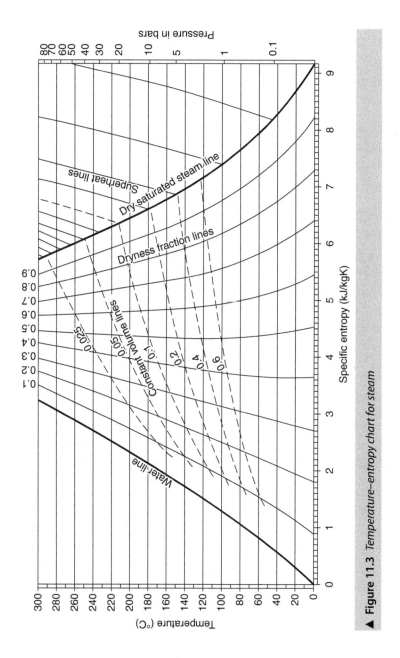

▲ **Figure 11.3** *Temperature–entropy chart for steam*

The WATER LINE gives the relation between the temperature of the water and its specific entropy. Curve AB on Figure 11.2 is part of this line.

DRY SATURATED STEAM LINE. This is produced by drawing a curve through points scaled off to the right from the water line, representing the increase in entropy for complete evaporation of 1 kg of water into steam at the various temperatures. Any point on this line represents dry saturated steam conditions at the given level of temperature. Point C on Figure 11.2 lies on this line. Inside this line the steam is wet, outside this line the steam is superheated.

SUPERHEAT LINES. These lines slope steeply upwards from points on the dry saturated steam line. These give the relation between the temperature of the steam and its specific entropy when the steam is superheated at constant pressure. CD on Figure 11.2 is one of these lines.

LINES OF CONSTANT DRYNESS FRACTIONS. These lines are drawn within the wet steam region, that is, between the two main boundary lines of the water and dry saturated steam. These are plotted by scaling off the values of xs_{fg} from the water line. For instance, if 0.9 of the values of specific entropy of complete evaporation for a series of temperatures were scaled off from the water line, the line drawn through these points is the dryness fraction line of 0.9. Lines for dryness fractions of 0.8, 0.7, etc., down to 0.1 are obtained in a similar manner.

CONSTANT VOLUME LINES. These are also included. Each constant volume line represents one particular volume occupied by 1 kg of saturated steam under varying conditions of quality from dry to very wet.

Taking a volume of 0.1 m³ as an example, we read from the tables that 0.1 m³ is the volume occupied by 1 kg of dry saturated steam when its temperature is 212.2°C. The curve for a volume of 0.1 m³ therefore begins at this point on the dry saturated steam line at the level of 212.2°C of temperature.

Now taking some other temperature, say 195°C, we note that the specific volume of dry saturated steam is given as 0.1408 m³/kg; hence, for 1 kg of steam of 195°C to occupy a volume of 0.1 m³ it must be wet, and its dryness fraction will be 0.1 ÷ 0.1408 = 0.7103. At another temperature, say 165°C, v_g = 0.2728 m³/kg for 1 kg of steam at 165°C to occupy a volume of 0.1 m³ the dryness fraction will be 0.1 ÷ 0.2728 = 0.3665. Plotting these three corresponding points of temperature and dryness fractions and drawing a curve through them, the constant volume line for 0.1 m³ is produced.

The above neglects the volume occupied by the water in the wet steam as being negligible, and obviously more than three plotted points are needed to obtain a true curve. Constant volume lines for various volumes are given covering such a range as likely to be met in practice.

Isothermal and Isentropic Processes

During an isothermal process the temperature remains constant; therefore, an isothermal operation is represented on a temperature–entropy diagram by a straight horizontal line.

During an adiabatic process, no heat transfer takes place to or from the surroundings, and a reversible adiabatic process is *isentropic*, that is, it takes place without change of entropy. Therefore, the entropy after the process is equal to the entropy before, and hence, an isentropic process appears on the temperature-entropy diagram as a straight vertical line. The processes in a Carnot cycle for a vapour, which is the same as for a gas, are shown in Figure 11.4.

EFFECT OF ISENTROPIC EXPANSION ON QUALITY. By drawing vertical lines down from the initial steam temperature on the temperature–entropy chart to represent isentropic expansion, it will be seen that dry saturated steam becomes wet during isentropic expansion, steam of normal wetness becomes wetter and steam which is very wet initially becomes slightly dryer. The T–S diagram of the Rankine cycle (Figure 12.11 of Chapter 12) illustrates this.

Example 11.5. Steam at a pressure of 14 bar expands isentropically to a pressure of 2.7 bar, find the dryness fraction at the end of expansion if the dryness of the steam at the beginning of expansion is (a) 1.0, (b) 0.8, (c) 0.6, (d) 0.4.

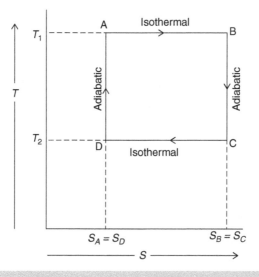

▲ **Figure 11.4** *T–S diagram for gases*

Tables page 4:

$$14 \text{ bar}, \quad s_f = 2.284 \quad s_{fg} = 4.185 \quad s_g = 6.469$$
$$2.7 \text{ bar}, \quad s_f = 1.634 \quad s_{fg} = 5.393$$

Entropy after expansion = Entropy before

(a) $1.634 + x \times 5.393 = 6.469$

$\qquad x \times 5.393 = 4.835$

$\qquad\qquad x = 0.8964 \quad$ Ans. (a)

(b) $1.634 + x \times 5.393 = 2.284 + 0.8 \times 4.185$

$\qquad x \times 5.393 = 3.998$

$\qquad\qquad x = 0.7415 \quad$ Ans. (b)

(c) $1.634 + x \times 5.393 = 2.284 + 0.6 \times 4.185$

$\qquad x \times 5.393 = 3.161$

$\qquad\qquad x = 0.5861 \quad$ Ans. (c)

(d) $1.634 + x \times 5.393 = 2.284 + 0.4 \times 4.185$

$\qquad x \times 5.393 = 2.324$

$\qquad\qquad x = 0.4309 \quad$ Ans. (d)

The above results can be checked graphically on the temperature–entropy chart by drawing a vertical line from the temperature level corresponding to 14 bar, which is 195°C, and the appropriate dryness, down to the temperature level corresponding to 2.7 bar, which is 130°C, and reading off the dryness fraction at the lower temperature.

Example 11.6. Superheated steam at a pressure of 20 bar and temperature 300°C is expanded isentropically. (i) At what pressure will the steam be just dry and saturated? (ii) What will be the dryness fraction if it is expanded to a pressure of 0.04 bar?

Tables page 7, 20 bar 300°C, $s = 6.768$

Tables page 3, 0.04 bar, $s_f = 0.422$, $s_{fg} = 8.051$

(i) Since entropy at 20 bar 300°C is 6.768, dry saturated steam at a lower pressure is to have the same entropy.

Tables page 4, gives $s_g = 6.761$ for 6 bar, therefore, pressure when steam is dry sat. = practically 6 bar. Ans. (i)

(ii) Final entropy = initial entropy

$$0.422 + x \times 8.051 = 6.768$$
$$x \times 8.051 = 6.346$$
$$x = 0.7881 \quad \text{Ans.(ii)}$$

T–S DIAGRAM FOR GASES provides a good illustration of isothermal and isentropic processes.

As an example, Figure 8.5 in Chapter 8 shows the pressure–volume diagram of the Carnot cycle for a gas. This on a temperature–entropy diagram appears as shown in Figure 11.4. Referring to this:

$$\text{Efficiency} = \frac{\text{Heat supplied} - \text{Heat rejected}}{\text{Heat supplied}}$$

$$\text{Efficiency} = 1 - \frac{\text{Heat rejected}}{\text{Heat supplied}}$$

$$= 1 - \frac{T_2(S_C - S_D)}{T_1(S_B - S_A)}$$

$$= 1 - \frac{T_2}{T_1} \text{ (as given in Chapter 8)}$$

Other ideal cycles can be similarly illustrated, for example, Figure 11.5 is for the Otto cycle (see the *p–V* diagram, Figure 8.1 of Chapter 8; also see for the Joule cycle the *T–S* diagram, Figure 12.17 of Chapter 12; there the lines of constant pressure are less steep than the lines of constant volume here).

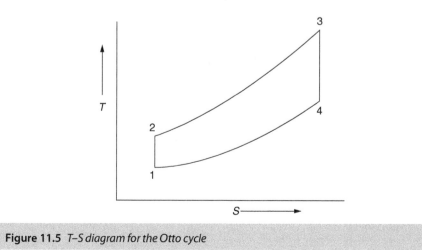

▲ **Figure 11.5** *T–S diagram for the Otto cycle*

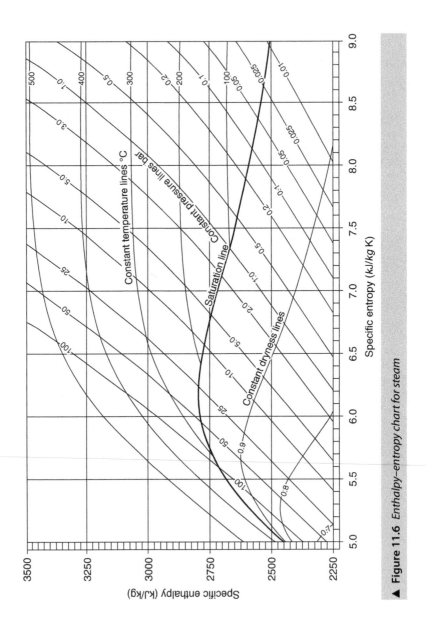

▲ **Figure 11.6** *Enthalpy–entropy chart for steam*

Enthalpy–Entropy (h-s) diagram

A diagram drawn with specific entropy of steam as the base, and the vertical ordinates representing specific enthalpy, is a most useful chart for solving problems or checking results of calculations. Figure 11.6 is a simplified chart. Complete full-size enthalpy–entropy *(h–s)* charts (also known as Mollier diagrams after its inventor Richard Mollier) drawn to scale can be purchased when more accuracy in calculations are required.

That part of the diagram under the saturation line (Shown in bold across the diagram in Figure 11.6) is referred to as the wet steam region, and above the line represents superheated conditions. Lines of constant dryness are plotted and drawn within the wet steam region (0.9, 0.8 and 0.7 dryness is shown in Figure 11.6), and constant temperature lines are drawn in the superheat region. Lines of constant pressure are drawn throughout the complete diagram.

The point of these diagrams is the simplicity of reading off the enthalpy drop and final condition when steam is expanded isentropically. This is due to isentropic expansion being represented by drawing a straight vertical line from the initial pressure and condition of steam, to the final pressure line. Throttling, being a constant enthalpy process, is represented by a straight horizontal line from the initial pressure and condition to the final pressure line.

Similar charts can be used for other vapours (e.g., refrigerants) and gases. The theoretical basis is also the same.

Test Examples 11

1. Find the entropy of 1 kg of wet steam at a pressure of 17 bar if the dryness fraction is 0.95.

2. Find the specific entropy of wet steam of temperature 195°C and dryness 0.9.

3. Dry saturated steam at 5.5 bar is expanded, isentropically, to 0.2 bar, find the dryness fraction of the steam at the end of expansion.

4. Superheated steam at 17 bar and 350°C is expanded in an engine and the final pressure is 1.7 bar. If the expansion is isentropic, find the dryness fraction of the expanded steam.

5. 1 kg of dry saturated steam at 22 bar is throttled to a pressure of 7 bar, then expanded at constant entropy to a pressure of 1.4 bar. Calculate (i) the degree of superheat at 7 bar, (ii) the increase in entropy and (iii) the dryness fraction at 1.4 bar.

12

TURBINES

A turbine is a machine for converting the heat energy in the working fluid (gas or steam) into mechanical energy at the output shaft. This series of events is reversed in the rotary compressor as the input drive gives increased velocity to the fluid in the blades which is converted to pressure rise in a fixed diffuser ring.

In the axial-flow turbine the rotor coupled to the shaft receives its rotary motion direct from the action of a high velocity jet impinging on blades fitted into grooves around the periphery of the rotor.

There are two types of turbine, the impulse and the 'reaction'. In both cases the fluid is allowed to expand from a high pressure to a lower pressure so that it acquires a high velocity at the expense of pressure, and this high velocity fluid is directed on to curved section blades which absorb some of its velocity; the difference is in the methods of expanding the working fluid.

In impulse turbines the fluid is expanded in nozzles in which the high velocity of the fluid is attained before it enters the blades on the turbine rotor, and the pressure drop and consequent increase in velocity therefore takes place in these nozzles. As the fluid passes over the rotor blades it loses velocity but there is no fall in pressure.

In 'reaction' turbines, expansion of the fluid takes place as it passes through the moving blades on the rotor as well as through the guide blades fixed to the casing. For a 50% reaction design (usual) the pressure (and enthalpy) ratios across fixed and moving blades are approximately equal.

The Impulse Turbine

As stated above, in the impulse turbine the high pressure fluid passes into nozzles wherein it expands from a high pressure to a lower pressure and thus the heat energy is converted into velocity energy (kinetic energy). The high velocity jet is directed on to blades fitted around the turbine wheel, the blades being of curved section so that the direction of the steam is changed thereby imparting a force to the blades to push the wheel around. The simplest form of an impulse turbine is where all the energy is converted in a single stage. To accomplish this the velocity of the steam converted into kinetic energy from the heat energy in the steam would need to match the turbine blade velocity, this would result in a very high speed of rotation for the turbine wheel.

Therefore, it is better to change only a portion of the steam entering the turbine into kinetic energy and thus have a lower speed of rotation. A number of such stages, arranged in series, can then be used to extract further kinetic energy from the steam, keeping the rotational speed relatively low. This process is called 'compounding'.

The best efficiency is obtained when the linear speed of the blades is half of the velocity of the fluid entering the blades; thus, when one set of nozzles is used to expand the fluid from its high supply pressure right down to the final low pressure, the resultant velocity of the fluid leaving the nozzles is very high, say about 1200 m/s. To obtain a high efficiency it means therefore that the wheel should run at a very high speed so that the linear velocity of the blades approaches 600 m/s, for example, in the case of a turbine wheel diameter of 1 m, the speed would be about 191 rev/s or 11,460 rev/min. Lower speeds, which are more suitable for practical machines, can be obtained by pressure-compounding, or velocity-compounding, or a combination of these termed pressure-velocity-compounding.

In the pressure-compounded impulse turbine, the drop in pressure is carried out in stages, each stage consisting of one set of nozzles and one bladed turbine wheel, the series of wheels being keyed to the one shaft with nozzle plates fixed to the casing between the wheels.

In the velocity-compounded impulse turbine, the complete drop in pressure takes place in one set of nozzles but the drop in velocity is carried out in stages, by absorbing only a part of the velocity in each row of blades on separate wheels and having guide blades fixed to the casing at each stage between the wheels to guide the fluid in the proper direction on to the moving blades.

The pressure-velocity-compounded turbine is a combination of these two design features.

CONSTRUCTION. In marine engines where the shaft is also required to run in the astern direction as well as ahead, a separate astern turbine is necessary. Figure 12.1 shows the principal parts of a pressure-velocity-compounded impulse steam turbine in which there are four pressure stages consisting of four sets of nozzles and four wheels in the ahead turbine, and two similar pressure stages in the astern turbine. Each wheel carries two rows of blades and there is one row of guide blades fixed to the casing protruding

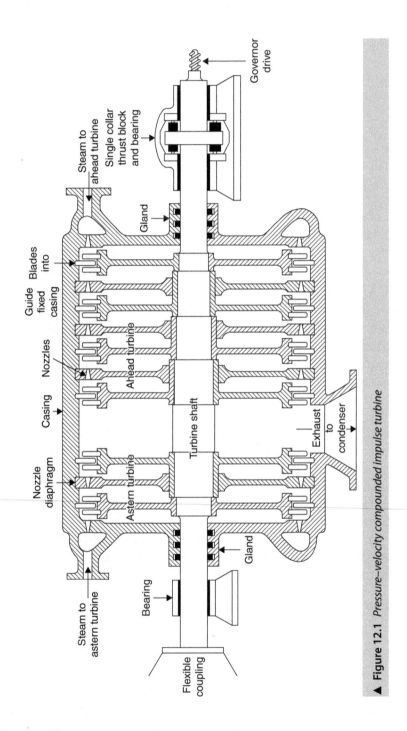

▲ **Figure 12.1** *Pressure–velocity compounded impulse turbine*

radially inwards between each row of moving blades, to drop the velocity in two steps from each set of nozzles.

The Reaction Turbine

The 'reaction' steam turbine is shown in Figure 12.2. The steam is expanded continuously through guide blades fixed to the casing and as it passes through the moving blades on the rotor on its way from the inlet end to the exhaust end of the turbine. There are no nozzles as in the impulse turbine. When the high pressure steam enters the reaction turbine, it is first passed through a row of guide blades in the casing through which the steam is expanded slightly, causing a little drop in pressure with a resulting increase in velocity, the steam being guided on to the blades in the first row of the rotor gives an impulse effect to these blades. As the steam passes through the rotor blades it is allowed to expand further so that the steam issues from them at a high relative velocity in a direction approximately opposite to the movement of the blades, thus exerting a further force due to reaction. This operation is repeated through the next pair of rows of guide blades and moving blades, then through the next and so on throughout a number of rows of guide and moving blades until the pressure has fallen to exhaust pressure. As explained, the action of the steam on the blades is partly impulse and partly reaction, and a more correct name for this type of turbine might be 'impulse-reaction', but it is generally known as a 'reaction' turbine to distinguish it from the pure impulse type.

Nozzles

The function of a nozzle is to produce a jet of high velocity which can be directed on to the blades of a turbine (Figure 12.3). The velocity is produced by allowing the working fluid to expand from a high pressure at the inlet to a lower pressure at the open exit; as no external work is done the decrease in enthalpy is converted into an increase in kinetic energy. The expansion is adiabatic because practically no heat energy is transferred in the very short time it takes the steam to pass along the nozzle. The ideal expansion would produce the maximum velocity at exit; this would be obtained if there were no friction, and the expansion was fully isentropic, which is not a practical option.

Let conditions at entrance be (Figure 12.3 - 1):

$$\text{Kinetic energy} = \frac{1}{2}mv_1^2$$
$$\text{Enthalpy} = H_1$$

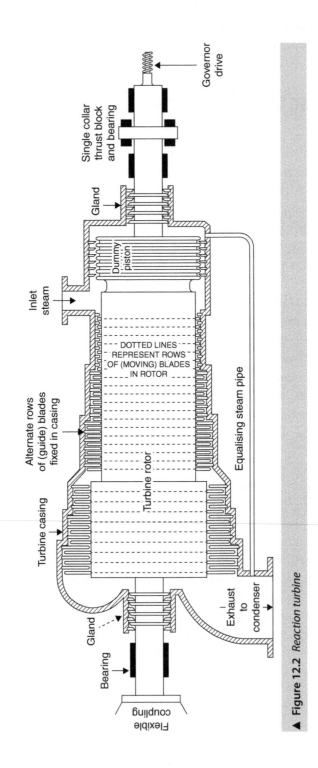

Governor drive

Single collar thrust block and bearing

Gland

Dummy piston

Inlet steam

DOTTED LINES REPRESENT ROWS OF (MOVING) BLADES IN ROTOR

Alternate rows of (guide) blades fixed in casing

Turbine casing

Turbine rotor

Equalising steam pipe

Gland

Bearing

Flexible coupling

Exhaust to condenser

▲ **Figure 12.2** *Reaction turbine*

and conditions at exit (Figure 12.3 - 2):

$$\text{Kinetic energy} = \frac{1}{2} m v_2^{\,2}$$
$$\text{Enthalpy} = H_2$$

then, working in joules of energy:

$$\text{Gain in kinetic energy} = \text{Enthalpy drop}$$
$$\frac{1}{2} m \, (v_2^2 - v_1^2) = H_1 - H_2$$

for one unit of mass:

$$\frac{1}{2} \, (v_2^2 - v_1^2) = h_1 - h_2$$

where h = specific enthalpy.

The inlet velocity is usually small compared with the exit velocity, and therefore V_1 is also very small, close to zero:

$$\frac{1}{2} v_2^2 = h_1 - h_2$$
$$v_2 = \sqrt{2 \, (h_1 - h_2)}$$
$$\therefore \; v \, [\text{m/s}] = \sqrt{2 \times \text{Spec. enthalpy drop [J/kg]}}$$

Since $\sqrt{2 \times 10^3} = 44.72$, the above may be written as
$$v = 44.72 \, \sqrt{\text{Spec. enthalpy drop [kJ/kg]}}$$

MASS AND VOLUME FLOW. At any point along the nozzle:

$$\text{Volume flow [m}^3\text{/s]} = \text{Area [m}^2\text{]} \times \text{Velocity [m/s]}$$
$$\text{Mass flow [kg/s]} = \frac{\text{Volume flow [m}^3\text{/S]}}{\text{Specific volume [m}^3\text{/kg]}}$$
$$\text{i.e. Area [m}^2\text{]} \times \text{Velocity [m/s]}$$
$$= \text{Mass flow [kg/s]} \times \text{Spec. vol. [m}^3\text{/kg]}$$

Example 12.1. Dry saturated steam enters a nozzle at 7 bar and leaves at 4 bar 0.98 dry. Find the velocity of the steam at exit. If the exit area of the nozzle is 300 mm², calculate the mass flow rate.

Tables page 4, at 7 bar, the specific enthalpy h_g = 2764 kJ/kg

$$4 \text{ bar, } h_f = 605 \; h_{fg} = 2134$$
$$\text{Spec. enthalpy drop} = 2764 - (605 + 0.98 \times 2134)$$
$$= 2764 - 2696$$
$$= 68 \text{kJ/kg} = 68 \times 10^3 \text{ J/kg}$$
$$v = \sqrt{2 \times 68 \times 10^3} = 368.7 \text{ m/s} \quad \text{Ans.}$$
$$\text{Volume flow [m}^3\text{s]} = \text{Area [m}^2\text{]} \times \text{Velocity [m/s]}$$
$$= 300 \times 10^{-6} \times 368.7$$
$$= 0.1106 \text{ m}^3/\text{s}$$

Tables page 4, at 4 bar the volume of the vapour v_g = 0.4623 m³/kg

$$\text{Spec. volume at 0.98 dry} = 0.98 \times 0.4623 = 0.453 \text{ m}^3/\text{kg}$$
$$\text{Mass flow} = \frac{0.1106}{0.453} = 0.2442 \text{ kg/s} \quad \text{Ans.}$$

Care must be taken to avoid confusion with the symbols: v usually represents velocity, however, here the same symbol is used to represent specific volume. In such cases where both velocity and specific volume would appear in the one expression or equation it is therefore advisable to write out the words fully or use an understandable abbreviation such as vol.

Another type of nozzle design used in the construction of steam turbines, is the convergent–divergent nozzle. Here the nozzle is arranged so that the entrance is slightly larger than the centre section and then the exit is much larger than the centre section. Therefore, the steam comes together in the centre section and expands out at the exit, as shown in Figure 12.3.

Example 12.2. Dry saturated steam enters a convergent–divergent nozzle at a pressure of 3.5 bar. The specific enthalpy drop between entrance and throat is 97 kJ/kg and the pressure there is 2.0 bar. The pressure at exit from the nozzle is 0.1 bar, the total possible specific enthalpy drop from entrance to exit is 534 kJ/kg and this is reduced by 12% due to the effect of friction in the divergent part of the nozzle. Calculate the nozzle area at (i) the throat and (ii) the mouth to pass 0.113 kg of steam per second.

Tables page 4, 3.5 bar, $h_g = 2732, v_g = 0.5241$

2 bar, $h_f = 505$ $h_{fg} = 2202$ $v_g = 0.8856$

page 3, 0.1 bar, $h_f = 192, h_{fg} = 2392, v_g = 14.67$

h of steam at 2 bar $= h_g$ at 3.5 bar $- 97$

$505 + x \times 202 = 2732 - 97$

$x \times 2202 = 2130$

Dryness at throat $x = 0.9674$

Spec. vol. of steam at throat

$= 0.9674 \times 0.8856 = 0.8567$ m³/kg

Velocity through throat $[\text{m/s}]$

$= \sqrt{2 \times \text{Spec. enthalpy drop [J/kg]}}$

$= \sqrt{2 \times 10^3 \times 97} = 440.4$ m/s

Area $[\text{m}^2] \times$ Velocity $[\text{m/s}] = $ Mass flow $[\text{kg/s}] \times$ Spec. vol. $[\text{m}^3/\text{kg}]$

$\therefore$ Area $[\text{mm}^2] = \dfrac{0.113 \times 0.8567}{440.4} \times 10^6$

$= 219.9$ mm² Ans. (i)

Spec. enthalpy drop between entrance and exit

$= 0.88 \times 534 = 469.9$ kJ/kg

h of steam at 0.1 bar $= H$ at 3.5 bar $- 469.9$

$192 + x \times 2392 = 2732 - 469.9$

$x \times 2392 = 2070$

Dryness at exit $x = 0.8654$

Spec. vol. of steam at exit$= 0.8654 \times 14.67 = 12.7$ m³/kg

Velocity at exit $= \sqrt{2 \times 10^3 \times 469.9}$

Area $[\text{m}^2] \times$ Velocity $[\text{m/s}] = $ Mass flow $[\text{kg/s}] \times$ Spec. vol. $[\text{m}^3/\text{kg}]$

$\therefore$ Area $[\text{mm}^2] = \dfrac{0.113 \times 12.7}{969.4} \times 10^6$

$= 1480$ mm² Ans. (ii)

CRITICAL PRESSURE RATIO. Figure 12.3 shows a convergent (entrance to throat)–divergent (throat to exit) nozzle. In the convergent part the fluid (steam, air or gas) is behaving according to the law $pV^n = $ constant, gaining in velocity as the area of the nozzle is reducing. Eventually a point is reached when any further reduction in nozzle area does not increase the velocity. The pressure ratio p_2/p_1, where p_2 (or p_T) is

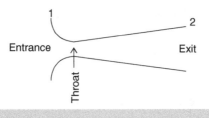

▲ **Figure 12.3** *A simple nozzle*

the pressure at the throat and p_1 is the inlet pressure, at which the minimum area of nozzle is reached is called the *critical pressure ratio*. At this point mass flow per unit area of nozzle is a maximum.

$$\text{Critical pressure ratio } \frac{p_T}{p_1} = \left(\frac{2}{n+1}\right)^{\frac{n}{n-1}}$$

This may vary from about 0.57 to 0.48 depending upon the value of n. If $n = \gamma = 1.4$ for air, then:

$$\frac{p_T}{p_1} = 0.528$$

Isentropic Efficiency

$$\text{Isentropic efficiency} = \frac{\text{Actual KE at nozzle exit}}{\text{Isentropic KE at nozzle exit}}$$

$$\text{Also, isentropic efficiency} = \frac{\text{Actual enthalpy change}}{\text{Isentropic enthalpy change}}$$

and for a perfect gas:

$$h_1 - h_2' = c_p(T_1 - T_2')$$

$$h_1 - h_2 = c_p\left(T_1 - T_2\right)$$

$$\text{Hence isentropic efficiency} = \frac{c_p(T_1 - T_2')}{c_p(T_1 - T_2)}$$

Example 12.3. Steam at 5.5 bar, 250°C expands isentropically to 1 bar through a convergent–divergent nozzle. The flow is in equilibrium throughout and the inlet velocity is negligible. Determine, if the critical pressure ratio is 0.546, the throat area of the nozzle for a mass flow rate of 0.15 kg/s and the condition of the steam at the exit.

$$p_T = 5.5 \times 0.546 = 3 \text{ bar}$$
from an $h - s$ chart
$$h_1 = 2960 \text{ kJ/kg}$$
$$h_2 = 2830 \text{ kJ/kg}$$

To determine h_1 and h_2 using tables: h_1 can be read directly, interpolation of entropy values is used for h_2, that is:

Isentropic expansion to throat $\therefore s_1 = s_2 = 7.227 \text{ kJ/kg K}$
If t = degrees of superheat,

$$\text{then } s_1 = s_{150} + \frac{t}{50}(s_{200} - s_{150}) \text{ at 3 bar}$$

$$\text{i.e. } 7.227 = 7.078 + \frac{t}{50}(7.312 - 7.078)$$

$$\therefore t = 50 \frac{7.227 - 7.078}{(7.312 - 7.078)} = 31.84°C$$

$$h_2 = h_{150} + \frac{t}{50}(h_{200} - h_{150})$$

$$= 2762 + \frac{31.84}{50}(2866 - 2762)$$

$$= 2828.2 \text{ kJ/kg}$$

$$\text{Velocity at throat} = \sqrt{2(h_1 - h_2)}$$

$$= \sqrt{2(2962 - 2828) \times 10^3}$$

$$= 513.8 \text{ m/s}$$

Specific volume at throat by interpolation (this is due to the actual values in the table not being recorded and intermediate values need to be calculated):

$$= v_{150} + \frac{t}{50}(v_{200} - v_{150})$$

$$= 0.6342 + \frac{31.84}{50}(0.7166 - 0.6342)$$

$$= 0.6867 \text{ m}^3/\text{kg}$$

$$\therefore \text{Nozzle area} \times 513.8 = 0.6867 \times 0.15$$

$$\text{Nozzle area} = 0.0002 \text{ m}^2 \text{ or 2 cm}^2 \quad \text{Ans.}$$

$$\text{Isentropic expansion to inlet} \therefore s_1 = s_2 = s_3$$

$$\text{At 1 bar, } s_1 = s_f + x s_{fg}$$

$$7.227 = 1.303 + x \times 6.056$$

$$x = 0.98 \text{ dry} \quad \text{Ans.}$$

Velocity Diagrams for Impulse Turbines

Figure 12.4 illustrates the vector diagram of velocities at the entrance side of the rotor blades ('moving' blades). v_1 represents the absolute velocity of the fluid (illustrated here as steam) directed towards the blades at an angle α_1, which is as close as practicable to the direction of their movement. The linear velocity of the blades is represented by u. Drawing the vectors of the steam velocity and blade velocity towards a common point, the vector joining these two, forming a closed diagram, is the velocity of the steam relative to the moving blades, represented by v_{r1}.

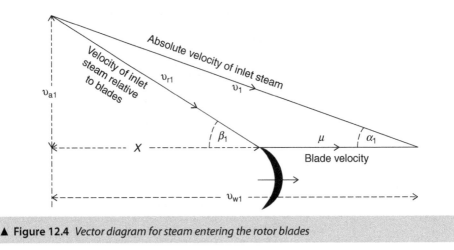

▲ **Figure 12.4** *Vector diagram for steam entering the rotor blades*

The relative direction of the steam to the blades is β_1 and in order that the steam should glide on to the blades without shock, the entrance edge of the blades must be in this direction. Therefore β_1 is also the entrance angle of the blades.

$v_{w1} = v_1 \cos \alpha_1$ is the component of the velocity of the steam jet in the direction of the blade movement, and is referred to as the *velocity of whirl at entrance*.

$v_{a1} = v_1 \sin \alpha_1$ is the axial component of the steam jet, that is, the component in the direction of the axis of the turbine.

Figure 12.5 is the vector diagram of velocities at the exit side of the moving blades. β_2 is the *exit angle of the blades* and the relative velocity of the exit steam vr_2 is in this direction. In impulse turbines, since there is no fall in steam pressure as it passes over the rotor blades, if friction is neglected, the relative velocity at exit is the same magnitude as the relative velocity at entrance, that is, $vr_2 = vr_1$. Friction between the steam and the blade surface reduces the velocity and, to take friction into account, $v_{r2} = kv_{r1}$, where the velocity coefficient k is in the region of 0.8–0.95. Loss of kinetic energy in the steam due to friction over the blade surfaces is converted into heat energy.

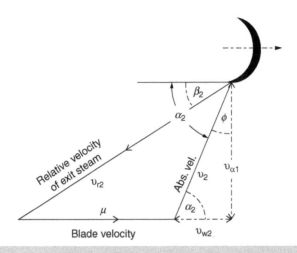

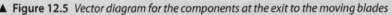

▲ **Figure 12.5** *Vector diagram for the components at the exit to the moving blades*

In the simple impulse turbine the blades are often symmetrical, which means that angle $\beta_1 = \beta_2$ and u is the vector of the blade velocity, v_2 is the vector of the absolute velocity of the exit steam. The direction of the exit steam is at α_2 to the direction of the blade movement, or at angle ø to the axis of the turbine.

$v_{w2} = v_2 \cos \alpha_2$ is the component of the exit steam velocity in the direction of blade movement and is referred to as the *velocity of whirl at exit*.

The axial component of the exit steam is $va_2 = v_2 \sin \alpha_2$.

Gas, for steam, would be treated in the same way.

Example 12.4. Gas at a velocity of 600 m/s from a nozzle is directed on to the blades at 20° to the direction of blade movement. (i) Calculate the inlet angle of the blades so that the gas will enter without shock when the linear velocity of the blades is 240 m/s. (ii) If the exit angle of the blades is the same as the inlet angle, find, neglecting blade friction, the magnitude and direction of the gas leaving the blades.

Referring to Figure 12.6:

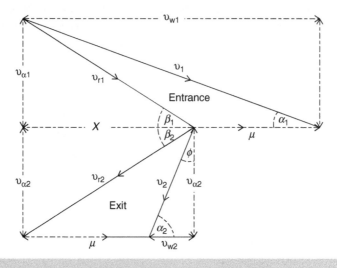

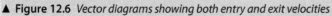

▲ **Figure 12.6** *Vector diagrams showing both entry and exit velocities*

$$V_{\alpha 1} = v_1 \sin \alpha_1 = 600 \times \sin 20° = 205.2 \text{ m/s}$$
$$V_{w1} = v_1 \cos \alpha_1 = 600 \times \sin 20° = 563.9 \text{ m/s}$$
$$x = V_{wl} - u = 563.9 - 240 = 323.9 \text{ m/s}$$
$$\tan \beta_1 = \frac{V_{\alpha 1}}{x} = \frac{205.2}{323.9}$$

Inlet angle of blades $\beta_1 = 32° \ 22'$ Ans. (i)

Neglecting friction across the blades, $v_{r2} = v_{r1}$ and since $\beta_2 = \beta_1$, then x is common to both entrance and exit diagrams, and $v_{a2} = v_{a1}$.

$$V_{w2} = x - u$$
$$= 323.9 - 240 = 83.9 \text{ m/s}$$
$$\tan \phi = \frac{V_{w2}}{V_{a2}} = \frac{83.9}{205.2} = 0.4089$$
$$\phi = 22° \ 14', \text{ and } \alpha_2 = 90° - 22° \ 14' = 67° \ 46'$$
$$v_2 = \frac{V_{w2}}{\sin \phi} = \frac{83.9}{0.3783} = 221.8 \text{ m/s}$$

$$\left. \begin{array}{l} \text{Abs. velocity of gas at exit} = 221.8 \text{ m/s} \\ \text{At } 67° \ 46' \text{ to blade movement} \\ \text{Or } 22° \ 14' \text{ to turbine axis} \end{array} \right\} \text{ Ans. (ii)}$$

Both entrance and exit velocity diagrams contain the vector of the blade velocity u, and therefore they can be combined together by using the blade velocity as a common base, as shown in Figure 12.7. This is a convenient diagram to solve either graphically or by calculation and will be used for all future problems of this type. A diagram is essential for reference in the calculations, students are advised to draw the diagram to scale rather than just a rough sketch, it will then serve as a check on the calculated results.

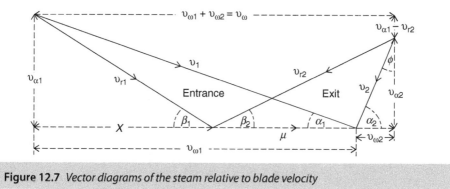

▲ **Figure 12.7** *Vector diagrams of the steam relative to blade velocity*

VELOCITY COMPOUNDING. Reference was made earlier to velocity compounding. This is a system of alternate moving and fixed blades which enables a large pressure and enthalpy drop to take place in the nozzles. This reduces pressure on the turbine casing and simplifies sealing at the glands, especially for high pressure steam turbines.

The fluid leaves the nozzles with a high velocity and enters a moving row of blades whose velocity is low compared with that of the fluid.

Upon leaving the moving blades at an angle of α_2 to the plane of rotation, the fluid enters a row of stationary blades whose entrance and exit angles are α_2, that is, symmetrical blades. Ideally, as the fluid passes through the fixed blades, there will be no alteration in the pressure, velocity or enthalpy, and the fluid will be guided without shock into the next row of moving blades.

Example 12.5. A velocity compounded impulse wheel has two rows of moving and one row of fixed blades, all of which are symmetrical.

Steam leaves the nozzles at 16° to the plane of rotation of the wheel with a velocity of 600 m/s. If the mean blade velocity is 125 m/s and the steam loses 10% of its inlet relative velocity in each of the moving blade passages determine: (i) the blade angles in each of the fixed and moving blades and (ii) die direction of the final velocity of the final exit steam (Figure 12.8).

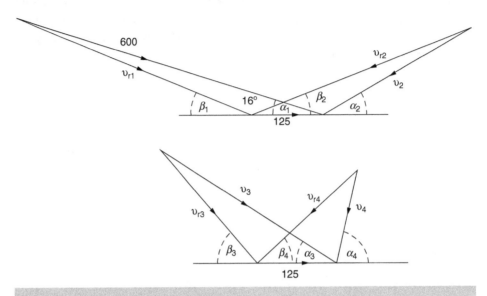

▲ **Figure 12.8** *Diagram for the example of a velocity compounded impulse wheel*

$$\tan \beta_1 = \frac{600 \sin 16°}{600 \cos 16° - 125}$$

$$\beta_1 = 20.1° \beta_1 = \beta_2 \quad \text{Ans.} (i)(a)$$

$$V_{r1} = \frac{600 \sin 16°}{\sin \beta_1} = 481.2 \text{ m/s}$$

$$V_{r2} = 0.9 \times 481.2 = 433.1 \text{ m/s}$$

$$\tan \alpha_2 = \frac{433.1 \sin 20.1°}{433.1 \cos 20.1° - 125}$$

$$\alpha_2 = 27.85° \; \alpha_2 = \alpha_3 \quad \text{Ans.} (i)(b)$$

$$V_2 = \frac{433.1 \sin 20.1°}{\sin 27.85°} = 318.6 \text{ m/s}$$

$$V_2 = V_3 \text{ as no friction in fixed blade (assumed)}$$

$$\tan \beta_3 = \frac{318.6 \sin 27.85°}{318.6 \cos 27.85° - 125}$$

$$\beta_3 = 43.52° \beta_3 = \beta_4 \quad \text{Ans.} (i)(c)$$

$$V_{r3} = \frac{318.6 \sin 27.85°}{\sin 43.52°} = 216 \text{ m/s}$$

$$V_{r4} = 0.9 \, v_{r3} = 194.5 \text{ m/s}$$

$$\tan \alpha_4 = \frac{194.5 \sin 43.52°}{194.5 \cos 43.52° - 125} \quad \alpha_4 = 83° \quad \text{Ans. (ii)}$$

Force on Blades

The effective velocity of the fluid causing motion of the rotor blades is the component in the direction of movement of the blades. This is the velocity of whirl. The velocity of whirl at entrance is $vw_1 = v_1 \cos \alpha_1$. The velocity of whirl at exit is $vw_2 = v_2 \cos \alpha_2$. The effective change of velocity of the fluid is the algebraic difference between vw_1 and vw_2; let this be represented by v_w. In the previous example, and in most cases, vw_2 is in the opposite direction to vw_1, then:

$$\text{Effective change of velocity} = v_{w1} - (-v_{w2})$$

$$\therefore v_w = v_{w1} + v_{w2}$$

If v_{w2} had been in the same direction as v_{w1}, then the effective change of velocity would be $v_{w1} - v_{w2}$

Since the motion of the fluid is changed as it passes over the blades, the *blades* must exert a force *on the fluid* to cause this change. From Newton's third law of motion which states that action and reaction are equal in magnitude and opposite in direction, this force is also the magnitude of the force exerted by the fluid on the *blades*.

$$\text{Force } [\text{N}] = \text{Mass } [\text{kg}] \times \text{Acceleration } [\text{m/s}^2]$$

$$= \text{Mass} \times \text{Change of velocity per second}$$

$$= \text{Change of momentum per second}$$

Therefore, if the mass rate of flow [kg/s] of the fluid is represented by $\dot{m}$, then:

$$\text{Force on blades } [\text{N}] = [\text{kg/s}] \times \text{Effective change of velocity}$$

$$= v_w$$

$$\text{Work done} = \text{Force} \times \text{Distance}$$

Since distance through which the force acts on the blades is the linear distance moved by the blades, we have:

Work done per second [Nm/s] = Force on blades [N] × Blade velocity [m/s]

Also:

Work done per second [Nm/s] = Power [J/s = W]

Therefore:

$$\text{Power } [W] = \dot{m}v_w u$$

The work supplied to the blades is the kinetic energy of the fluid jet:

$$\text{Work supplied per second} = \frac{1}{2}\dot{m}V_1^2$$

Hence:

$$\text{Diagram efficiency} = \frac{\text{Work done on blades [J/s]}}{\text{Work supplied [J/s]}}$$
$$= \frac{\dot{m}v_w u}{1/2\dot{m}v_1^2} = \frac{2uv_w}{v_1^2}$$

This is sometimes called the blade efficiency.

The *axial* force on the blades is due to the difference between the axial components of the fluid velocities at entrance and exit, thus:

$$\text{Axial thrust} = \dot{m}\left(v_{a1} - v_{a2}\right)$$

Example 12.6. Steam leaves the nozzles of a single stage impulse turbine at a velocity of 670 m/s at 19° to the plane of the wheel, and the steam consumption is 0.34 kg/s. The mean diameter of the blade ring is 1070 mm. Find (i) the inlet angle of the blades to suit a rotor speed of 83.3 rev/s. If the velocity coefficient of the steam across the blades is 0.9 and the blade exit angle is 32°, find (ii) the force on the blades, (iii) the power given to the wheel and (iv) the diagram efficiency.

Referring to Figure 12.7:

$$\text{Linear velocity of blades} = \text{Mean circumference} \times \text{rev/s}$$
$$u = \pi \times 1.07 \times 83.3 = 280 \text{ m/s}$$
$$v_{a1} = v_1 \sin \alpha_1 = 670 \times \sin 19° = 218.1 \text{ m/s}$$
$$v_{w1} = v_1 \cos \alpha_1 = 670 \times \cos 19° = 633.6 \text{ m/s}$$
$$\tan \beta_1 = \frac{v_{a1}}{x} = \frac{218.1}{353.6} = 0.6169$$
$$\tan \beta_1 = \frac{v_{a1}}{x} = \frac{218.1}{353.6} = 0.6169$$
$$\text{Entrance angle} = 31° \ 40' \quad \text{Ans.} \ (\text{i})$$

$$v_{r1} = \frac{v_{a1}}{\sin \beta_1} = \frac{218.1}{\sin 31° 40'} = 415.6 \text{ m/s}$$
$$v_{r2} = 0.9 v_{r21} = 0.9 \times 415.6 = 374 \text{ m/s}$$
$$v_{w2} = v_{r2} \cos \beta_2 - u = 374 \cos 32° - 280$$
$$= 317.1 - 280 = 37.1 \text{ m/s}$$
$$\text{Effective change of velocity} = v_w = v_{w1} + v_{w2}$$
$$\text{Force on blades} [N] = m \ [\text{kg/s}] \times v_w \ [\text{m/s}]$$
$$= 0.34 \times 670.7$$
$$= 228.1 \text{ N} \quad \text{Ans.} \quad (\text{ii})$$

$$\text{Power} \ [W = J/s = N\,m/s] = \text{Force} \ [N] \times \text{Linear velocity} \ [\text{m/s}]$$
$$= 228.1 \times 280$$
$$= 63870 \text{ W or } 63.87 \text{ kW} \quad \text{Ans.} \quad (\text{iii})$$

$$\text{Diagram efficiency} = \frac{\text{Work done on blades [J/s]}}{\text{Work supplied [J/s]}}$$
$$= \frac{\dot{m} v_w u}{1/2 \dot{m} v_1^2} = \frac{2 u v_w}{v_1^2}$$
$$= \frac{2 \times 280 \times 670.7}{670^2}$$
$$= 0.8368 \text{ or } 83.68\% \quad \text{Ans.} \quad (\text{iv})$$

Example 12.7. An impulse turbine has a row of nozzles set at angle α_1 to the plane of motion of the moving blades and issues steam at a velocity of C. The blades have a tangential velocity of u and are symmetrical. Neglecting the effects of friction show that the diagram efficiency is given by:

$$\eta = \frac{4u\,(C\cos\alpha_1 - u)}{C^2}$$

$$v_w = v_{r1}\cos\beta_1 + v_{r2}\cos\beta_2$$

$$v_{r1} = v_{r2}\ \text{as no blade friction}$$

$$\beta_1 = \beta_2\ \text{as blades symmetrical}$$

$$v_{r1}\cos\beta_1 = C\cos\alpha_1 - u$$

$$P = \dot{m}\,v_w u$$

$$= \dot{m}\times 2\,(C\cos\alpha_1 - u)\times u$$

$$\text{Work supplied} = \dot{m}\,1/2\,C^2$$

$$\eta = \frac{2\dot{m}u\,(C\cos\alpha_1 - u)}{0.5\,\dot{m}C^2}$$

$$= \frac{4u\,(C\cos\alpha_1 - u)}{C^2} \quad \text{Ans.}$$

Velocity Diagrams for Reaction Turbines

It has been seen that, in the impulse turbine, the fluid expands while passing through fixed nozzles, and there is no expansion on its passage through the channels between the moving blades. All generation of velocity takes place in the nozzles.

In the reaction turbine, expansion of the fluid takes place during its passage through the fixed (guide) blades which take the place of nozzles, and it also expands as it passes through the moving blades. Therefore the velocity of the fluid is increased as it passes through the fixed blades, and the relative velocity of the fluid to the moving blades is increased as it passes through the moving blades, so that v_{r2} is greater than v_{r1}.

In the reaction turbine, the fixed and moving blades are often of the same section and reversed in direction. In such cases, the entrance and exit angles of the fixed blades are the same as those of the moving blades, and the velocity vector diagram at entrance is identical with the velocity vector diagram at exit, and therefore the combined diagram is symmetrical (see Figures 12.9 and 12.10). Thus the relative velocity of the fluid at exit from the moving blades is equal to the absolute velocity at entrance, $v_{r2} = v_1$, and the absolute velocity of the fluid at exit from the moving blades is equal to the relative velocity at entrance, $v_2 = v_{r1}$. Hence, $\beta_2 = \alpha_1$ and $\alpha_2 = \beta_1$.

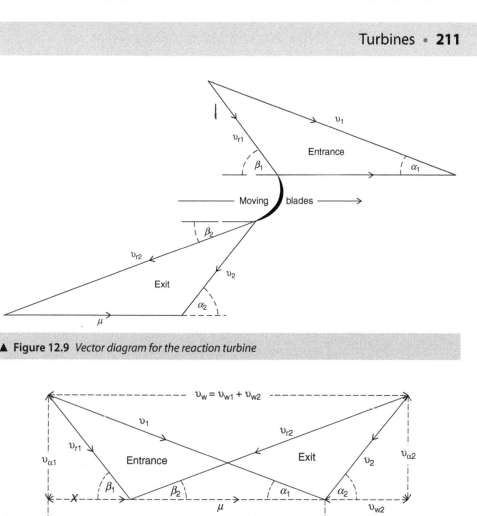

▲ **Figure 12.9** *Vector diagram for the reaction turbine*

▲ **Figure 12.10** *Combined vector diagram for the reaction turbine*

Example 12.8. At one stage of a reaction turbine the velocity of the steam leaving the fixed blades is 90 m/s and the exit angle is 20°. The linear velocity of the moving blades is 60 m/s and the steam consumption is 1.1 kg/s. Assuming the fixed and moving blades to be of identical section, calculate (i) the entrance angle of the blades, (ii) the force on the blades and (iii) the stage power.

$$v_{a1} = v_1 \sin \alpha_1 = 90 \times \sin 20° = 30.78 \text{ m/s}$$
$$v_{w1} = v_1 \cos \alpha_1 = 90 \times \sin 20° = 84.57 \text{ m/s}$$
$$x = v_{w1} - u = 84.57 - 60 = 24.57 \text{ m/s}$$
$$\tan \beta_1 = \frac{30.78}{24.57} = 1.253$$

Entrance angle $= 51° \ 24'$ Ans. (i)

Effective change of velocity $= v_w = v_{w1} + v_{w2}$

$\left(\text{note that } v_{w2} = x\right)$

$v_w = 84.57 + 24.57 = 109.14$ m/s

Force on blades $[N] = [kg/s] \times v_w [m/s]$

$= 1.1 \times 109.19$

$= 120$ N   Ans. (ii)

Power $[W = J/s = N\,m/s] = \text{Force}[N] \times \text{Velocity}[m/s]$

$= 120 \times 60$

$= 7200$ W or 7.2 kW   Ans. (iii)

Ideal Cycles

The Carnot cycle has been described for a gas and a vapour (see Figure 8.5 (pressure–volume diagram) of Chapter 8 and Figure 11.4 (temperature–entropy) of Chapter 11). The wider the range of temperature the more efficient the cycle. The lowest practical condensing temperature is governed by the coolant temperature and the highest practical temperature is governed by the limits of temperature and pressure that can be withstood by the materials that make up the internal components of the turbine.

There are two major reasons why the Carnot cycle is not used as the theoretical basis for vapour (steam) cycles. First, it has a low work ratio, and second it is difficult to compress a wet vapour efficiently; it is easier to fully condense the vapour and compress the liquid to boiler pressure using a feed pump. The resulting cycle is known as the Rankine cycle. It is important for marine engineers to know that as a consequence the feed pump is working at the limit of its capability and, if all the temperatures and pressures of the fluid around the system are not kept correctly, then the feed pump can have a tendency to 'gas up'. This is where the water turns into steam inside the pump and it cannot compress the 'vapour' enough to introduce a fresh charge of water to the boiler. The water level inside the boiler could then reduce to a dangerous level.

The Rankine cycle is illustrated in Figure 12.11 on a pV diagram and on a $T–S$ diagram and consists of:

(i)   a to b. Feed water pumped into the boiler and receiving heat energy as it is increased in pressure from p_2 to p_1 and in temperature from T_2 to T_1.

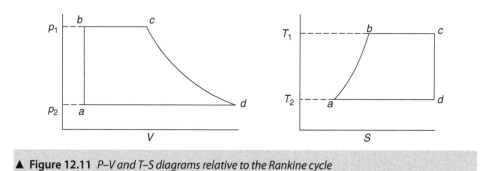

▲ **Figure 12.11** *P–V and T–S diagrams relative to the Rankine cycle*

(ii)   *b* to *c*. The water is completely evaporated into steam in the boiler at constant pressure p_1 and constant temperature T_1 and the steam is supplied to the engine as it is generated.

(iii)   *c* to *d*. The steam supply to the engine expands isentropically from the highest to the lowest limits of pressure and temperature.

(iv)   *d* to *a*. The steam is exhausted from the engine into the condenser and condensed into water at constant pressure p_2 and constant temperature T_2.

In the ideal engine there would be no energy losses due to friction or heat transfer, no leakage of steam and no undercooling of the condensate in the condenser, that is, the condensate would be at the saturation temperature corresponding to the condenser steam pressure.

Therefore, the work which would be done in the ideal engine is equal to the heat energy given up by the steam on its passage through the engine. This is, per unit mass flow and equal to the enthalpy drop, the difference between the enthalpy of the supply steam (h_1) and the enthalpy of the exhaust steam (h_2).

The heat energy supplied per unit mass flow to the feed water in the boiler to produce the steam is the difference between the enthalpy of the supply steam (h_1) and the enthalpy of the feed water (hf_2), the feed water temperature being equal to the condensate temperature.

Thus, neglecting the work done by the feed pump in compressing the water as being comparatively small, the efficiency of the Ranking cycle is:

$$\text{Rankine efficiency} = \frac{h_1 - h_2}{h_1 - h_{f2}}$$

A cycle operating with steam in the wet region will be limited to a maximum at the critical state (22.1 bar, 374.15°C) which limits efficiency. Superheating raises the

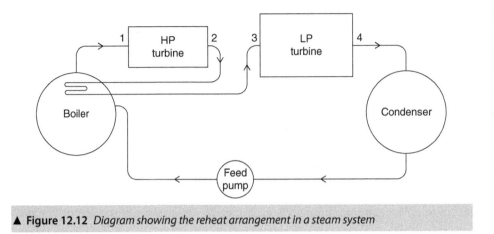

▲ **Figure 12.12** *Diagram showing the reheat arrangement in a steam system*

temperature (above saturation) without raising the boiler pressure and this increases the ideal cycle efficiency.

It is also common practice to reheat steam between stages which reduces wetness at low pressures and increases power output for a given size of units. Figure 12.12 and the example following illustrate both applications and a resulting Rankine cycle. Examples are often solved more quickly using an h–s chart.

$$\text{Rankine efficiency} = \frac{\text{Heat energy given up by steam through engine}}{\text{Heat energy supplied by boiler and reheater}}$$
$$= \frac{(h_1 - h_2) + (h_3 - h_4)}{(h_1 - h_{f4}) + (h_3 - h_2)}$$

Example 12.9. In an ideal steam reheat cycle the steam is expanded in the first stage of a turbine from 40 bar, 400°C to 3.0 bar. At this pressure the steam is passed through a reheater and its temperature is raised to 300°C at constant pressure. It then passes through the remainder of the turbine and is expanded to 0.75 bar. Calculate the Rankine efficiency.

$$h_1 \text{ at 40 bar, 400°C} = 3214$$

Isentropic expansion from 40 bar, 400°C to 3 bar:

$$\text{Entropy after expansion} = \text{Entropy before}$$
$$1.672 + x \times 5.321 = 6.769$$
$$x \times 5.321 = 5.097$$
$$x = 0.9579$$

h_2 at 3 bar, 0.9579 dry

$$= 561 + 0.579 \times 2164 = 2634$$

h_3 at 3 bar 300°C $= 3070$

Isentropic expansion from 3 bar, 300°C to 0.075 bar:

$$\text{Entropy after expansion} = \text{Entropy before}$$
$$0.57 + x \times 7.674 = 7.702$$
$$x \times 7.674 = 7.126$$
$$x = 0.9287$$

h_4 at 0.075 bar, 0.9287 dry
$$= 169 + 0.9287 \times 2405 = 2403$$

h_{f4} at 0.075 bar $= 169$

$$\text{Rankine efficiency} = \frac{(h_1 - h_2) + (h_3 - h_4)}{(h_1 - h_{f4}) + (h_3 - h_2)}$$
$$= \frac{(3214 - 2634) + (3070 - 2403)}{(3214 - 169) + (3070 - 2634)}$$
$$= \frac{580 + 667}{3045 + 436} = \frac{1247}{3481}$$
$$= 0.3583 \text{ or } 35.83\% \quad \text{Ans.}$$

Actual Steam Cycles

An estimate of the actual vapour (steam) cycle efficiency can be arrived at by amending the isentropic process of expansion (and compression) using an efficiency factor. Net work is the turbine (expansion), which is actual work minus the actual compressor (compression) or pump work.

ISENTROPIC EFFICIENCY. When an actual process is compared with an isentropic process the resulting efficiency is called 'isentropic efficiency'.

If we consider steam passing through, for example, a turbine, the ratio of the actual enthalpy drop to the isentropic enthalpy drop is the isentropic efficiency. This can be easily shown on a h–s chart.

$$h = \text{Specific enthalpy kJ/kg}; \; s = \text{Specific entropy kJ/kg K}$$
$$\text{Isentropic efficiency} = \frac{\text{Actual enthalpy drop}}{\text{Isentropic enthalpy drop}} \times 100\%$$
$$= \frac{h_a}{h_s} \times 100\%$$

Example 12.10. Steam at 50 bar and 400°C is expanded in a turbine to 2 bar and 0.96 dry. It is then reheated at constant pressure to 250°C and finally expanded to 0.06 bar

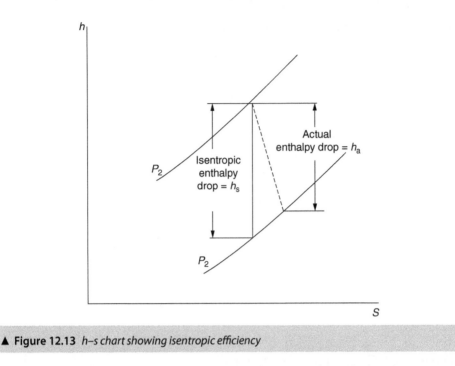

▲ **Figure 12.13** *h–s chart showing isentropic efficiency*

with isentropic efficiency of 0.8. Using the *h–s* chart, find (i) the isentropic efficiency of the first expansion (ii) the total power developed for a mass flow of 3 kg/s.

If the reheater was by-passed, determine, if its isentropic efficiency was unaltered, the loss of power in the second stage.

Specific enthalpy values read from the chart (Figure 12.14) as follows:

$$h_1 = 3195 \text{ kJ/kg}, h_3 = 2970 \text{ kJ/kg}, h_s = 2120 \text{ kJ/kg}$$
$$h_2' \ 2615 \text{ kJ/kg} \ h_4' = ?$$
$$h_2 = 2512 \text{ kJ/kg}$$
$$h_4 = 2370 \text{ kJ/kg}$$

$$\text{Isentropic efficiency of first stage} = \frac{h_1 - h_2'}{h_1 - h_2}$$
$$= \frac{3195 - 2615}{3195 - 2515} \times 100\%$$
$$= 85\%$$

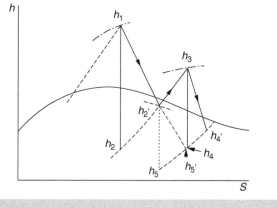

▲ **Figure 12.14** *Specific enthalpy values*

Actual enthalpy drop in second stage = Isentropic enthalpy drop × Isentropic efficiency

$$= \left(h_3 - h_4\right) \times 0.8$$
$$= \left(2970 - 2370\right) \times 0.8$$
$$= 480 \text{ kJ/kg}$$

Total power developed $= \dot{m}[(h_1 - h_2') + (h_3 - h_4')]$

$$= 3\left[580 + 480\right]/10^3$$
$$= 318 \text{ MW}$$

With reheater by-passed:

Actual enthalpy drop in second stage $= \left(h_2' - h_5\right) \times 0.8$
$$= \left(2615 - 2120\right) \times 0.8$$
$$= 396 \text{ kJ/kg}$$

Power loss in second stage = Original power − New power

$$= 3\left[480 - 396\right]$$
$$= 252 \text{ kW}$$

Thermal Efficiency

The thermal efficiency of an engine is the ratio of the heat energy converted into work in the engine to the heat energy supplied. For a steam engine this is the heat energy that is transferred to the steam in the boilers, starting from the feed water at condenser condensate temperature. Thus:

Energy [kJ] supplied = Steam consumption [kg] $\times (h_1 - hf_2)$ [kJ/kg]

where h_1 = Enthalpy per kg of supply steam

hf_2 = Enthalpy per kg of water at saturation temperature

corresponding to condenser pressure.

Hence, on a time basis of one second:

$$\text{Thermal efficiency} = \frac{\text{Power [kW = kJ/s]}}{\text{Steam consumption [kg/s]} \times (h_1 - h_{f2})}$$

or, on a time basis of 1 h (3600 seconds):

$$\text{Thermal efficiency} = \frac{\text{Power [kW]} \times 3600}{\text{Steam consumption [kg/h]} \times (h_1 - h_{f2})}$$

or, on a basis of 1 kWh:

$$\text{Thermal efficiency} = \frac{3600}{\text{Specific steam cons. [kg/kWh]} \times (h_1 - h_{f2})}$$

Gas Turbine Cycles

Gas turbines have become much more common in cruise ships and superyachts. However, as hybrid technology starts to develop, the 'fuel hungry' gas turbines are falling out of favour again.

Marine type gas turbines work on the ideal constant pressure (Joule) cycle. Figure 12.15 is a diagrammatic sketch of a simple open cycle gas turbine plant which consists of three essential parts – air compressor, combustion chamber and turbine. Referring to Figure 12.15, air is drawn in from the atmosphere and compressed from $p_1 V_1 T_1$ to the higher pressure, smaller volume and higher temperature $p_2 V_2 T_2$. The compressed air is delivered to the combustion chamber. Some of this air is used for burning the fuel which is admitted through the burner into the combustion chamber, the remainder of the air passes through the jacket surrounding the burner housing, mixes with the products of combustion and is heated at constant pressure while the volume and temperature increases, the conditions now being $p_3 V_3 T_3$. The mixture of hot air and gases now passes through the turbine where it expands to $p_4 V_4 T_4$ as it does work in driving the rotor. Finally, the gases exhaust at constant pressure.

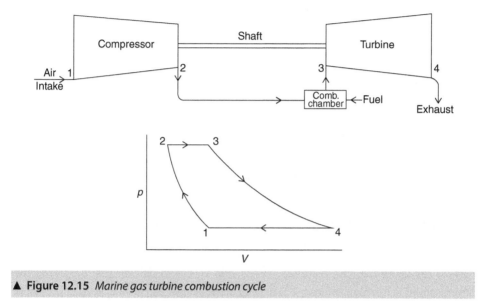

Of the power developed in the turbine, some is absorbed in driving the compressor, the remainder being available for external use such as for propulsion or driving an electric generator. A starting motor is fitted at the opposite end of the shaft to that for the external drive.

In the ideal cycle, compression and expansion are isentropic between the same pressures p_2 and p_1 following the law $pVr = a$ constant. Referring to Figure 12.15:

$$\text{Ideal thermal efficiency} = \frac{\text{Heat energy converted into work}}{\text{Heat energy supplied}}$$

$$= \frac{\text{Heat supplied} - \text{heat rejected}}{\text{heat supplied}}$$

$$= 1 - \frac{\text{Heat rejected}}{\text{Heat supplied}}$$

$$= 1 - \frac{m \times c_p \times (T_4 - T_1)}{m \times c_p \times (T_3 - T_2)}$$

$$= 1 - \frac{T_4 - T_1}{T_3 - T_2} \qquad \text{(i)}$$

$$\text{Let } r_p = \text{Pressure ratio} = \frac{P_2}{P_1} = \frac{P_3}{P_4}$$

$$\frac{T_2}{T_1} = \left\{\frac{P_2}{P_1}\right\}^{\frac{\gamma-1}{\gamma}} \qquad \therefore T_2 = T_1 \times r_p^{(\gamma-1)/\gamma}$$

$$\frac{T_3}{T_4} = \left\{\frac{P_3}{P_4}\right\}^{\frac{\gamma-1}{\gamma}} \qquad \therefore T_3 = T_4 \times r_p^{(\gamma-1)/\gamma}$$

$$T_3 - T_2 = r_p^{(\gamma-1)/\gamma}(T_4 - T_1)$$

Substituting this value of $T_3 - T_2$ into (i):

$$\text{Ideal thermal efficiency} = 1 - \frac{1}{r_p^{(\gamma-1)/\gamma}} \qquad \text{(ii)}$$

Showing that the ideal thermal efficiency depends upon the pressure ratio.

Example 12.11. In a simple gas turbine plant working on the ideal constant pressure cycle, air is taken into the compressor at 1 bar, 16°C and delivered at 5.5 bar. If the temperature at turbine inlet is 700°C, calculate (i) the temperature at the end of compression, (ii) temperature at exit from the turbine and (iii) the ideal thermal efficiency. Take $\gamma = 1.4$.

$$\text{Pressure ratio } r_p = \frac{P_2}{P_1} = \frac{P_3}{P_4} = \frac{5.4}{1} = 5.4$$

$$\frac{\gamma-1}{\gamma} = \frac{0.4}{1.4} = \frac{2}{7}$$

$$r_p^{(\gamma-1)/\gamma} = 5.4^{2/7} = 1.619$$

$$\frac{T_2}{T_1} = \left\{\frac{P_2}{P_1}\right\}^{\frac{\gamma-1}{\gamma}}$$

$$T_2 = 289 \times 1.619 = 467.8 \text{ K}$$

Temperature at end of compression
$$= 194.8°C \quad \text{Ans.} \quad \text{(i)}$$

$$\frac{T_4}{T_3} = \left\{\frac{P_4}{P_3}\right\}^{\frac{\gamma-1}{\gamma}}$$

$$T_4 = \frac{973}{1.619} = 601 \text{ K}$$

Temperature at end of expansion
$$= 328°C \quad \text{Ans.} \quad \text{(ii)}$$

$$\text{Ideal thermal efficiency} = 1 - \frac{T_4 - T_1}{T_3 - T_2}$$

$$= 1 - \frac{601 - 289}{973 - 467.8}$$

$$= 0.3824 \text{ or } 38.24\% \quad \text{Ans.(iii)}$$

Alternatively:

$$\text{Ideal thermal efficiency} = 1 - \frac{1}{r_p^{(\gamma-1)/\gamma}}$$

$$= 1 - \frac{1}{1.619} = 0.3823$$

HEAT EXCHANGER. By including a heat exchanger, some of the heat energy in the exhaust gases can be utilized by transferring it to the air before it enters the combustion chamber, resulting in less fuel being needed to raise the temperature of the air to the required turbine inlet temperature and therefore increasing the thermal efficiency. A typical arrangement is shown diagrammatically in Figure 12.16.

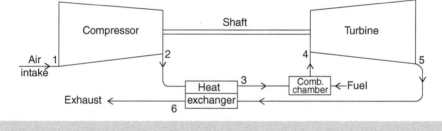

▲ **Figure 12.16** *System for pre-heating the air before combustion*

The lowest temperature entering the heat exchanger is that of the compressed air T_2 and the highest temperature is that of the exhaust gases from the turbine T_5. The difference $T_5 - T_2$ is the overall temperature range for the exchanger. The air, on its passage through the exchanger, is increased in temperature from T_2 to T_3, that is, an increase of $T_3 - T_2$. The ratio of the increase in air temperature to the overall temperature range is termed the *thermal ratio* or *effectiveness* of the heat exchanger. Thus:

$$\text{Thermal ratio of heat exchanger} = \frac{T_3 - T_2}{T_5 - T_2}$$

ACTUAL CYCLE EFFICIENCY. The actual processes of compression and expansion can be estimated by applying the isentropic efficiency to the ideal isentropic processes. Isentropic efficiency is determined from enthalpy change, which, for a perfect gas, is given by:

$$\Delta h = mc_p \Delta T$$

The processes and cycle for gas turbines are best illustrated, as in Figure 12.17, by a T–S diagram.

Example 12.12. An open cycle gas turbine unit has a pressure ratio of 6:1 and a maximum cycle temperature of 760°C. The isentropic efficiency of the compressor and turbine is 0.85. The minimum cycle temperature is 15°C.

(i) Sketch the cycle on a temperature–entropy diagram.

(ii) Calculate the thermal efficiency of the unit.

For the working substance $c_p = 1005$ J/kg K; $\gamma = 1.4$.

Refer to Figure 12.17. Ans. (i)

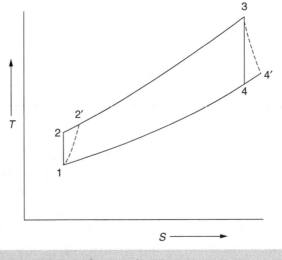

▲ **Figure 12.17** *Typical T–S diagram for gas turbine engine*

$$\frac{T_2}{T_1} = \left\{ \frac{p_2}{p_1} \right\}^{\frac{\gamma-1}{\gamma}}$$

$$T_2 = 288 \times 6^{\frac{0.4}{1.4}}$$

$$= 480.4 \text{ K}$$

$$T_2 - T_1 = 480.4 - 288$$

$$= 192.3 \text{ K}$$

$$T_2' - T_1 = \frac{192.3}{0.85}$$

$$= 226.3 \text{ K}$$

$$T_2' = 226.3 + 288$$

$$= 514.3 \text{ K}$$

$$\frac{T_3}{T_4} = \left\{ \frac{p_3}{p_4} \right\}^{\frac{\gamma-1}{\gamma}}$$

$$\frac{1033}{T_4} = 6^{\frac{0.4}{1.4}}$$

$$T_4 = 619.3 \text{ K}$$

$$T_3 - T_4 = 1033 - 619.3$$

$$= 413.7 \text{ K}$$

$$T_3 - T_4' = 0.85 \times 413.7$$

$$= 351.6 \text{ K}$$

$$T_4' = 681.4 \text{ K}$$

$$\text{Thermal efficiency} = \frac{\text{Network}}{\text{Heat supplied}} \times 100$$

$$= \frac{351.6 - 226.3}{1033 - 514.3} \times 100$$

$$= 24.16\% \quad \text{Ans. (ii)}$$

Note: c_p and $\dot{m}$ are common to each term in the above expression for thermal efficiency and cancel out, leaving temperatures. c_p, although given, is not required.

Test Examples 12

1. Dry saturated steam at 8 bar is expanded in turbine nozzles to a pressure of 5 bar, 0.97 dry. Find (i) the velocity at exit. If the area at exit is 14.5 cm², find (ii) the mass flow of the steam in kg/s.

2. The velocity of the steam from the nozzles of a turbine is 450 m/s, the angle of the nozzles to blade motion is 20° and the blade entrance angle is 33°. Find (i) the linear velocity of the blades so that the steam enters without shock, and (ii) the rotational speed of the rotor in rev/s if the mean diameter of the blade ring is 660 mm.

3. At a certain stage of a reaction turbine, the steam leaves the guide blades and enters the moving blades at an absolute velocity of 243 m/s at an angle of 23° to the plane of rotation, and the blade velocity is 159 m/s. Fixed and moving blades have the same inlet and exit angles and the steam flow is 0.9 kg/s. Calculate (i) the inlet angle of the blades, (ii) the force on the blades and (iii) the stage power.

4. A single stage impulse steam turbine has symmetrical rotor blades with inlet and outlet angles of 35°. The mean blade diameter is 600 mm, the turbine runs at 100 rev/s and the steam leaves the rotor in the axial direction:

    (i)   Determine:

        (a) the nozzle angle;

        (b) the velocity of steam leaving the nozzle;

        (c) the power developed per kg of steam flow.

    (ii) State the major assumption made to solve this problem.

5. Dry saturated steam at 14 bar is expanded in a turbine nozzle to 10 bar, expansion following the law pVn = constant, where the value of n is 1.135. Calculate:

    (i)   the dryness fraction of the steam at exit;

    (ii) the enthalpy drop through the nozzle per kg of steam;

    (iii) the velocity of discharge;

    (iv) the area of nozzle exit in mm² per kg of steam discharged per second.

6. A perfect gas expands from 7 bar, 150°C to atmospheric pressure isentropically through a convergent–divergent nozzle. If the mass flow rate of the gas through the nozzle is 0.25 kg/s determine:

    (i)   the velocity of the gas at the nozzle throat;

    (ii) the throat area.

    Note: $\gamma = 1.67$, $R = 2078.5$ J/kg K and $c_p = 833.9$ J/kg K for the gas.

    Also $\dfrac{p_T}{p_1} = \left[\dfrac{2}{\gamma+1}\right]^{\frac{\gamma-1}{\gamma}}$ is the critical pressure ratio for a gas expanding in a convergent–divergent nozzle according to the law pV = constant, where p_T and p_1 are the pressures at throat and inlet, respectively.

7. In an impulse turbine the theoretical enthalpy drop of the steam through the nozzles is 312.5 kJ/kg and 10% of this is lost in friction in the nozzles. The nozzle angle is 20°, the inlet angle of the blades is 35° and the absolute velocity of the steam leaving the blades is 204 m/s in the direction of the axis of the turbine. Calculate on the basis of 1 kg of steam supplied per second:

   (i)   blade velocity so that there is no shock at steam entry;

   (ii)   blade angle at exit;

   (iii) energy lost due to friction of the steam across the blades;

   (iv)   axial thrust;

   (v)   power supplied;

   (vi)   efficiency of the blading.

8. An engine is supplied with steam at a pressure of 15 bar and temperature 250°C, and the pressure of the exhaust is 0.16 bar. Assuming isentropic expansion, find (i) the dryness fraction of the steam after expansion and (ii) the Rankine efficiency.

9. Dry saturated steam at 20 bar is throttled to supply a turbine at 10 bar, and this is followed by isentropic expansion to 1.4 bar. It is then reheated at constant pressure to 200°C and finally expanded isentropically to 0.4 bar. Using the $h-s$ chart determine per kg of steam,

   (i)   changes in enthalpy during each stage;

   (ii)   the overall change in entropy;

   (iii) the condition of the steam at the end of expansion.

10. Air is drawn into a gas turbine working on the constant pressure cycle, at 1 bar, 21°C and compressed to 5.7 bar. The temperature at the end of heat supply is 680°C. Taking expansion and compression to be adiabatic where $c_v = 0.718$ kJ/kg K, $c_p = 1.005$ kJ/kg K, calculate (i) temperature at end of compression, (ii) temperature at exhaust, (iii) heat energy supplied per kg at constant pressure, (iv) increase in internal energy per kg from inlet to exhaust and (v) ideal thermal efficiency.

13

COMBUSTION IN BOILERS AND IC ENGINES

There are two types of boiler: (i) the fire-tube boiler, in which the hot gases from the combustion of the fuel in the furnace pass through the tubes, while the water is around the outside of the tubes; and (ii) the water-tube boiler, in which the water flows through the tubes while the hot gases pass around the outside of the tubes.

Capacity and Equivalent Evaporation

The *capacity* of a boiler is the mass of steam that it can produce in a given time, which is usually within one hour. However, since the feed water temperature, and pressure and temperature of the steam vary with different plants, it is necessary for purposes of comparison to refer the evaporative capacity to a common standard. This reference, standard condition, assumes that the feed water temperature is at 100°C and will be converted into dry saturated steam at 100°C. This will then give a direct comparison of the steam evaporated by a unit mass of fuel burned in the furnace. This comparison is termed the *equivalent evaporation, per kg of fuel, from and at* 100°C. For clarity the 'per kg of fuel' is usually omitted.

Thus, if, for example, a boiler produces 60 Mg of steam per hour at 40 bar and 450°C, from feed water at 130°C, when 4800 kg of fuel are burned per hour, then:

Tables   page   7,   Steam 40 bar 450°C, $h = 3330$

Tables   page   4,   Water at 130°C, $h = 546$

Heat energy transferred to each kg of steam

$$= 3330 - 546 = 2784 \text{ kJ}$$

Actual evaporative capacity is 60×10^3 kg/h.

Therefore, heat energy transferred to steam

$$= 2784 \times 60 \times 10^3 \text{ kJ/h}$$

Now suppose the temperature of the feed water was 100°C and dry saturated steam at 100°C was produced, the heat energy required per kilogram of steam would be h_{fg} at 100°C which is 2256.7 kJ (Steam Tables page 2 – please see the full reference on page 159 of this book), and therefore the mass of steam that would be produced under these conditions of 'from and at 100°C' would be:

$$= \frac{\text{Total heat energy transfered to steam per hour}}{\text{Heat energy required for each kg}}$$

$$= \frac{2784 \times 60 \times 10^3}{2256.7} = 74 \times 10^3 \text{ kg/h or 74 Mg/h}$$

This is the capacity of the boiler from and at 100°C.

Since 4800 kg of fuel are burned per hour, the mass of steam that would be produced by each kg of fuel is

$$\frac{74 \times 10^3}{4800} = 15.42 \text{ kg steam kg fuel}$$

This is the equivalent evaporation, per kg of fuel, from and at 100°C.

Summing up,

if m_s = actual mass of steam generated per unit mass of fuel burned

h_1 = enthalpy per kg of steam

h_w = enthalpy per kg of feed water

h_{fg} 100°C = enthalpy of evaporation per kg at 100°C,

then, equivalent evaporation from and at 100°C

$$= \frac{m_s(h_1 - h_w)}{h_{fg}\,100°C}$$

Boiler Efficiency

The efficiency of a boiler is the ratio of the heat energy transferred to the feed water when converting it into steam, to the heat energy supplied to the boiler by the combustion of the fuel.

The heat energy transferred to the water to produce steam is the difference between the enthalpy of the steam leaving the boiler and the enthalpy of the feed water entering the boiler, thus:

$$\text{mass of steam generated} \times (h_1 - h_w)$$

The heat energy supplied to the boiler is the energy released during combustion of the fuel, which is the product of the mass of fuel burned and its calorific value. The calorific value, as previously explained (see Chapter 7), is the heat energy given off during complete combustion of a unit mass of the fuel. Therefore:

$$\text{Boiler efficiency} = \frac{\text{Heat energy transfered to water and steam}}{\text{Heat energy supplied by fuel}}$$

$$= \frac{\text{Mass of steam} \times (h_1 - h_w)}{\text{Mass of fuel} \times \text{Calorific value}}$$

or, if we let m_s represent the mass of steam generated per unit mass of fuel burned:

$$\text{Boiler efficiency} = \frac{m_s(h_1 - h_w)}{\text{Calorific value of the fuel}}$$

Example 13.1. An oil-fired boiler working at a pressure of 15 bar generates 14.5 kg of steam per kg of fuel burned. The feed water temperature is 95°C and the steam leaves the boiler 0.98 dry. If the calorific value of the oil is 42 MJ/kg, calculate the thermal efficiency of the boiler and the equivalent evaporation from and at 100°C.

$$\text{Tables page 4,} \quad \text{steam 15 bar,} \quad h_f = 845 \quad h_{fg} = 1947$$

$$\text{Tables page 2,} \quad \text{water 95°C,} \quad K = 398$$

$$\text{Boiler steam,} \quad h_1 = h_f + xh_{fg}$$

$$= 845 + 0.98 \times 1947 = 2753$$

Heat energy transferred to steam per kg of fuel burned

$$= m_s(h_1 - h_w)$$
$$= 14.5(2753.398)$$
$$= 14.5 \times 2355 \text{ kJ}$$

Heat energy supplied to boiler per kg of fuel burned

$$= 42 \times 10^3 \text{ kJ}$$

$$\text{Boiler efficiency} = \frac{14.5 \times 2355}{42 \times 10^3}$$

$$= 0.8132 \text{ or } 81.32\% \quad \text{Ans.} (i)$$

Equivalent evaporation, per kg of fuel, from and at 100°C

$$= \frac{m_s (h_1 - h_w)}{h_{fg} \, 100°C}$$

$$= \frac{14.5 \times 2355}{2256.7}$$

$$= 1513 \text{ kg steam/kg fuel} \quad \text{Ans.} (ii)$$

Feed Water

Every precaution must be taken to maintain water as pure as possible. There are a couple of sources of contamination. One is from sea water (where sea water is still used as a primary source of cooling) and the other is from oil as the steam will be used as the heating source for oil heaters. Eventually joints can leak. However, modern materials and manufacturing techniques are much better than they were and contamination is far less common than in the past.

There is always a certain amount of fluid loss in the steam circuit, and this loss must be made up to maintain the working level of the water in the boiler. Make-up feed water may be taken direct from the reserve fresh water tanks, or distilled water may be produced by evaporating sea water in evaporators. Due to the purity of feed water required, it is usual to pass all make-up feed water through evaporators.

The quantity of dissolved solids present in water is expressed in parts of solids per million parts of water, abbreviated to parts per million and represented by ppm. This is the same ratio as the grams of solids in 1 million grams of water. Note that 10^6 g = 10^3 kg = 1 tonne, and this is the mass of 1 m^3 of pure water. For example, if there are 80 g of dissolved solids in 1 m^3 of water, it is expressed as 80 ppm. As a further example, if 5 tonnes of water has 150 ppm dissolved solids, then the total mass of dissolved solids is:

$$5 \times 150 = 750 \text{ g}$$

Equations dealing with the amount of dissolved solids in evaporators and boilers are considered on this basis. Thus, in boilers with a contaminated feed water, the equation is built upon the simple principle:

Initial mass [g] of solids in boiler	+	Mass [g] of solids put in with feed water	=	Final mass [g] of solids in boiler

Mass [t] of water in boiler	×	Initial ppm	+	Mass [t] of feed	×	Feed ppm	=	Mass [t] of water in boiler	×	Final ppm

In fresh water generators (evaporators) where it is common to have a constant blow-down to maintain a steady pre-determined density of the water in the machine, the equation would be based upon:

Mass of solids put in = Mass of solids blown out

Mass of feed × Feed ppm = Mass blown out × Blow out ppm

If the evaporated water contains a given amount of solids, this would be included on the right-hand side of the above equation.

Example 13.2. A boiler initially contains 4 tonne of water of 80 ppm. If the evaporation rate is 500 kg/h and the feed water contains 150 ppm of dissolved solids, calculate (i) the density (ppm) of the boiler water after 12 h; and (ii) the time for the water to reach 2000 ppm from the time of initial condition.

$$\text{Amount of feed} = \text{Evaporation rate}$$
$$= 500 \text{ kg/h} = 0.5 \text{ tonne/h}$$
$$= 0.5 \times 12 \text{ tonne in 12 h}$$

Solids in initially + Solids put in = Solids in finally

Water in boiler	×	Initial ppm	+	Amount of feed	×	Feed ppm	=	Water in boiler	×	Final ppm

$$4 \times 80 + 0 - 5 \times 12 \times 150 = 4 \times \text{Final density}$$
$$320 + 900 = 4 \times \text{Final density}$$
$$\text{Final density} = \frac{1220}{4} = 305 \text{ ppm} \quad \text{Ans. (i)}$$

Let t = time in hours for boiler water to reach 2000 ppm; then feed in t hours = 0.5t tonne:

Water in boiler	×	Initial ppm	+	Amount of feed	×	Feed ppm	=	Water in boiler	×	Final ppm

$$4 \times 80 + 0.5_t \times 150 = 4 \times 2000$$
$$75t = 7680$$
$$t = 102.4 \text{ h} \quad \text{Ans.} \quad \text{(ii)}$$

If an analysis of the feed water is given so that a distinction can be made with regard to the solids that remain in solution and those that precipitate to form scale when the water is heated, then the density in the evaporator or boiler is taken as that due to permanently soluble solids.

The density of sea water and the composition of the dissolved solids vary throughout different parts of the world. A typical sample could be taken as dissolved solids 32 280 ppm and mass analysis of the solid matter would show as:

Sodium chloride   79.3%

Magnesium chloride   10.2%

Magnesium sulfate   6.1%

Calcium sulfate   3.8%

Calcium bicarbonate   0.6%

Scale is formed from magnesium sulfate, calcium sulfate and calcium bicarbonate. Scale is a very poor conductor of heat and therefore it will reduce the transfer of heat to a cooling medium/surface. The result is overheating of the metal heating surfaces with consequent loss of strength and possibility of collapse. The remainder of the solids in the sea water may be regarded as permanently soluble solids.

Principles of Combustion (Applicable to Boilers and Internal Combustion Engines)

Combustion is the rapid oxidation of a fuel which releases heat. The fuel can be a solid, liquid or gas. To start the combustion of the fuel, oxygen and a high temperature are required, then additional heat energy and some light are released throughout the process. The heat energy release comes from the atoms breaking and reforming into a new arrangement. For example, in a perfect reaction, fossil fuel and oxygen becomes carbon dioxide and water as the heat is released.

Liquid and gaseous marine fuels are a carbon and hydrogen mix, in different percentages of composition, and in most marine applications controlled combustion will take place in the furnaces of boilers and the cylinders of internal combustion engines.

The oxygen required for combustion will be taken from the air that is supplied to the machinery space by powerful fans. This 'atmospheric' air is made up of nitrogen, oxygen, argon, carbon dioxide and traces of other chemicals such as neon, helium and krypton. Due to the different molecular weights of the chemicals there is a slightly different percentage composition by volume and by mass. See table below:

Chemical	% (Volume)	% (Mass)	K-mol
Nitrogen	78.09	75.47	28.02
Oxygen	20.95	23.2	32
Argon	0.933	1.28	39.94
Carbon dioxide	0.026	0.046	44.01
Others	0.001	0.004	

In the modern world there is considerable focus on the use of different fuels. Within the air entering the engine, oxygen is the active element required for combustion; however, the attention has now shifted to nitrogen. This is due to the fact that modern diesel engines produce very high peak temperatures and at these temperatures nitrogen (N_2) oxidises into oxides of nitrogen (NOx), which are released into the atmosphere.

Metals can be used as fuel but are not generally considered in marine applications due to their very high ignition temperatures. It is important to note this point as metal fires have been known to occur on-board ship. These are so hot that if water is introduced into them the water is broken down into hydrogen and oxygen that can then work to add more fuel to the fire.

The difficulty comes when a physical machine needs to be built to carry out the theoretical process. With combustion the oxygen from the air must be intimately mixed with the fuel to give as close a homogeneous mix as possible so that the oxidation process (burning) can occur. Therefore, the amount of fuel that can be burned, at any one time, depends upon the quantity of oxygen (air) that is supplied.

The fuel and the oxygen need to interact at the molecular level; however invariably they are being introduced into a 'combustion space' from different starting points and in large quantities, making mixing more difficult.

The fuel and the oxygen are then required to mix microscopically in a fraction of a second. Advances in material science have enabled better mechanical components and this, together with the introduction of electronic control, has enabled the mixing to become more efficient. Despite these developments, the combustion process in the real world is still not absolutely complete. It is only gas or vaporized mixtures that can achieve true homogeneity.

Excess air, over and above the theoretical minimum quantity of air for complete combustion, is always necessary as the fuel and air must find each other at a molecular level in the combustion space, and this is difficult to achieve in the time allowed.

The amount of excess air depends upon the design of the combustion space and conditions under which the fuel is burned. If there is an insufficient air supply, combustion will not be complete, one indication of this being black smoke and the soot produced. If too much air is supplied, an unnecessary amount of heat energy will be carried away to waste. Each case represents a loss of efficiency.

Air in excess of that required to complete the combustion process will also carry oxygen that will then be able to react with the nitrogen and produce more NOx, which will enter the exhaust. Exhaust gas recirculation (EGR) used in modern diesel engines reduces the 'excess' oxygen as well as having a higher specific heat capacity than air, and thus lowers the thermal NOx that ends up in the exhaust. (See page 244 on the environmental improvement of ships' exhausts.)

Approximating the composition of air by mass as 23% oxygen and 77% nitrogen/other gases, then 23 kg of oxygen will be obtained from 100 kg of air. Therefore, applying the same proportion, to obtain 1 kg of oxygen, the supply of air must be:

$$\frac{100}{23}\text{kg}$$

Students will be able to deduce that the theoretical minimum mass of air required for a theoretical complete combustion will be:

$$100/23 \times \text{the mass of } O_2 \text{ required}$$

Elements. These are pure substances they consist of only *one* type of atom and cannot be broken down by non-nuclear processes into simpler substances. They are represented by their own symbol, usually the first letter of their names except when necessary to distinguish between two whose names have the same first letter. Hydrogen is an element since it cannot be broken down further, is oxygen. Water, however, is not an element as it can be broken down into hydrogen and oxygen. The molecules of some elements consist of single atoms, such as carbon (C) and sulphur (S). Other elements consist of molecules of two atoms each and these are distinguished by the subscript 2, examples of these being hydrogen (H_2), nitrogen (N_2) and oxygen (O_2).

Compounds. These are chemical combinations of different elements and are denoted by placing together the symbols of the elements which constitute the compound. Thus, each molecule of water (and steam) is composed of two atoms of hydrogen and one atom of oxygen and is therefore written as H_2O. Carbon dioxide is represented by

CO_2 which signifies that each molecule of carbon dioxide is composed of one atom of carbon and two atoms of oxygen. The number of molecules, when more than one, is represented by placing that number in front, thus $3CO_2$ represents three molecules of carbon dioxide.

Atomic number. This is a number assigned to a substance that denotes the number of 'protons' that are part of the makeup of the nucleus in the atom of the substance. Hydrogen has one proton that makes up its nucleus and therefore its *atomic number* is 1. Oxygen has 8 protons and therefore has an atomic number of 8. It also means that all atoms with 8 protons are oxygen atoms. Normal atoms are electrically balanced, and therefore the atomic number is also representative of the number of electrons present in an atom.

Isotopes. The majority of the atoms inside each element conform to a standard format which is that the number of protons are balanced by the same number of neutrons and electrons. In this configuration the element is stable. However, if the nucleus contained extra neutrons, then the element would still be stable, but it would have extra components, that is, the neutrons. Such elements, which have a different number of neutrons in their atoms, are termed 'isotopes'.

Atomic weight. The atomic weight of a substance is the averaged sum of the atomic weights, including all of its isotopes, of the atoms of which the molecules are composed. A list of the masses of the substances involved in the calculations on the combustion of fuels is given in the following table. The actual weights are calculated to 3 decimal

Substance	Symbol	Atomic weight	Molecular weight
Elements			
Hydrogen	H_2	1	$2 \times 1 = 2$
Carbon	C	12	$1 \times 12 = 12$
Nitrogen	N_2	14	$2 \times 14 = 28$
Oxygen	O_2	16	$2 \times 16 = 32$
Sulphur	S	32	$1 \times 32 = 32$
Compounds			
Water, steam	H_2O		$2 \times 1 + 1 \times 16 = 18$
Carbon dioxide	CO_2		$1 \times 12 + 2 \times 16 = 44$
Sulphur dioxide	SO_2		$1 \times 32 + 2 \times 16 = 64$
Carbon monoxide	CO		$1 \times 12 + 1 \times 16 = 28$

places. However, the numbers are close to whole numbers and the whole numbers are sufficiently accurate for practical calculations.

STOICHIOMETRY. This is the term given to the calculation of the exact proportion of elements that are required to make pure chemical compounds. In the examples that follow the expression *stoichiometric air requirement* is sometimes used. The stoichiometric air requirement is the exact amount of air required to carry the correct quantity of oxygen for complete combustion of the fuel being used for combustion.

LAMBDA READING. The air-to-fuel ratio is given as a lambda reading for a given fuel. 1.0 is the stoichiometric reading for any given fuel. This is the air-to-fuel ratio required for complete combustion of the fuel and is 14.7:1 (by mass) for gasoline (petrol) engines. This is an important development as it is crucial to modern engine control technology in smaller engines, whereas the larger marine engines tend to use 'knock sensors' to inform the combustion control system.

CALORIFIC VALUES. The calorific value (cv) of a substance is the amount of heat energy released during complete combustion of unit mass of that substance and, for fuels, it is usually expressed in megajoules of energy per kilogram of mass [MJ/kg]. When hydrogen burns it combines with oxygen to form steam and the heat energy released is about 144 MJ/kg of hydrogen.

Carbon, if supplied with sufficient oxygen, will burn completely to carbon dioxide and, in doing so, about 33.7 MJ of heat energy is released per kilogram of carbon. If there is a deficiency of oxygen, some or all of the carbon will burn to carbon monoxide and the heat energy released by each kilogram of carbon is then only about one-third of that when carbon dioxide is formed. Thus there is a great loss when carbon monoxide is produced due to an insufficient air supply to the fuel.

Also by looking up the calorific values of substances we see that with the burning of sulphur, it chemically combines with oxygen to form sulphur dioxide and about 9.3 MJ of heat energy is released per kilogram of sulphur.

Students can see that taken at face value hydrogen seems to show potential as a possible 'clean' fuel. Some interesting observations are explored below.

CHEMICAL EQUATIONS. They enable us to calculate the proportions in which the elements combine and then by substituting the atomic weights, the relative mass of oxygen required for the burning of each combustible element can be calculated.

For combustion of hydrogen to water plus heat (steam) we need the hydrogen and oxygen in the correct proportions; therefore:

$$2H_2 + O_2 = 2H_2O$$

Inserting the atomic weights gives:

$$2 \times (2 \times 1) + 2 \times 16 = 36$$
$$4 \text{ kg } H_2 + 32 \text{ kg } O_2 = 36 \text{ kg } H_2O$$
$$1 \text{ kg } H_2 + 8 \text{ kg } O_2 = 9 \text{ kg } H_2O$$

That is, 1 kg of hydrogen requires 8 kg of oxygen to burn it completely, and this produces 9 kg of steam.

Note: The above reaction is a complete 'oxidation' process where heat is liberated in significant quantities. Alternatively, hydrogen and oxygen might react in the following way, called 'oxygen reduction':

$2H_2 + O_2 = H_2O_2$ (hydrogen peroxide or hydroxide): this happens inside some hydrogen fuel cells.

Combustion of carbon combined with oxygen produces carbon dioxide + heat:

$$C + O_2 = CO_2$$
$$1 \times 12 + 2 \times 16 = 44$$
$$12 \text{ kg } C + 32 \text{ kg } O_2 = 44 \text{ kg } CO_2$$
$$1 \text{ kg } C + 2\tfrac{2}{3} \text{ kg } O_2 = 3\tfrac{2}{3} \text{ kg } CO_2$$

Thus, 1 kg of carbon requires $2\tfrac{2}{3}$ kg of oxygen to burn it completely and this produces $3\tfrac{2}{3}$kg of carbon dioxide.

Combustion of carbon to carbon monoxide (incomplete combustion):

$$2C + O_2 = 2CO$$
$$2 \times 12 + 2 \times 16 = 2 \times 28$$
$$24 \text{ kg } C + 32 \text{ kg } O_2 = 56 \text{ kg } CO$$
$$1 C + 1\tfrac{1}{3} O_2 = 2\tfrac{1}{3} CO$$

Thus, 1 kg of carbon if supplied with $1\tfrac{1}{3}$ kg of oxygen gives incomplete combustion which produces $2\tfrac{1}{3}$ kg of carbon monoxide.

Combustion of sulphur to sulphur dioxide:

$$S + O_2 = SO_2$$
$$1 \times 32 + 2 \times 16 = 64$$
$$32 \text{ kg } S + 32 \text{ kg } O_2 = 64 \text{ kg } SO_2$$
$$1 \text{ kg } S + 1 \text{ kg } O_2 = 2 \text{ kg } SO_2$$

Hence, 1 kg of sulphur requires 1 kg of oxygen to burn it and 2 kg of sulphur dioxide is produced. Students will notice that reducing the sulphur content of the fuel will have a knock-on effect and reduce the sulphur dioxide content of the exhaust gas.

The analysis of some fuels shows that a little oxygen is present; in these cases, it is usual to assume that this oxygen is combined with some of the hydrogen in the form of water and therefore this portion of hydrogen cannot be burned. We have seen above that hydrogen and oxygen combine in the proportion of 1:8, and hence the amount of hydrogen which is not available for combustion is one-eighth of the mass of oxygen present in the fuel. The remaining hydrogen in the fuel is referred to as *available hydrogen*, and thus:

$$\text{Available hydrogen} = H_2 - \frac{O_2}{8}$$

From the foregoing notes we can now write the expressions:

$$\text{Calorific value} = 33.7\,C + 144\left(H_2 - \frac{O_2}{8}\right) + 9.3\,S \text{ MJ/kg}$$

$$\text{Oxygen required} = 2^{2/3}C + 8\left(H_2 - \frac{O_2}{8}\right) + S \text{ kg } O_2/\text{kg fuel}$$

$$\begin{aligned}\text{Minimum air reqd.} \\ \text{(stoichiometric)}\end{aligned} = \frac{100}{23} \times \text{Oxygen required}$$

$$= \frac{100}{23}\left\{2^{2/3}C + 8\left(H_2 - \frac{O_2}{8}\right) + S\right\} \text{kg air/kg fuel}$$

Example 13.3. An oil fuel is composed of 86% carbon, 11% hydrogen, 2% oxygen and 1% impurities. Calculate the calorific value and stoichiometric mass of air required to burn 1 kg of the fuel, taking the calorific values of carbon and hydrogen as 33.7 and 144 MJ/kg, respectively.

$$\text{Available hydrogen} = H_2 - \frac{O_2}{8}$$

$$= 0.11 - \frac{0.02}{8} = 0.1075 \text{ kg}$$

Heat energy from 0.86 kg of carbon
$$= 0.86 \times 33.7 = 28.98 \text{ MJ}$$
Heat energy from 0.1075 kg of hydrogen
$$= 0.1075 \times 144 = 15.48 \text{ MJ}$$
Total heat energy per kg of fuel
$$= 28.98 + 15.48 = 44.46 \text{ MJ/kg} \quad \text{Ans. (i)}$$

Oxygen required to burn the carbon

$= 2\frac{2}{3}$ kg $\times 0.86 = 2.293$ kg

Oxygen required to burn the available hydrogen

$= 8 \times 0.1075 = 0.86$ kg

Total oxygen $= 2.293 + 0.86 = 3.153$ kg

Air required (stoichiometric) $= \dfrac{100}{23} \times 3.153 = 13.71$ kg air/kg fuel   Ans. (ii)

HIGHER AND LOWER CALORIFIC VALUES. H_2O formed by the hydrogen in the fuel cannot exist as water in the high temperatures of boiler flue gases or the exhaust gases from internal combustion engines. It exists in the form of steam, and this steam passing away in the waste gases carries enthalpy of evaporation which is not available as heat energy to the boiler or engine and therefore will contribute to the 'Waste Heat' within the exhaust.

Therefore, the theoretical calorific value of a fuel containing hydrogen, as calculated in the above example, is termed the *higher* (or *gross*) *calorific value* and, when the unavailable heat energy is subtracted from this, it is termed the *lower* (or *net*) *calorific value*. It has been recommended that the amount of heat energy to be considered as not available should be 2.442 MJ/kg of steam in the products of combustion. The amount of steam, as shown above, is 9 times the mass of hydrogen in the fuel.

In the previous example the mass of hydrogen in each kilogram of fuel is 0.11 kg, therefore there will be $9 \times 0.11 = 0.99$ kg of steam formed. The unavailable heat energy is therefore to be taken as $0.99 \times 2.442 = 2.418$.

Hence the lower calorific value of this fuel is:

$$44.46 - 2.418 = 42.042 \text{ MJ/kg}$$

The calorific values of solid and liquid fuels can be found experimentally by means of a bomb calorimeter. In this device a known weight of fuel is combusted and the rise in temperature of the container and its surrounding coolant is measured. The masses and specific heat values of the container and the coolant are known and therefore this, together with the temperature rise, will give the information to calculate the heat energy transferred.

Composition of Flue/Exhaust Gases

An estimate of the analysis of the flue gases can be calculated from the composition of the fuel and mass of supply air as demonstrated in Example 13.4.

Example 13.4. An oil fuel is composed of 85% carbon, 12% hydrogen, 2% oxygen and 1% incombustible solid matter. If the air supply is 50% in excess of the stoichiometric, find the mass of each product of combustion per kg of fuel burned and the percentage mass analysis of the flue gases. Under international regulations, in 2022, the sulphur should be less than 0.5% by mass.

$$\text{Stoichiometric air/kg fuel} = \frac{100}{23}\left\{2^{2/3}C + 8\left(H_2 - \frac{O_2}{8}\right)\right\}$$

$$= \frac{100}{23}\left\{2^{2/3}C + 8\,H_2 - O_2\right\}$$

$$= \frac{100}{23}\left\{2^{2/3} \times 0.85 + 8 \times 0.12 - 0.02\right\}$$

$$= \frac{100}{23} \times 3.207 = 13.94 \text{ kg air/kg fuel}$$

$$\text{Excess air} = 0.5 \times 13.94 = 6.97 \text{ kg}$$

$$\text{Actual air supply} = 13.94 + 6.97 = 20.91 \text{ kg air/kg fuel}$$

Mass of each product of combustion, Ans. (i):

Mass of CO_2 formed = $\frac{2}{3} \times 0.85 = 3.117$ kg

Mass of H_2O formed = $9 \times 0.12 = 1.08$ kg

Mass of O_2 = excess of oxygen from the excess air = 23% of 6.97 = 1.603 kg

Mass of N_2 = mass of nitrogen in all the air supply = 77% of 20.91 = 16.1 kg

Total products of combustion = $3.117 + 1.08 + 1.603 + 16.1 = 21.9$ kg/kg fuel

Alternatively, the total mass of products of combustion per kg fuel = Mass of air supplied + (Mass of fuel – Incombustibles)

$$= 20.91 + (1 - 0.01)$$

$$= 20.91 + 0.99 = 21.9 \text{ kg/kg fuel}$$

$$CO_2 = \frac{3.117}{21.9} \times 100 = 14.23\%$$

$$H_2O = \frac{1.08}{21.9} \times 100 = 4.931\%$$

$$O_2 = \frac{1.603}{21.9} \times 100 = 7.319\%$$

$$N_2 = \frac{16.1}{21.9} \times 100 = 73.52\%$$

Exhaust Gas Analysis

ORSAT APPARATUS (now obsolete but retained here to give the students a background appreciation of the original diagnostic principles). Originally it was found that exhaust gas composition could be analysed by using the apparatus first constructed by Louis

Orsat. This enables a volumetric analysis of the dry products of combustion to be made. It operated on the principle of passing a sample of the flue gases through a series of three bottles containing different chemical solutions. The chemicals in each of the bottles absorb one of the constituents in the gas, and the remaining gas is then measured each time. The reduction in volume of the sample each time represents the volume of the constituent absorbed.

The chemical solutions used for absorbing the different parts of the gas could have been:

    (i) caustic potash which absorbs the CO_2;

    (ii) pyrogallic acid which then absorbs the CO_2 and O_2;

    (iii) cuprous chloride which can absorb CO_2, O_2 and CO.

The modern systems have the ability to measure the composition of boiler flue gas or diesel engine exhaust, directly and in real time. The technology used is infrared spectrometry which uses an infrared beam projected through the exhaust gas. The way that the different molecules in the gas react with the beam will change its 'reflected' characteristics.

By measuring the profile of the reflection, the exhaust gas composition can be determined and decisions made based on the results. The output could be used to inform a computerised control system or an independent 'compliance' monitoring system.

If a volumetric analysis is used as obtained by the Orsat apparatus, then the results had to be converted into a mass analysis. By Avogadro's law, equal volumes of any gas at the same temperature and pressure contain the same number of molecules, therefore, although the *mass* of a molecule of one gas is different from the mass of a molecule of another gas, the *volume* of each molecule is the same at the same temperature and pressure.

These calculations can now be completed by a simple processor to give a reading based on mass rather than volume. Students will be able to study the logic behind these calculations and thus understand the methods of computation.

Hence the ratio of the volumes of each gas in the mixture of flue gases (N mols) can be converted into a ratio of masses by multiplying by the molecular weight of the gas (M), and from this the percentage composition by mass (m%) can be calculated.

Example 13.5. In a test on a sample of funnel gases the percentage composition by volume was found to be: $CO_2 = 8.5\%$, $O_2 = 9.5\%$, $CO = 3\%$ and the remainder 79% was assumed to be N_2. Convert these quantities into a mass analysis. See the table below where the columns are:

    DFG = Dry Flue Gas component

    N = the number of molecules in the gas

    M = the molecular weight of the gas

m = the composition by mass

m% = the percentage composition

DFG	N	M	m	m%
CO_2	8.5	44	$8.5 \times 44 = 374$	$\dfrac{374}{2974} \times 100 = 12.58\%$
O_2	9.5	32	$9.5 \times 32 = 304$	$\dfrac{304}{2974} \times 100 = 10.22\%$
CO	3.0	28	$3 \times 28 = 84$	$\dfrac{84}{2974} \times 100 = 2.82\%$
N_2	79.0	28	$79 \times 28 = 2212$	$\dfrac{2212}{2974} \times 100 = 74.38\%$
Total			2974	100.00%

In effect the N_2 actually interacts with any spare O_2 to form nitrogen oxide and nitrogen dioxide. Students can now see how combusting fossil fuels is so environmentally unfriendly.

Example 13.6. The analysis of a sample of coal burned in the furnace of a boiler is 80% carbon, 5% hydrogen, 4% oxygen and the remainder ash etc. Calculate (i) the stoichiometric air required per kg of coal, (ii) the actual mass of air if it is supplied with 70% excess and (iii) the percentage mass analysis of the products of combustion.

$$\text{Stoichiometric air} = \frac{100}{23}\left\{2^{2/3}C + 8\left(H_2 - \frac{O_2}{8}\right)\right\}$$

$$= \frac{100}{23}\left\{2^{2/3}C + 8\,H_2 - O_2\right\}$$

$$= \frac{100}{23}\left\{2^{2/3} \times 0.8 + 8 \times 0.05 - 0.04\right\}$$

$$= \frac{100}{23} \times 2.439 = 10.84 \text{ kg air/kg fuel} \quad \text{Ans. (i)}$$

Excess air $= 0.7 \times 10.84 = 7.59$

Actual air $= 10.84 + 7.59 = 18.43$ kg air/kg fuel   Ans. (ii)

Mass products of combustion per kg of coal:

$$CO_2 = 3\tfrac{2}{3}C = 3\tfrac{2}{3} \times 0.8 = 2.933 \text{ kg}$$

$$9H_2 = 9 \times 0.05 = 0.45$$

$$O_2 = 23\% \text{ of excess air} = 0.23 \times 7.59 = 1.746$$

$$N_2 = 77\% \text{ of all air} = 0.77 \times 18.43 = 14.19$$

Total $= 19 - 319$ kg

% mass analysis   Ans. (iii)

$$CO_2 = \frac{2.933}{19.319} \times 100 = 15.18\%$$

$$H_2O = \frac{0.45}{19.319} \times 100 = 2.33\%$$

$$O_2 = \frac{1.746}{19.319} \times 100 = 9.04\%$$

$$N_2 = \frac{14.19}{19.319} \times 100 = 73.45\%$$

Dry flue products are the total products less the amount of steam

$$= 19.319 - 0.45 = 18.869 \text{ kg}$$

CONVERSION FROM MASS TO VOLUMETRIC ANALYSIS. This is sometimes required and can be obtained by reversing the above procedure. To illustrate, determine the percentage analysis of the dry flue gases by volume from the previous worked example.

DFG	m%	M	N	N%
CO_2	15.18	44	$15.18 \div 44 = 0.345$	$\frac{0.345}{3.2505} \times 100 = 10.61\%$
O_2	9.04	32	$9.04 \div 32 = 0.2825$	$\frac{0.2825}{3.2505} \times 100 = 8.69\%$
N_2	73.45	28	$73.45 \div 28 = 2.623$	$\frac{2.623}{3.2505} \times 100 = 80.7\%$
Total			3.2505	100.00%

HYDROCARBON FUELS CxHy. The group of fuels which have a chemical formula made up of primarily carbon and hydrogen CxHy are called hydrocarbon fuels. Some examples are:

Methane CH_4

Propane C_3H_8

Butane C_4H_{10}

Benzene C_6H_6

Naphthalene $C_{10}H_8$

More information can be found on the composition of hydrocarbon fuel in *Reeds Volume* 8, Chapter 2.

The make-up of marine fuel is not straightforward as it contains several of the different examples shown above. These different chemicals have different points in the combustion process when they interact with oxygen. Students will be able to see why it is difficult to achieve a complete and clean combustion of fossil fuels.

If we consider 1 kg of hydrocarbon fuel CxHy it will be necessary to obtain the masses of carbon and hydrogen contained in the 1 kg in order to evaluate the air required for combustion.

$$\text{The mass of one mol of the fuel} = 12 \times x + 1 + y \text{ kg}$$
$$M = 12 \times x + 1 \times y$$
$$\text{The fraction of hydrogen by mass} = \frac{1 \times y}{M} \text{ kg}$$
$$\text{And the fraction of carbon by mass} = \frac{12 \times x}{M} \text{ kg}$$

Example 13.7. Determine the amount of air required for the complete combustion of 1 kg of propane, formula C_3H_8.

$$\text{Mass of 1 mol of the fuel} = 12 \times 3 + 1 \times 8 = 44 \text{ kg}$$
$$\text{Fraction of hydrogen by mass} = \frac{8}{44} = 0.1818$$
$$\text{Fraction of carbon by mass} = \frac{12 \times 3}{44} = 0.8182$$
$$= \Sigma 1.0000$$

Hence 1 kg of C_3H_8 consists of 0.8182 kg of carbon and 0.1818 kg of hydrogen.

$$\text{Stoichiometric air} = \frac{100}{23}[0.8182 \times 2^{2/3} + 0.1818 \times 8]$$
$$= 15.81 \text{ kg/kg of fuel}$$

Note: In 1 kg of a pure hydrocarbon fuel, if x is the mass of carbon then (1 − x) is the mass of hydrogen.

Liquid natural gas (LNG) is gaining popularity as a marine fuel. LNG is nearly completely made up of methane CH_4. Therefore, the exothermic reaction for complete combustion becomes:

$$CH_4 + 2O_2 \Rightarrow CO_2 + 2H_2O$$

INCOMPLETE COMBUSTION. It is difficult to comprehend that any fuel introduced into an extremely hot combustion space will survive the process without being consumed.

However, in reality it is very difficult to design a machine to achieve a complete combustion process. Incomplete combustion will occur when insufficient air, or variation in conditions, internal or external to the equipment leads to unsatisfactory combustion. This is especially true when a diesel engine is changing its load conditions.

Where there is incomplete combustion the exhaust gas analysis will include carbon monoxide as the carbon atom has only combined with one oxygen atom instead of the two oxygen atoms required for carbon dioxide. Balancing of the fuel–gas combustion equation allows analysis of the constituents and dry gas and air amount.

Engine design is a key component on the way to achieving complete combustion. The features affecting design are:

Improving swirl

Promoting squash

Compression ratio

Reducing peak temperatures by using Millar timing

Stratified fuel injection

Speed of the flame front

Exhaust gas recirculation

Exhaust gas balancing equations, starting with the carbon balance:

$$\frac{x}{12}C + \cdots \rightarrow aCO_2 + bCO + \cdots$$

$$\text{C balance is } \frac{x}{12} = a + b$$

Similarly for oxygen, hydrogen etc. balances, these relations can be used to complete an otherwise unknown combustion equation; the mass of dry gases can be found from a carbon balance.

Relative mass of carbon in fuel

$$= \frac{12}{44}\left\{\text{Rel. mass (m) } CO_2\right\} + \frac{12}{28}\left\{\text{Rel. mass (m) } CO\right\} \text{ in gases}$$

$$= \frac{12}{44}\left\{\text{Rel. vol (N) } CO_2 \times 44\right\} + \frac{12}{28}\left\{\text{Rel. vol. (N) } CO \times 28\right\}$$

$$= \left\{\text{Rel. vol. (N) } CO_2 + \text{Rel. vol. (N) } CO\right\}$$

Example 13.8. The exhaust gas from an engine has a dry analysis by volume as follows:

$$CO_2 \; 8.8\%, \quad CO \; 1.25\%, \quad O_2 \; 6.9\%, \quad N_2 \; 83.05\%$$

The fuel supplied to the engine has a mass analysis as follows:

$$C \; 84\%, \; H_2 \; 14\%, \; O_2 \; 2\%$$

Determine the mass of air supplied per kg of fuel burned.

Note: Air contains 20.7% oxygen by volume and 23% oxygen by mass. Atomic mass relationships: hydrogen = 1, carbon =12, nitrogen = 14, oxygen =16.

DFG	N	M	m
CO_2	8.8	44	387.2
CO	1.25	28	35
O_2	6.9	32	220.8
N_2	83.05	28	2325.4
Total			2968.4

$$\text{Relative gas mass} = 2968.4$$
$$\text{Relative C mass} = 12(8.8 + 1.25)$$
$$= 120.6$$

i.e. 120.6 kg of C in 2968.4 kg of dry flue gases

$$0.84 \text{ kg of C in} \frac{2968.4 \times 0.84}{120.6} \text{kg dry gases}$$

$$\text{Dry flue gases} = 20.68 \text{ kg}$$
$$\text{Water vapour} = 9 \times 0.14$$
$$= 1.26 \text{ kg}$$
$$\text{Total gases} = 21.94 \text{ kg}$$
$$\text{Mass of air supplied} = 20.94 \text{ kg/kg fuel} \quad \text{Ans.}$$

ENVIRONMENTAL ISSUES: The composition of exhaust gases is now a central part of Marpol Annex VI. Ships sailing worldwide and into emission control areas (ECAs) need to comply with the tight regulations governing their exhaust composition. To ensure that this is the case ships will now need to comply with the requirements limiting the release of oxides of nitrogen, sulphur oxides and carbon particulate levels in the exhaust gas.

The sulphur content of the exhaust gas can be controlled by limiting the sulphur content within the fuel and by removing the sulphur from the exhaust gas by means of exhaust gas cleaning (EGC) technology.

Nitrogen is drawn into the combustion process from the atmospheric air that is used to carry the oxygen required for combustion. The heat from combustion oxidises some of the nitrogen and produces nitrogen oxide (NO) and nitrogen dioxide (NO_2), generally shown as NOx. This can be limited by 'engine technology' and design or it can be removed by the exhaust gas cleaning (scrubbing) technology.

Ships are also required to become more efficient year on year. This means becoming more energy efficient and therefore less power is required to achieve the same job as before, which reduces the CO_2 levels in the exhaust gas.

Using LNG as a fuel means that the combustion equation for methane comes under close inspection. As the LNG is less complicated in its make-up the combustion is closer to being complete and therefore much less CO, CO_2, NOx and SOx are produced. However, due to its global warming potential (GWP) being much higher than CO_2, complete combustion of methane is very important.

Reducing 'methane slip' down to the lowest value possible is a very important aim for engine designers.

Another strategy that is gaining in popularity is the use of bio-fuel. The main aim here is to reduce the CO_2 produced from the fuel. However, it does rely on the production of the fuel producing less CO_2 than the fossil fuel extraction and production process, rather than producing much less CO_2 during the actual combustion process. In addition, the bio-fuel has only recently extracted carbon from the atmosphere, unlike the fossil fuel that has had its carbon stored in the ground for millions of years and is now being released in a fraction of that time.

If hydrogen is used in a fuel cell to produce electricity, rather than combusted in an internal combustion engine or a boiler, then there are no resulting pollutants or global warming gasses from the activity.

The problem with using hydrogen is that it must be extracted from other substances and currently that process does produce pollutants. If the hydrogen is produced using energy from renewable sources then the hydrogen fuel cell will be a 'clean' fuel.

Handling and storage of hydrogen also needs careful consideration, especially where pipes and fixings run through enclosed spaces. Effective crew training is essential.

Test Examples 13

1. Steam at 30 bar, 375°C is generated in a boiler at the rate of 30 000 kg/h from feed water at 130°C. The fuel has a calorific value of 42 MJ/kg and the daily consumption is 53 tonne. Calculate (i) the boiler efficiency, (ii) the equivalent engine power if the overall efficiency of the plant is 0.13, (iii) the evaporative capacity of the boiler in kg/h from and at 100°C and (iv) the equivalent evaporation per kg of fuel from and at 100°C.

2. A boiler contains 3.5 tonne of water, initially having 40 ppm dissolved solids, and after 24 h the dissolved solids in the water measure 2500 ppm. If the feed rate is 875 kg/h, find the ppm of dissolved solids contained in the feed water.

3. At its rated output a waste heat evaporator produces 10 tonne/day of fresh water containing 250 ppm dissolved solids from sea water containing 31 250 ppm. The dissolved solid content of the brine in the evaporator shell is maintained at 78 125 ppm. Calculate the mass flow per day of:

   (i) the sea water feed;

   (ii) the brine discharge.

4. The constituents of a fuel are 85% carbon, 13% hydrogen and 2% oxygen. When burning this fuel in a boiler furnace the air supply is 50% in excess of the stoichiometric, inlet temperature of the air being 31°C and funnel temperature 280°C. Calculate (i) the calorific value of the fuel, (ii) mass of air supplied per kg of fuel and (iii) the heat energy carried away to waste in the funnel gases expressed as a percentage of the heat energy supplied, taking the specific heat of the flue gases as 1.005 kJ/kg K.

5. A fuel consists of 84% carbon, 13% hydrogen, 2% oxygen and the remainder incombustible solid matter. Calculate (i) the calorific value, (ii) the stoichiometric air required per kg of fuel and (iii) an estimate of the mass analysis of the flue gases if 22 kg of air are supplied per kg of fuel burned.

6. The analysis of a fuel oil is 85.5% carbon, 11.9% hydrogen, 1.6% oxygen and 1% impurities. Calculate the percentage CO_2 in the flue gases when (i) the quantity of air supplied is the stoichiometric, and when the excess air over the stoichiometric is (ii) 25%, (iii) 50% and (iv) 75%.

7. The mass analysis of a fuel is 84% carbon, 14% hydrogen and 2% ash. It is burned with 20% excess air relative to stoichiometric requirement. If 100 kg/h of this fuel is burned in a boiler, determine (i) the volumetric analysis of the dry flue

gas and (ii) the volumetric flow rate of the gas if its temperature and pressure are 250°C and 1 bar, respectively.

Note: Air contains 23% oxygen by mass. $R_o = 8.3143$ kJ/kg mol K and atomic mass relationships are: oxygen 16, carbon 12, hydrogen 1.

8. The fuel oil supplied to a boiler has a mass analysis of 86% carbon, 12% hydrogen and 2% sulphur. The fuel is burned with an air-to-fuel ratio of 20:1. Calculate:

   (i)   the mass analysis of the wet flue gases;

   (ii) the volumetric analysis of the wet flue gases.

9. Benzene fuel C_6H_6 is burned in a boiler furnace under stoichiometric conditions of combustion. Calculate:

   (i)   the mass of air required for 1 kg of benzene;

   (ii) the mass analysis of the flue gases;

   (iii) the volumetric analysis of the dry flue gas.

10. The volumetric analysis of a dry flue gas is: carbon dioxide 10.8%, oxygen 7.2%, carbon monoxide 0.8% and the remainder nitrogen. Calculate:

   (i)   the analysis of the dry flue gas by mass;

   (ii) the mass percentage of carbon and hydrogen in the fuel, assuming it to be a pure hydrocarbon.

14

REFRIGERATION

As we have seen the natural transfer of heat energy is from a hot body to a colder one. The function of a refrigeration plant is to act as a heat pump and reverse this process.

Refrigeration units can be divided into two classes: (i) those which require a supply of mechanical work and use a vapour compression system, and (ii) those which require a heat supply and work on the absorption system. The absorption system tends to be an older design that was more suited to small domestic use. However, even these operate on the more effective vapour compression system which is described here. (More details of the absorption type systems can be found in *Reeds Vol. 8,* Chapter 7.)

The design and construction of refrigeration systems has just come through a major period of change. The substance used as a refrigerant needs to have the ability to carry heat from one place in a refrigeration circuit to another in different part of the circuit.

To achieve this, the substance must also be able to exist as a gas in one part of the circuit, and as a liquid in another part, at the operating temperatures and pressures designed for the system.

During the last half of the past century gases from the chlorine, fluorine, carbon (CFC) family were developed and used in refrigeration systems. It is now known that these refrigerant gases were very damaging to the environment. As a consequence, they have now been phased out of use. New CFC gases cannot be purchased and replacements are being used in new systems.

Fluorinated gasses are also up to 23,000 times stronger as a greenhouse gas than carbon dioxide. Therefore, there is a need to remove these gases from use, which has a specific impact upon refrigerants.

The problem has led administrations to develop rules around both the use and handling of these refrigerants. The regulations are referred to as the 'F gas' regulations.

Ozone Depleting Substances (ODSs)

Ozone is made up of three oxygen atoms joined together. This is slightly different to the normal two oxygen atoms that form the bulk of the free oxygen that exists in the atmosphere. The bulk of atmospheric ozone lies in the stratosphere, which is approximately 10–25 miles (15–40 km) above the surface of the earth. It carries out the important function of reflecting back into space some of the ultra-violet (UV) radiation that has come from the sun. Any reduction in this protective layer would mean that more harmful radiation would otherwise reach the surface of the earth.

Ozone is made and destroyed by the natural processes that are part of nature; however, since the development of and subsequence release into the atmosphere of CFCs the natural production of ozone has not been able to keep up and the total amount of ozone has reduced. The CFC is a very stable chemical and is not broken down by the water in the lower part of the atmosphere. Therefore, CFCs remain intact until they reach the stratosphere where they are struck by intense radiation from the sun. The CFC finally decomposes under the sunlight and the chlorine released interacts with the ozone leading to its destruction. The chemicals that are most potent at reducing the ozone in the atmosphere are CFCs, carbon tetrachloride, methyl bromide, methyl chloroform and halons. Hydrochlorofluorocarbon (HCFC), which is a compound composed of hydrogen, chlorine, fluorine and carbon atoms, are being used as a replacement. The introduction of the hydrogen means that the compound breaks down in the lower atmosphere before it is able to reach the stratosphere where the ozone is situated.

The boiling and condensation points of a liquid depend upon the pressure exerted upon it. For example, if water is under atmospheric pressure, it will evaporate at 100°C; if the pressure is 7 bar, the water will not change into steam until its temperature is 165°C; at 14 bar the boiling point is 195°C; and so on.

The ideal refrigerant must be one that will evaporate at very low temperatures. The boiling point of carbon dioxide at atmospheric pressure, for example, is about −78°C (*note*: *minus* 78). By increasing the pressure, the temperature at which liquid CO_2 will evaporate (or CO_2 vapour will condense) is raised accordingly so that any desired evaporation and condensation temperature can be attained, within certain limits, by subjecting the substance to the appropriate pressure.

Refrigerants

Desirable properties of a refrigerant:

- low boiling point (otherwise operation at a high vacuum becomes necessary);
- low condensing pressure (to avoid a heavy machine, plant scantlings and to reduce the leakage risk);
- high specific enthalpy of vaporisation (to reduce the quantity of refrigerant in circulation and lower machine speeds, sizes, etc.);
- low specific volume in the vapour phase (this reduces size and increases efficiency);
- high critical temperature (temperature above which vapour cannot be condensed by isothermal compression);
- non-corrosive and non-solvent (pure or mixed);
- stable under working conditions;
- non-flammable and non-explosive (pure or mixed);
- low interaction with oil (the fact that most refrigerants are miscible may be advantageous, i.e., removal of oil films, lowering pour point, etc., provided separators are fitted);
- easy leak detection;
- non-toxic (non-poisonous and non-irritating);
- cheap, easily stored and obtained.

Refrigerant gases

Given the situation described about the effect that some refrigerants have on the atmosphere world leaders have taken action to restrict their use. In July 2020, the 100th signatory to the Kigali Amendment to the Montreal Protocol was reached. This amendment has now obligated UN members to cut the use of Hydrofluorocarbons (HFCs) by up to 80% in the next 25–30 years. The changeover from using CFCs in new systems has been completed as ODSs are not prohibited for use in systems that were built before May 2005, which is the date that Marpol Annex VI came into force. New installations that operate on HCFC were permitted until 1 January 2020. However, already in force – from 1 January 2010 – is the EU/UK regulation that prohibits the use of virgin HCFC in the maintenance of any system.

Initially, only recycled gases were permitted to be used and from 1st January 2015 the use of all HCFC gases for maintenance purposes has not been allowed. HCFCs can still be used in existing instillations so long as there is no maintenance required to the system. Recycling means that recovered HCFCs have been placed through a filtering or moisture removal process. The recycled gas must not then be placed on the open market; they can only be used by a recycling company for the maintenance of existing systems.

With chlorine being the negative substance for ODS and fluorine being the most potent GWP substance, refrigerants still in existence are:

- CFCs: R11, R12 and R502;
- HCFCs: R22;
- HCFC blends including: R401a, R402a, R403a and R406a;
- CFC = chlorine–fluorine–carbon;
- HCFC = Hydrogen–chlorine–fluorine–carbon.

The common trading names are: Arcton, Freane, Freon, Isceon, Solkane and Suva. Due to the restrictions in the use of HCFC as a refrigerant one of the alternatives is ammonia (NH_3) because it is environmentally friendly. It is not, however, completely human friendly and great care must be taken when handling ammonia or using an ammonia refrigeration plant. An ammonia machine should be built in a compartment of its own so that it can be sealed off in the case of a serious leakage; water will absorb ammonia and therefore a water spray is a good combatant against a leakage. Ammonia will corrode copper and copper alloys and therefore parts in contact with it should be made of such metals as nickel steel and monel metal. Its natural boiling point is about −39°C, and therefore the pressures required are much lower than those required in a CO_2 machine. Smaller plants also have problems finding compatible lubricating oils.

Carbon dioxide (CO_2) is an inert vapour, non-poisonous, odourless and without any corrosive action on the metals. Its natural boiling point is very low, which means that a CO_2 plant must be run at very high pressures to bring it to the conditions where it will evaporate and condense at the normal temperatures of a refrigerating machine. A further disadvantage is that its critical temperature is about 31°C which falls within the range of coolant temperatures. At its critical temperature and above, it is impossible to liquefy the vapour no matter to what pressure it is subjected. Despite the problems with CFC refrigerant gases, CO_2 has not made a comeback as a popular choice for a refrigerant gas, though it is still sometimes used in the fishing industry.

Freon R12 and R22 were the two gases that were developed for universal use in marine refrigeration systems 30–40 years ago. Their replacements have the designated numbers R-134a, R-404A and R-407C. Common gases found in newer systems are R-410A and R-507.

Information about the most popular refrigerants in marine systems are shown in the table below.

Name	Chemical formula/Blend	Critical point	Boiling point
R134a	CF_3CH_2F	101.1°C at 40.59 bar	−26.3°C
R404a	R-125/143a/134a at percentages of (44/52/4)	72.5°C at 37.8 bar	−51.8°C
R407c	R-32/125/134c at percentages of (23/25/52)	86.1°C at 46.3 bar	−43.6°C
R410a	R-32/125 at percentages of (50/50)	72.0°C at 47.7 bar	−48.5°C
Ammonia	NH_3	132.3°C at 113 bar	−33.34°C

TEMPERATURE GLIDE. Modern refrigerants such as R-134a are made up of a blend of different gases. The problem with this is that each gas in the blend has its own boiling point, as well as other properties. The difference in these temperatures is referred to as the 'temperature glide'.

Working Cycle

Figure 14.1 illustrates diagrammatically the essential components of the vapour compression system, which consists of the compressor driven by an electric motor, the condenser which is circulated by the water, the expansion valve (which may be referred

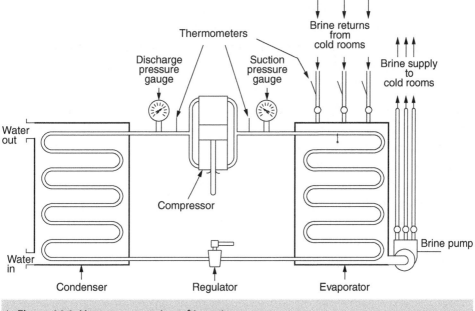

▲ **Figure 14.1** *Vapour compression refrigeration system*

to as the regulator) and the evaporator which in this case is circulated by very salty water called brine.

This system is a completely enclosed circuit with the same quantity of refrigerant passing continually around the system. Systems only require charging when there are losses due to leakage. Refrigeration plant may be designed using any of the refrigerants described previously, with the only difference in their working being the pressures and temperatures throughout the system. The different pressures will, of course, dictate different equipment design to accommodate the different gas used. Therefore during the examples in the following 'general' description words such as 'refrigerant', 'the liquid', 'the vapour', etc. may be used.

Starting our repeating cycle at the compressor, the refrigerant is drawn into the compressor, as a vapour at low pressure, from the evaporator. Here it is compressed to a high pressure and delivered into the coils of the condenser in the state of a superheated vapour. As it passes through the condenser coils/tubes/plates, the vapour is cooled and condensed into a liquid at approximately condenser coolant inlet water temperature, the latent heat energy given up being absorbed by the coolant surrounding the coils/tubes/plates. The liquid, still at a high pressure, passes along to the expansion valve (regulator), which is a valve just partially open to limit the flow through it.

The pipeline from the discharge side of the compressor to the expansion valve is under high pressure, but from the expansion valve to the suction side of the compressor it is at a lower pressure due to the expansion valve restricting the flow and acting to 'throttle' the refrigerant. As the liquid passes through the expansion valve from a region of high pressure to a region of low pressure, the throttling occurs. Some of the liquid automatically changes itself into a vapour absorbing the required amount of heat energy to do so from the remainder of the liquid and causing it to fall to a low temperature. This temperature is regulated by the pressure so that the refrigerant enters the evaporator at a temperature lower than that of the secondary refrigerant, which in this case is brine.

The liquid (more correctly a mixture of liquid and vapour) now passes through the coils/tubes/plates of the evaporator where it receives heat to evaporate it before being taken into the compressor to go through the cycle again. The required heat absorbed by the liquid refrigerant in the coils/tubes of the evaporator to cause evaporation is extracted from the brine surrounding the coils/tubes, resulting in the brine being cooled to a low temperature. The brine is a 'secondary' refrigerant that is used to cool a cargo hold or other compartment that is usually remote from the primary refrigerant circuit. In other systems the evaporator will be situated to cool a direct flow of air from the room or compartment that is to be cooled.

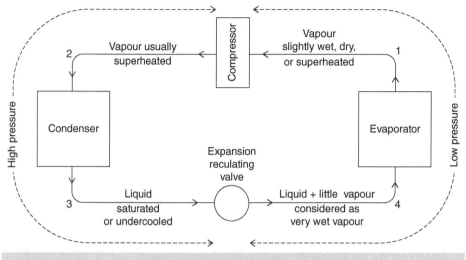

▲ **Figure 14.2** *Typical sketch of a simple refrigeration/heat pump system*

The Circuit of the Refrigerant

It is necessary to have a clear picture in mind when performing calculations involving the changes that take place on the refrigerant as it flows through the components of the circuit, and therefore a simple diagrammatic sketch such as Figure 14.2 should be made with points in the circuit suitably marked. Reference should be made to Figure 14.2 when reading the following notes and examples.

Properties of some refrigerants are given in *Thermodynamic and Transport Properties of Fluids: SI Units*, 5th edition, arranged by G.F.C. Rogers and Y.R. Mayhew. This booklet was referred to in previous chapters dealing with properties of steam. In examples given here, extracts from the tables in this booklet will be used when appropriate, stating the page number.

Principles and calculations involving enthalpy of the refrigerant, dryness fraction, volume, effect of throttling etc, follow the same pattern as for other vapours. However, whereas the enthalpy and entropy of steam are measured from the datum of water at 0°C, it is more convenient to choose a lower datum temperature for refrigerants to avoid negative quantities within the normal range of refrigeration. This datum is − 40°C, and thus the enthalpy and entropy of the saturated liquid refrigerant are taken as 0 at this temperature.

Note that when the value of h_{fg} is not given in the tables, it is found by subtraction: $h_{fg} = h_g - h_f$. Similarly for the value of s_{fg}.

The most common gases in modern systems are:

R134A

R404A

R407C

R410A.

Referring to Figure 14.2: The work done on the vapour in the compressor increases the specific enthalpy of the refrigerant from h_1 to h_2. Therefore the effective work done per second [kJ/s], which is the effective power [kW], is equal to the product of the mass flow [kg/s] of the refrigerant and its gain of specific enthalpy $(h_2 - h_1)$ [kJ/kg].

In the condenser, heat transfer takes place from the refrigerant to the condenser circulating water, causing condensation of the refrigerant from a vapour to a liquid. The heat rejected by the refrigerant per kilogram is the change of specific enthalpy $h_2 - h_3$.

The effect of throttling the refrigerant through the expansion valve is to reduce the pressure of the saturated liquid with consequent reduction in temperature and the formation of a very small amount of vapour. During a throttling process there is no change of enthalpy, therefore h_4 is equal to h_3.

In passing through the evaporator coils, heat transfer takes place from the brine (or other surrounds) to the refrigerant, which cools the brine and causes the refrigerant to evaporate. The gain in specific enthalpy of the refrigerant is $h_1 - h_4$ and this is the *refrigerating effect* or *cooling effect* per kilogram of the refrigerant flowing through the circuit.

Example 14.1. In an NH_3 refrigerator, the ammonia leaves the evaporator and enters the compressor as dry saturated vapour at 2.68 bar, it leaves the compressor and enters the condenser at 8.57 bar with 50°C of superheat, it is condensed at constant pressure and leaves the condenser as saturated liquid. If the rate of flow of the refrigerant through the circuit is 0.45 kg/min, calculate (i) the compressor power, (ii) the heat rejected to the condenser cooling water in kJ/s and (iii) the refrigerating effect in kJ/s.

From tables page 14, NH_3:

$$2.68 \text{ bar}, \quad h_g = 1430.5$$
$$8.57 \text{ bar}, \quad h_f = 275.1 \qquad h \text{ supht } 50° = 1597.2$$
$$\text{Mass flow of refrigerant} = \frac{0.45}{60} = 0.0075 \text{ kg/s}$$

Enthalpy gain per kg of refrigerant in compressor

$$= h_2 - h_1$$
$$= 1597.2 - 1430.5$$
$$= 166.7 \text{ kJ/kg}$$

Enthalpy gain per second

$$= \text{Mass flow } [\text{kg/s}] \times 166.7 \, [\text{kJ/kg}]$$
$$= 0.0075 \times 166.7 = 1.25 \text{ kJ/s}$$

kJ/s = kW, therefore useful compressor power

$$= 1.25 \text{ kW} \quad \text{Ans.(i)}$$

Enthalpy drop per kg of refrigerant through condenser

$$= h_2 - h_3$$
$$= 1597.2 - 275.1 = 1322.1 \text{ kJ/kg}$$

Heat rejected $= 0.0075 \times 1322.1$

$$= 9.915 \text{ U/s} \quad \text{Ans.(ii)}$$

Enthalpy gain per kg of refrigerant through evaporator

$$= 1430.5 - 275.1 = 1155.4 \text{ kJ/kg}$$

(*Note*: The value of h_4 is the same as h_3 because there is no change of enthalpy in the throttling process through the expansion valve between the condenser exit and the evaporator inlet.)

$$\text{Refrigerating effect} = 0.0075 \times 1155.4$$
$$= 8.666 \text{ kg/s} \quad \text{Ans.(iii)}$$

Example 14.2. In a refrigerator using refrigerant 134a, the refrigerant leaves the condenser as a saturated liquid at 20°C, and the evaporator temperature is −10°C and it then leaves the evaporator as a vapour 0.97 dry. Calculate (i) the dryness fraction at the evaporator inlet, (ii) the cooling effect per kg of refrigerant and (iii) the volume flow of refrigerant entering the compressor if the mass flow is 0.1 kg/s.

From tables page 15, refrigerant 134a:

$$20°C, \quad h_f = 227.45$$
$$-10°C, \quad h_f = 186.71 \quad h_g = 392.51 \quad v_g = 0.0995$$
$$h_{fg} = h_g - h_f = 392.51 - 186.71 = 205.80$$

Throttling process through expansion valve:

$$\text{Enthalpy after throttling } (h_4) = \text{Enthalpy before } (h_3)$$
$$h_{fg} + x_4 h_{fg4} = h_{f3}$$
$$392.51 + x_4 \times 205.80 = 227.45$$
$$x_4 \times 205.80 = 28$$
$$x_4 = 0.1792 \quad \text{Ans. (i)}$$

Enthalpy gain of refrigerant through evaporator = Cooling effect
$$= h_1 - h_4 = (26.87 + 0.97 \times 156.32) - (26.87 + 0.1792 \times 156.32)$$
$$= (0.97 - 0.1792) \times 156.32$$
$$= 0.7908 \times 156.32 = 123.6 \text{ kJ/kg} \quad \text{Ans. (ii)}$$

Alternatively, since $h_4 = h_3$:

$$\text{Cooling effect} = (26.87 + 0.97 \times 156.32) - 54.87$$
$$= 123.6 \text{ kJ/kg}$$

Specific volume of refrigerant leaving evaporator

$$= 0.97 \times 0.0766 \text{ m}^3 / \text{kg}$$

Volume of refrigerant entering compressor per second $\left[\text{m}^3 / \text{kg} \right]$
$$= 0.1 \times 0.97 \times 0.0766$$
$$= 7.43 \times 10^{-3} \text{ m}^3 / \text{s}$$
or 7.43 l/s   Ans. (iii)

Capacity and Performance

The capacity of a refrigerating plant is usually expressed in terms of the tonnes of ice at 0°C that could be made in 24 h from water at 0°C.

The *coefficient of performance* is the ratio of the refrigerating effect to the net work done on the refrigerant in the compressor:

$$\text{Coeff. of performance} = \frac{\text{Heat extracted by refrigerant}}{\text{Work done on refrigerant}}$$
$$= \frac{h_1 - h_4}{h_2 - h_1}$$

In Chapter 8, the reversed Carnot cycle was explained as the ideal cycle for a refrigerator and the coefficient of performance of this cycle given as:

$$\frac{T_2}{T_1 - T_2}$$

where T_1 and T_2 are the highest and lowest absolute temperatures in the cycle. This is referred to as the *Carnot coefficient of performance*. Note, however, that the notation is different to that used in the practical circuit of Figure 14.1; to avoid confusion write subscripts H for highest and L for lowest:

$$\frac{T_L}{T_H - T_L}$$

Example 14.3. A refrigerating system, using refrigerant R134a, operates on the ideal vapour compression cycle between the limits −15°C and 25°C. The vapour is dry saturated at the end of isentropic compression and there is no undercooling of the condensate in the condenser. Calculate (i) the dryness fraction at the suction of the compressor, (ii) the refrigerating effect per kg of refrigerant, (iii) the coefficient of performance and (iv) the Carnot coefficient of performance.

Information from R134a data:

$$25°C, \quad h_f = 234.6 \quad h_g = 412.6 \quad s_g = 1.7175$$
$$-15°C, \quad h_f = 180.20 \quad h_g = 389.80$$
$$h_{fg} = 389.80 - 180.20 = 209.60$$
$$s_f = 0.9257 \quad s_g = 1.7379$$
$$s_{fg} = 1.7379 - 0.9257 = 0.8122$$

For isentropic compression in the compressor:

$$\text{Entropy before } (s_1) = \text{Entropy after } (s_2)$$
$$0.9257 + x_1 \times 0.8122 = 1.7175$$
$$\text{Therefore, } x_1 = 0.9749 \quad \text{Ans. (i)}$$
$$\text{To find } h_1 = 180.20 + (0.9749 \times 209.60)$$
$$h_1 = 384.54$$
$$\text{Dry vapour at } h_2 = h_g = 412.60$$

Since there is no undercooling, the condensate leaves the condenser as saturated liquid and therefore $h_3 = h_f$ at 25°C = 234.60. In addition, as there is no change in enthalpy during the throttling process $h_4 = h_3 = 234.60$. Therefore:

$$\text{Refrigerating effect} = h_1 - h_4 = 384.54 - 234.60$$
$$= 149.94 \text{ kJ/kg} \quad \text{Ans.(ii)}$$

$$\text{Work done in compressor/kg} = h_2 - h_1$$
$$= 412.60 - 384.54 = 28.06 \text{ kJ/kg}$$

$$\text{Coefficient of performance} = \frac{\text{Refrigerating effect}}{\text{Work transfer}}$$

$$= \frac{h_1 - h_4}{h_2 - h_1} = \frac{384.53 - 234.60}{412.60 - 384.54}$$

$$= \frac{149.94}{28.06} = 5.34 \quad \text{Ans.(iii)}$$

$$\text{Carnot coeff. of performance} = \frac{T_L}{T_H - T_L}$$

$$\text{where } T_H = 25 + 273 = 298 \text{ K,}$$
$$T_L = -15 + 273 = 258 \text{ K}$$

$$\therefore \text{ Carnot c.o.p.} = \frac{258}{298 - 258} = 6.45 \quad \text{Ans. (iv)}$$

T–S and p–h diagrams illustrate, and allow analysis of, the refrigeration circuit (see Figure 14.3).

Example 14.4. An ammonia refrigeration plant operates between temperature limits of 20°C and 32°C. The refrigerant leaves the evaporator as saturated vapour and enters the expansion valve as saturated liquid. Assuming the isentropic efficiency of the compressor is 0.85, determine the coefficient of performance of the plant and sketch the cycle on p–h and T–S diagrams.

From tables:

$$h_2 = 1469.9 \text{ kJ/kg}$$
$$h_3 = h_4 = 332.8 \text{ kJ/kg}$$
For isentropic compression $s_1 = s_2 = 4.962 \text{ kJ/kg K}$

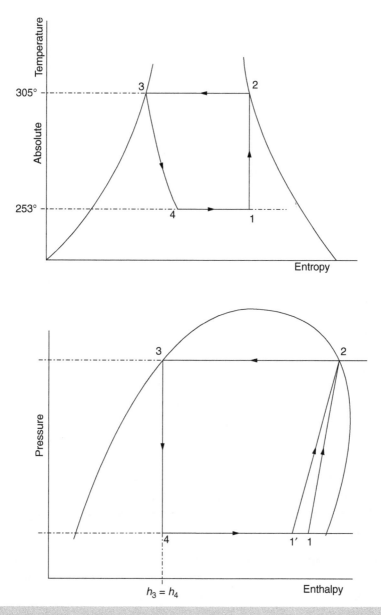

▲ **Figure 14.3** *Typical p–h and T–S diagrams for a refrigeration cycle*

$$at - 20°C$$

$$s_1 = s_f + x \left(s_g - s_f \right)$$

$$4.962 = 0.368 + x \left(5.623 - 0.368 \right)$$

$$x = 0.874 \text{ dry}$$

also at − 20°C

$$h_1 = h_f + x \left(h_g - h_f \right)$$

$$= 89.8 + 0.874 \left(1420 - 89.8 \right)$$

$$= 1252.4 \text{ kJ/kg}$$

Isentropic work done in compressor $= h_2 - h_1$

$$= 1469.9 - 1252.4$$

$$= 217.5 \text{ kJ/kg}$$

Actual work done in compressor × Isentropic efficiency = Isentropic work done in compressor

i.e. Actual work done in compressor $= \dfrac{217.5}{0.85} = 255.9 \text{ kJ/kg}$

$$\therefore h_1' = h_2 - 255.9$$

$$= 1469.9 - 255.9 = 1214 \text{ kJ/kg}$$

$$\text{c.o.p.} = \frac{h_1' - h_4}{h_2 - h_1} = \frac{1214 - 332.8}{1469.9 - 1214}$$

$$= 3.443 \quad \text{Ans.}$$

HEAT PUMPS. If in Figure 14.2 we consider the condenser as a heater where air, or water, is heated as the refrigerant condenses, the heated air or water being subsequently used for domestic purposes in accommodation spaces, we are then using the plant as a *heat pump*.

The refrigerant is expanded through the expansion valve and the temperature drops. Heat is then collected through the evaporator due to its higher temperature. Further heat is gained as the gas passes through the compressor. The refrigerant is passed through the heat exchanger, that in the refrigeration circuit is called the condenser, before the refrigerant again reaches the expansion valve. In this system the heat from the condenser is used to heat up the space that it is situated in.

The coefficient of performance of the heat pump (or heat pump coefficient) is defined as the ratio:

$$\frac{\text{Heat transfer in condenser}}{\text{Work done on refrigerant}}$$

i.e. $\text{c.o.p.} = \dfrac{h_2 - h_3}{h_2 - h_1}$ for heat pump

Test Examples 14

1. In a CO_2 refrigerating plant, the specific enthalpy of the refrigerant as it leaves the condenser is 135 kJ/kg, and as it leaves the evaporator it is 320 kJ/kg. If the mass flow of the refrigerant is 5 kg/min, calculate the refrigerating effect per hour.

2. Refrigerant R134a leaves the condenser of a refrigerating plant as a saturated liquid at 571.62 kPa. The evaporator pressure is 132.72 kPa and the refrigerant leaves the evaporator at this pressure and at a temperature of −10°C. Calculate (i) the dryness fraction of the refrigerant as it enters the evaporator and (ii) the refrigerating effect per kg.

   Additional information given from tables relating to R134a page 15 of *Thermodynamic and Transport Properties of Fluids: SI Units*, 5th edition, arranged by G.F.C. Rogers and Y.R. Mayhew.

   At 571.62 kPa $t = 20°C$ $h_f = 227.45$ $h_g = 409.62$

   and at 132.72 kPa $t = -20°C$ $h_f = 173.67$ $h_g = 386.44$

3. A refrigerating machine uses ammonia as the working fluid. It leaves the compressor as dry saturated vapour at 8.57 bar, passes through the condenser at this pressure and leaves as saturated liquid. The pressure in the evaporator is 1.902 bar and the ammonia leaves the evaporator 0.96 dry. If the rate of flow of the refrigerant through the circuit is 2 kg/min, calculate (i) the heat rejected in the condenser in kJ/min, (ii) the refrigerating effect in kJ/min and (iii) the volume taken into the compressor in m³/min.

4. The dryness fractions of the CO_2 entering and leaving the evaporator of a refrigerating plant are 0.28 and 0.92, respectively. If the specific enthalpy of evaporation (h_{fg}) of CO_2 at the evaporator pressure is 290.7 kJ/kg, calculate (i) the refrigerating effect per kg, (ii) the mass of ice at −5°C that would theoretically be made per day from water at 14°C when the mass flow of CO_2 through the machine is 0.5 kg/s. Take the values:

   Specific heat of water = 4.2 kJ/kg K

   Specific heat of ice = 2.04 kJ/kg K

   Enthalpy of fusion of ice = 335 kJ/kg

5. A refrigerating machine is driven by a motor of output power 2.25 kW and 2.5 tonne of ice at −7°C is made per day from water at 18°C. Calculate the coefficient of performance of the machine and express its capacity in terms of tonnes of ice per 24 hours from and at 0°C, taking the following values:

Specific heat of water = 4.2 kJ/kg K

Specific heat of ice = 2.04 kJ/kg K

Enthalpy of fusion of ice = 335 kJ/kg

6. The refrigerant leaves the compressor and enters the condenser of a refrigerating plant running on R404a, at 1097.7 kPa and 50°C, and leaves the condenser as saturated liquid at the same pressure. At compressor suction the pressure is 611.1 kPa and temperature 0°C. Calculate the coefficient of performance.

   Information given from suitable Thermodynamic tables for R404a

   At 1097.7 kPa $t = 20$°C $h_f = 229.9$ $h_g = 378.3$ 30°K of Superheat $h = 384.3$

   At 611.1 kPa $t = -15$°C $h_f = 179.4$ $h_g = 359.6$ 15°K of Superheat $h = 365.3$

7. In an ammonia refrigerating plant the discharge from the compressor is 14.7 bar, 63°C. The refrigerant leaves the condenser at this pressure as liquid with no undercooling. The suction pressure at the compressor is 2.077 bar and compression is isentropic. Calculate for a mass flow of refrigerant of 0.15 kg/s (i) the refrigerating effect, (ii) the work transfer in the compressor and (iii) the coefficient of performance.

8. A refrigeration plant using R134a operates between 488.33 kPa and 84.35 kPa. The refrigerant is dry saturated on entry to the compressor and no undercooling takes place in the condenser. If compression is isentropic and losses are neglected, determine:

   (i) the c.o.p;

   (ii) power input to compressor if the flow rate of refrigerant entering the compressor is 0.15 m³/s.

   Sketch the cycle on temperature–entropy and pressure–enthalpy axes. Use the thermodynamic data presented on page 15 of Rogers and Mayhew.

9. A heat pump using ammonia operates between saturation temperature limits of −16°C and 40°C. The ammonia leaves the compressor after isentropic compression as dry saturated vapour and is then used to heat air from ambient, 13°C–20°C which is circulated in an insulated room of internal volume 300 m³. If the room requires four changes of air per hour determine the c.o.p. of the heat pump and the mass flow rate of ammonia.

   Take for the air $c_p = 1005$ J/kg, $R = 287$ J/kg K and pressure = 1.013 bar.

10. In a refrigeration plant the refrigerant R134a leaves the condenser as a liquid at 25°C. The refrigerant leaves the evaporator as a dry saturated vapour at − 15°C and leaves the compressor at 665.25 kPa and 35°C. The cooling load is 70.5 kW.

    Calculate:

    (i) the mass flow rate of refrigerant;

    (ii) the power of the compressor;

    (iii) the condition of the refrigerant after the expansion valve.

    Using the table on page 15 of Rogers and Mayhew.

SOLUTIONS TO TEST EXAMPLES

TEST EXAMPLES 1

1. Mass of water lifted $= 50 \times 10^3$ kg/h

   Weight of water lifted $= 50 \times 10^3 \times 9.81$ N/h

   $$\text{Power} \left[\text{W} = \text{J/s} = \text{N m/s} \right]$$
   $$= \text{Weight lifted per second} \left[\text{N/s} \right] \times \text{Height} \left[\text{m} \right]$$
   $$= \frac{50 \times 10^3 \times 9.81 \times 8}{3600}$$
   $$= 1090 \text{ W} = 109 \text{ kW} \quad \text{Ans. (i)}$$
   $$\text{Input power} = \frac{\text{Output power}}{\text{Efficiency}}$$
   $$= \frac{1.09}{0.69} = 1.58 \text{ kW} \quad \text{Ans. (ii)}$$
   $$\text{Energy} \left[\text{kWh} \right] = \text{Power} \left[\text{kW} \right] \times \text{Time} \left[\text{h} \right]$$
   $$= 1.58 \times 2 = 3.16 \text{ kWh} \quad \text{Ans. (iii)}$$
   $$\text{Energy} \left[\text{MJ} \right] = \text{Energy} \left[\text{kWh} \right] \times 3.6 \left[\text{MJ/kWh} \right]$$
   $$= 3.16 \times 3.6 = 11.38 \text{ MJ} \quad \text{Ans. (iv)}$$

2. A column of water 1 mm high exerts a pressure of 9.81 N/m², therefore 20 mm water is equivalent to:

   $$20 \times 9.81 = 196.2 \text{ N/m}^2 \quad \text{Ans. (i)(a)}$$
   $$1 \text{ mbar} = 100 \text{ N/m}^2$$
   $$\therefore 1962.2 \text{ N/m}^2 = 1.962 \text{ mbar} \quad \text{Ans. (ii)(a)}$$

   A column of mercury 1 mm high exerts a pressure of 133.3 N/m², therefore 750 mmHg is equivalent to:

$$750 \times 133.3 = 1 \times 10^5 \text{ N/m}^2$$
$$= 100 \text{ kN/m}^2 \quad \text{Ans. (i)(b)}$$
$$= 1 \text{ bar} \quad \text{Ans. (ii)(b)}$$

3. Abs. press. = (Barometer mmHg − Vacgauge mmHg) × 133.3

$$= (757 - 715) \times 133.3$$
$$= 42 \times 133.3$$
$$= 5.6 \times 10^3 \text{ N/m}^2$$
$$= 5.6 \text{ kN/m}^2$$
$$= 0.056 \text{ bar} \quad \text{Ans.}$$

4.
$$C = (140 - 32) \times \frac{5}{9}$$

$$= 108 \times \frac{5}{9}$$
$$= 60° \quad \text{Ans. (i)}$$

$$C = (5 - 32) \times \frac{5}{9}$$
$$= -27 \times \frac{5}{9}$$
$$= -15°C \quad \text{Ans. (ii)}$$

$$C = (-31 - 32) \times \frac{5}{9}$$
$$= -63 \times \frac{5}{9}$$
$$= -35°C \quad \text{Ans. (iii)}$$

$$C = (-40 - 32) \times \frac{5}{9}$$
$$= -72 \times \frac{5}{9}$$
$$= -40°C \quad \text{Ans. (iv)}$$

5. Area of tube bore $= 0.7854 \times 0.032 \text{ m}^2$

Area of 16 tubes $= 0.7854 \times 0.03^2 \times 16 \text{ m}^2$
Volume flow $[\text{m}^3/\text{s}] = \text{Area } [\text{m}^2] \times \text{Velocity } [\text{m/s}]$
$$= 0.7854 \times 0.032 \times 16 \times 2$$
$$= 0.02261 \text{ m}^3/\text{s}$$

$$1 \text{ m}^3 = 10^3 \text{ l}$$

i.e. Volume flow $[\text{1/s}] = 0.02261 \times 10^3$

$$= 22.61 \text{ l/s} \quad \text{Ans. (i)}$$

Density $= 0.85 \text{ g/ml} = 0.85 \text{ kg/l}$

Mass flow $[\text{kg/min}] = Vol. \text{ flow} [\text{l/min}] \times \text{Density} [\text{kg/l}]$

$$= 22.61 \times 60 \times 0.85$$

$$= 1153 \text{ kg/min} \quad \text{Ans. (ii)}$$

TEST EXAMPLES 2

1. Mechanical energy converted into heat energy per minute:

Energy $[\text{kJ} = \text{Power} [\text{kW} = \text{kJ/s}] \times \text{Time} [\text{s}]$

$$= 70 \times 60$$

$$= 4200 \text{ kJ} = 4.2 \text{ MJ} \quad \text{Ans. (i)}$$

All the heat energy being transferred to water:

$$Q[\text{kJ}] = m [\text{kg}] \times A [\text{kJ/kg K}] \times (T_2 - T_1) [\text{K}]$$

$$4200 = m \times 4.2 \times 10$$

$$m = 100 \text{ kg/min} \quad \text{Ans. (ii)}$$

2. Friction force at effective radius of pads

$$= \mu \times \text{force between surfaces}$$

$$= 0.025 \times 240 = 6 \text{ kN}$$

Work to overcome friction in one revolution

$$= \text{Friction force} [\text{kN}] \times \text{Circumference} [\text{m}]$$

$$= 6 \times 2\pi \times 0.23 \text{ kNm} = \text{kJ}$$

Power loss $[\text{kW} = \text{kJ/s}] = \text{Work per rev} [\text{kJ}] \times \text{rev/s}$

$$= 6 \times 2\pi \times 0.23 \times \frac{93}{60}$$

$$= 13.45 \text{ kW} \quad \text{Ans. (i)}$$

Mechanical energy converted into heat energy per hour:

$$\text{Energy} \left[kW\,h \right] = \text{Power} \left[kW \right] \times \text{Time} \left[h \right]$$
$$= 13.45 \times 1 = 13.45 \text{ kW h}$$
$$13.45 \text{ kW h} \times 3.6 = 48.42 \text{ MJ} \quad \text{Ans. (ii)}$$

Quantity of heat energy transferred to oil per hour:

$$Q \left[kJ \right] = m \left[kg \right] \times A \left[kJ/kg\,K \right] \times \left(T_2 - T_1 \right) \left[K \right]$$
$$48.42 \times 10^3 = m \times 2 \times 20$$
$$m = 1210.5 \text{ kg/h} \quad \text{Ans. (iii)}$$

3. Let θ = initial temperature of the copper = temperature of the flue gases

$$\text{Heat lost by copper} = \text{Heat gained by water}$$
$$m_c \times c_c \times \text{temp. fal}_c = m_w \times c_w \times \text{Temp. rise}_w$$
$$1.8 \times 0.395 \times \left(\theta - 37.2 \right) = 2.27 \times 4.2 \times \left(37.2 - 20 \right)$$
$$0.711\theta - 26.44 = 1641$$
$$0.711\theta = 190.44$$
$$\theta = 267.9°C \quad \text{Ans.}$$

4. Mass of 2.3 l of water = 2.3 kg

Let c [kJ/kg K] = specific heat of the iron

$$\text{Heat lost by iron} = \text{Heat gained by water and vessel}$$
$$2.15 \times c \times \left(100 - 244 \right) = \left(2.3 + 0.18 \right) \times 4.2 \times \left(244.17 \right)$$
$$2.15 \times c \times 75.6 = 2.48 \times 4.2 \times 7.4$$
$$c = \frac{2.48 \times 4.2 \times 7.4}{2.15 \times 75.6}$$
$$= 0.4742 \text{ kJ/kg K} \quad \text{Ans.}$$

5. Heat transferred to the ice to raise all of it in temperature from −5°C to 0°C

$$= 0.5 \times 2.04 \times 5 = 5.1 \text{ kJ}$$

Heat transferred from the water and vessel in cooling from 17°C to 0°C

$$= \left(1.8 + 0.148 \right) \times 4.2 \times 17$$
$$= 1.948 \times 4.2 \times 17 = 139.1 \text{ kJ}$$

Therefore, heat from water available to melt some of the ice

$$= 139.1 - 5.1 = 134 \text{ kJ}$$

Each kilogram of ice requires 335 kJ to melt it; therefore mass of ice melted by 134 kJ

$$= \frac{134}{335} = 0.4 \text{ kg}$$

Hence, $\left.\begin{array}{l} \text{Final mass of water} = 1.8 + 0.4 = 2.2 \text{ kg} \\ \text{Final mass of ice} = 0.5 - 0.4 = 0.1 \text{ kg} \end{array}\right\}$ Ans.

TEST EXAMPLES 3

1. Free expansion $= \alpha \times l \times (\theta_2 - \theta_1)$

$$= 1.25 \times 10^{-5} \times 3.85 \times (260 - 18)$$
$$= 0.01165 \text{ m}$$
$$= 11.65 \text{ mm} \quad \text{Ans.}$$

2.
$$= \frac{\pi}{6} d^3 = \frac{\pi}{6} \times 0.15^3 \text{ m}^3$$

Mass $[\text{kg}] = $ Volume $[\text{m}^3] \times$ Density $[\text{kg/m}^3]$

$$= \frac{\pi}{6} \times 0.15^3 \times 7.21 \times 10^3$$
$$= 12.75 \text{ kg}$$

$$Q[\text{kJ}] = m[\text{kg}] \times A[\text{kJ/kg K}] \times (T_2 - T_1)[\text{K}]$$
$$2110 = 12.75 \times 0.54 \times (T_2 - T_1)$$
$$\text{Temp. rise} = \frac{2110}{12.75 \times 0.54} = 306.7 \text{ K}$$
$$= 306.7^\circ\text{C}$$

Increase in diameter $= \alpha \times d \times (\theta_2 - \theta_1)$
$$= 1.12 \times 10^{-3} \times 150 \times 306.7$$
$$= 0.5152 \text{ mm} \quad \text{Ans.}$$

3. Free expansion $= \alpha \times l \times (\theta_2 - \theta_1)$

Heated length $= l + \alpha l (\theta_2 - \theta_1) = l\{1 + \alpha(\theta_2 - \theta_1)\}$

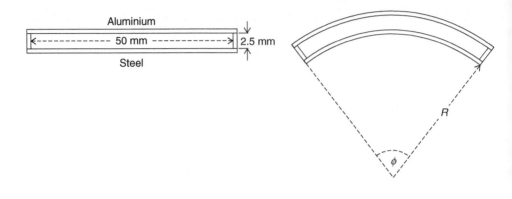

Heated length of aluminium strip

$$= 50\left(1 + 2.5 \times 10^{-5} \times 200\right)$$
$$= 50 \times 1.005 = 50.25 \text{ mm}$$

Heated length of steel strip

$$= 50\left(1 + 1.2 \times 10^{-5} \times 200\right)$$
$$= 50 \times 1.0024 = 50.12 \text{ mm}$$

Heated length of brass distance pieces

$$= 2.5\left(1 + 2 \times 10^{-5} \times 200\right)$$
$$= 2.5 \times 1.004 = 2.51 \text{ mm}$$
$$\text{Let } R = \text{ radius of curvature of steel}$$
$$R + 2.51 = \text{radius of curvature of aluminium}$$

Subtended angle φ is common to both

$$\varphi \text{ [rad]} = \frac{\text{Arc length [mm]}}{\text{Radius [mm]}}$$
$$\therefore \frac{50.25}{R + 2.51} = \frac{50.12}{R}$$
$$50.25R = 50.12\left(R + 2.51\right)$$
$$50.25R = 50.12R + 125.8$$
$$0.13R = 125.8$$

$$R = 967.7 \text{ mm}$$
$$\text{Radius of steel} = 967.7 \text{ mm} \quad \text{Ans.}$$
$$\text{Radius of aluminium} = 967.7 + 2.51$$
$$= 970.21 \text{ mm} \quad \text{Ans.}$$

4. Internal volume of pipe (= vol. of oil in pipe)

$$= 0.7854 \times 0.032 \times 13.7 \text{ m}^3$$

Co-efficient of cubical expansion of steel pipe

$$\beta_p = 3 \times 1.2 \times 10^{-5} = 3.6 \times 10^{-5} /°C$$

Co-efficient of cubical expansion of oil

$$\beta_o = 9 \times 10^{-4} = 90 \times 10^{-5} /°C$$

Increase in volume of pipe $= \beta_p \times V \times (\theta_2 - \theta_1)$
Increase in volume of oil $= \beta_o \times V \times (\theta_2 - \theta_1)$

$$\text{Oil overflow} = \beta_o V \left(\theta_2 - \theta_1\right) - \beta_p V \left(\theta_2 - \theta_1\right)$$
$$= \left(\beta_o - \beta_p\right) \times V \times \left(\theta_2 - \theta_1\right)$$
$$= (90 - 3.6) \times 10^{-5} \times 0.7854 \times 0.032 \times 13.7 \times 27$$
$$= 2.259 \times 10^{-4} \text{ m}^3$$

$$1 \text{ m}^3 = 10^3 \text{ l}$$
$$\text{i.e. } 2.259 \times 10^{-4} \times 10^3 = 0.2259 \text{ l} \quad \text{Ans.}$$

5. Free expansion $= \alpha \times l \times (\theta_2 - \theta_1)$
    If fully restricted,

$$\text{Strain} = \frac{\text{Change of length}}{\text{Original length}} = \frac{\alpha l (\theta_2 - \theta_1)}{l} = \alpha (\theta_2 - \theta_1)$$

$$\text{Stress} = \text{Strain} \times E$$
$$= 1.12 \times 10^{-5} \times (220 - 17) \times 206 \times 10^9$$
$$= 1.12 \times 203 \times 206 \times 10^4$$
$$= 4.684 \times 10^8 \text{ N/m}^2 = 4684 \text{ MN/m}^2$$

This compressive stress is to be relieved by exerting an initial tensile stress so that the compressive stress does not exceed 350 MN/m²:

$$\text{Initial tensile stress} = 4684 - 350$$
$$= 1184 \text{ MN/m}^2$$

TEST EXAMPLES 4

1.
$$Q[J] = \frac{k[W/mK] \times A[m^2] \times t[s] \times (T_1 - T_2)[K]}{S[m]}$$
$$= \frac{150 \times 0.7854 \times 0.127^2 \times 60 \times 5}{0.019}$$
$$= 3 \times 10^4 \text{ J} = 30 \text{ kJ} \quad \text{Ans.}$$

2. Total surface area of walls, ceiling and floor

$$= 2(4.5 \times 2.5 + 4 \times 2.5 + 4.5 \times 4)$$
$$= 78.5 \text{ m}^2$$
$$Q = \frac{kAt(T_1 - T_2)}{S}$$
$$= \frac{5.8 \times 10^{-2} \times 78.5 \times 3600 \times 20}{0.15}$$
$$= 2.185 \times 10^6 \text{ J}$$
$$= 2.185 \text{ MJ or } 2185 \text{ kJ} \quad \text{Ans.}$$

3. For each thickness:

$$Q = \frac{kAt \times \text{Temp.diff.}}{S} \qquad\qquad \therefore \text{Temp. diff.} = \frac{QS}{kAt}$$

For each three thicknesses:

$$T_D = \frac{Q}{AT}\left\{\frac{S_1}{k_1} + \frac{S_2}{k_2} + \frac{S_3}{k_3}\right\}$$

Heat transfer per second per square metre of area:

$$T_D = \frac{Q}{At}\Sigma\left\{\frac{S}{k}\right\}$$
$$23 = \frac{Q}{1 \times 1} \times 4.325$$
$$Q = \frac{23}{4.325} = 5.318 \text{ J} \quad \text{Ans. (i)}$$

Heat transfer per hour through whole side:

$$Q = 5.318 \times 6 \times 3.7 \times 3600$$
$$= 4.25 \times 10^5 \text{ J}$$
$$= 425 \text{ kJ} \quad \text{Ans. (ii)}$$

Temperature drop across outer wood wall

$$= \frac{QS_1}{k_1 At}$$
$$= 0.7977 \text{ K, say } 0.8°\text{C}$$

Temperature at interface of cork and outer wood wall

$$= 20 - 0.8 = 19.2°\text{C} \quad \text{Ans. (iii)(a)}$$

Temperature drop across cork

$$= \frac{QS_2}{k_2 At}$$
$$= \frac{5.318 \times 0.168}{0.042} = 21.27 \text{ K} = 21.27°\text{C}$$

Temperature at interface of cork and inner wood wall

$$= 19.2 - 21.27 = -2.07°\text{C} \quad \text{Ans. (iii)(b)}$$

4. $Q = 5.67 \times 10^{-11} At \left(T_1^4 - T_2^4 \right)$

$$T_1 = 488 \text{ K}$$
$$T_2 = 318 \text{ K}$$
$$T_1^4 - T_2^4 = 4.647 \times 10^{10}$$
$$Q = 5.67 \times 10^{-11} \times 0.7854 \times 0.5^2 \times 3600 \times 4.647 \times 10^{10}$$
$$= 1862 \text{ kJ} \quad \text{Ans.}$$

5. $\dfrac{1}{U} = \dfrac{1}{h_G} + \dfrac{S_p}{k_p} + \dfrac{1}{h_A}$

$$= \frac{1}{31.5} + \frac{0.01}{50} + \frac{1}{32}$$
$$= 0.03175 + 0.0002 + 0.03125$$
$$= 0.0632$$

$U = 1/0.0632 = 15.82 \text{ W/m}^2\text{ K}$ Ans. (i)

$Q = UAtT_D$

$Q = 15.82 \times 1 \times 60 \times (280 - 35)$

$= 2.327 \times 10^5 \text{ J} = 232.7 \text{ kJ}$ Ans. (ii)

6. Heat flow from outside atmosphere to exposed surface of outer wood wall, and also from exposed surface of inner wood wall to room atmosphere:

$Q = hAt \times$ temp. difference per m² of surface area per second,

$42 = 15 \times 1 \times 1 \times$ temp. difference

$$\text{Temp. difference} = \frac{42}{15} = 2.8°C$$

Temperature of exposed surface of outer wood

$$= 25 - 2.8 = 22.2°C \quad \text{Ans.} \quad (i)(a)$$

Temperature of exposed surface of inner wood

$$= -20 + 2.8 = -17.2°C \text{ Ans.} \quad (i)(b)$$

Heat flow through each wood wall:

$$Q = \frac{kAt \times \text{Temp. difference}}{0.03}$$

Temperature difference across each wood wall

$$= \frac{42 \times 0.03}{0.2} = 6.3°C$$

Interface temperature of outer wood and cork

$$= 22.2 - 6.3 = 15.9°C \quad \text{Ans.} \quad (ii)(a)$$

Interface temperature of inner wood and cork

$$= -17.2 + 6.3 = -10.9°C \quad \text{Ans.} \quad (ii)(b)$$

Temperature difference across cork

$$= 15.9 - (-10.9) = 26.8°C$$

Heat flow through cork:

$$Q = \frac{kAt \times \text{temp. difference}}{S}$$

$$S = \frac{0.05 \times 1 \times 1 \times 26.8}{42}$$

$$= 0.0319 \text{ m} = 31.9 \text{ mm} \quad \text{Ans.} \quad \text{(iii)}$$

7. Let T_1 = abs. temp. of shell before lagging

   = 503 K

   T_2 = abs. temp. of surrounds before lagging

   = 324 K

   T_3 = abs. temp. of cleading after lagging

   = 342 K

   T_4 = abs. temp. of surrounds after lagging

   = 300 K

Heat radiated from shell before lagging

$$= 5.67 \times 10^{-11} \left(T_1^4 - T_2^4 \right) \text{ kJ/m}^2\text{s}$$

Heat radiated from cleading after lagging

$$= 5.67 \times 10^{-11} \left(T_3^4 - T_4^4 \right) \text{ kJ/m}^2$$

Heat saved by lagging

$$= 5.67 \times 10^{-11} \left\{ \left(T_1^4 - T_2^4 \right) - \left(T_3^4 - T_4^4 \right) \right\} \text{ kJ/m}^2\text{s}$$

$$= 4.74 \times 10^{10}$$

Total surface area = Area of ends + Area of cylinder

$$= \pi d^2 + \pi d l$$
$$= \pi d (d + l)$$
$$= \pi \times 1.22 (1.22 + 4.78)$$
$$= \pi \times 1.22 \times 6$$
$$\text{Lagged area} = 0.75 \times \pi \times 1.22 \times 6$$
$$= 17.26 \text{ m}^2$$

Total heat saved by lagging per hour

$$= 5.67 \times 10^{-11} At \left\{ \left(T_1^4 - T_2^4 \right) - \left(T_3^4 - T_4^4 \right) \right\}$$
$$= 5.67 \times 10^{-11} \times 17.26 \times 3600 \times 4.74 \times 10^{10}$$
$$= 1.67 \times 10^5 \text{ kJ}$$
$$= 1.67 \text{ MJ} \quad \text{Ans.}$$

8.　　For lagging $Q = \dfrac{2\pi k (T_1 - T_2)}{\ln(D/d)}$ W/m

For surface film $Q = \dfrac{hA (T_2 - T_3)}{L}$ W/m

$$= h\pi D \left(T_2 - T_3 \right)$$

Q for lagging $= Q$ for surface film

$$\frac{2\pi k \times 0.05(T_1 - T_2)}{\ln(D/d)} = h\pi D(T_2 - T_3)$$

$$\therefore \frac{2\pi \times 0.05(350 - T_2)}{\ln\left(\dfrac{340}{200}\right)} = 10 \times \pi \times \frac{340}{10^3}(T_2 - 15)$$

From which $T_2 = 32.6°C$　Ans. (i)

$$\therefore Q = hA\left(T_2 - T_3 \right)$$

$$= 10 \times \pi \times \frac{340}{10^3} \times 20 \ (32.6 - 15)/10^3$$

$$= 3.76 \text{ kJ/s} \quad \text{Ans. (ii)}$$

9.　Let T = temperature of the cooling water at outlet

Therefore $\theta_1 = 410 - 10, \ \theta_2 = 130 - T$
$$= 400$$

$$\therefore \theta_m = \frac{400 - (130 - T)}{\ln\left(\dfrac{400}{130 - T}\right)} \qquad (1)$$

For the gas $Q = 0.4 \times 1130 \times (410-130)/10^3$

$$= 126.56 \text{ kW}$$

For the water $Q = 0.5 \times 4190 \times (T-10)/10^3$

$$126.56 = 0.5 \times 4190 \times (T-10)/10^3$$
$$= \text{From which } T = 70.4°C \text{ substitute in (1)}$$
$$\therefore \theta_m = \frac{400 - (130 - 70.4)}{\ln\left(\dfrac{400}{130 - 70.4}\right)}$$
$$= 178.8°C$$

also $Q = UAt\theta_m$

$$A = \frac{Q}{UAt\theta_m}$$
$$= \frac{123.56 \times 10^3}{140 \times 1 \times 178.8}$$
$$= 5.06 \text{ m}^2 \quad \text{Ans.}$$

10. *Note*: This question illustrates one fluid being at constant temperature, that is, the condensing steam.

$$\text{Water mass flow} = 0.7854 \times 0.0252 \times 0.6 \times 10^3$$
$$= 0.2945 \text{ kg/s}$$
$$Q = 0.2945 \times 4.18 \times 12$$
$$= 14.77 \text{ kW}$$
$$\theta_m = \frac{27 - 15}{\ln\left(\dfrac{27}{15}\right)}$$
$$= 20.41°C$$

$$Q = UAt\theta_m$$
$$U = \frac{41770}{\pi \times 0.025 \times 2.75 \times 1 \times 20.41}$$

Overall heat transfer coefficient

$$= 3350 \text{ W/m}^2 \text{ K} \quad \text{Ans. (i)}$$

$$\frac{1}{U} = \frac{1}{h_0} + \frac{1}{h_i}$$

$$\frac{1}{3350} = \frac{1}{17000} + \frac{1}{h_i}$$

Inside surface heat transfer coefficient (i.e., between water and tube) is h_i:

$$= 4172 \text{ W/m}^2 \text{ K} \quad \text{Ans. (ii)}$$

TEST EXAMPLES 5

1.
$$p_1 V_1 = p_2 V_2$$

$$120 \times 1 = 960 \times V_2$$

$$V_2 = \frac{120 \times 1}{960} = 0.125 \text{ m}^3 \qquad \text{Ans. (i)}$$

2. $\quad p_1 V_1 = p_2 V_2$

$$1.05 \times V_1 = 42 \times 5.6$$

$$V_1 = \frac{42 \times 5.6}{1.05} = 224 \text{ m}^3 \quad \text{Ans. (ii)}$$

$$\frac{p_1 V_1}{T_1} = \frac{p_2 V_2}{T_2}$$

$$\frac{1350 \times 0.2}{450} = \frac{250 \times 0.9}{T_2}$$

$$T_2 = \frac{450 \times 250 \times 0.9}{1350 \times 0.2} = 375 \text{ K}$$

$$= 102°C \quad \text{Ans.}$$

3. $\quad pV = mRT$

$$m_1 = \frac{pV}{RT}$$

$$= \frac{10 \times 100 \times 2}{287 \times 293}$$

$$m_2 = 0.92 \times 0.02378$$

$$= 0.02188 \text{ kg}$$

$$T = \frac{pV}{mR}$$

$$= \frac{20 \times 100 \times 2}{0.02188 \times 287}$$

$$= 636.98 \text{ K}$$

Temperature = 363.98°C Ans.

4. By Charles' law, when volume is constant:

$$\frac{p_1}{p_2} = \frac{T_1}{T_2}$$

$$p_2 = \frac{(3200 + 100) \times 308}{289}$$

$$= \frac{3200 \times 308}{289} = 3518 \text{ kN/m}^2$$

or 3518 − 100 = 3418 kN/m² gauge   Ans. (i)

$$Q[\text{kJ}] = m \text{kJ} \times c [\text{kJ/kg K}] \times (T_2 - T_1)[\text{K}]$$

$$= 20 \times 0.718 \times (35 - 16)$$

$$= 272.9 \text{ kJ   Ans. (ii)}$$

5. Volume of saloon = 12 × 16.5 × 4 = 792 m³

Volume of air to be supplied per hour = 2 × 792 = 1584 m³

Volume varies as the absolute temperature when the pressure is constant, therefore density varies inversely as the absolute temperature.

Density at 0°C and atmos. press. = 1.293 kg/m³

i.e., density at 21°C and atmos. press.

$$= 1.293 \times \frac{273}{294} = 1.2 \text{ kg/m}^3$$

Mass of air to be supplied per hour

$$= \text{Volume } [\text{m}^3/\text{h}] \times \text{Density } [\text{kg/m}^3]$$

$$= 1584 \times 1.2 = 1901 \text{ kg/h}$$

Heat extracted per hour

$$Q\left[kJ/h\right] = m\left[kg/h\right] \times c_p\left[kJ/kg\ K\right] \times \left(T_1 - T_2\right)\left[K\right]$$
$$= 1901 \times 1.005 \times \left(30 - 21\right)$$
$$= 17190\ kJ/h\ or\ 17.19\ MJ/h\quad Ans.\ (i)$$

Power [kW] = kilojoules per second

$$= \frac{17190}{3600} = 4.7775kW\quad Ans.\ (ii)$$

6. Negligible velocity, so kinetic energies are zero Steady Flow Energy Equation:

$$h_1 + q = h_2 + w\ per\ unit\ mass$$
$$w_{OUT} = h_1 - h_2 + q_{IN}$$
$$w_{OUT} = 1.7 \times 0.9 \times 480 - 1.7 \times 10$$
$$= 717.4\ kW\quad Ans.$$

7. At constant volume:

$$Q = mc_v\left(T_2 - T_1\right)$$
$$= 1.36 \times 0.718 \times \left(468 - 40\right)$$
$$= 417.8\ kJ\quad Ans.\ (i)(a)$$

External work done = nil   Ans. (i) (b)

From the energy equation:

Heat energy supplied = Increase in internal energy + External work done

---

$$417.8 = Increase\ in\ internal\ energy + 0$$
$$\therefore Increase\ in\ internal\ energy = 417 - 8\ kJ\quad Ans.\ (i)(c)$$

At constant pressure:

$$Q = mc_p\left(T_2 - T_1\right)$$
$$= 1.36 \times 1.005 \times \left(468 - 40\right)$$
$$= 584.8\ kJ\quad Ans.\ (ii)(a)$$

Internal energy depends only upon the temperature and is independent of changes in pressure and volume. Therefore the increase in internal energy is the same as before

$$= 417.8\ kJ\quad Ans.\ (ii)(b)$$

Heat energy supplied = Increase in internal energy + External work done

---

$$584.8 = 417.8 + \text{Work done}$$
$$\text{Work done} = 584.8 - 417.8$$
$$= 167 \text{ kJ} \quad \text{Ans.} \ (\text{ii})(\text{c})$$

Note also:

$$R = c_p - c_v = 1.005 - 0.718 = 0.287 \text{ kJ/kg K}$$
$$\text{Work done} = mR(T_2 - T_1)$$
$$= 1.36 \times 0.287 \times 428$$
$$= 167 \text{ kJ} \ (\text{as above})$$

8. Ratio of partial pressures is equal to the ratio of partial volumes.

$$\left.\begin{array}{l}\text{partial press. of} \ldots CO_2 = 0.1 \times 1.015 = 0.1015 \text{ bar}\\ \text{partial press. of} \ldots O_2 = 0.08 \times 1.015 = 0.0812 \text{ bar}\\ \text{partial press. of} \ldots N_2 = 0.82 \times 1.015 = 0.8323 \text{ bar}\end{array}\right\} \text{Ans.}$$

$$pV = mRT \qquad \therefore m = \frac{pV}{RT}$$

where p = partial pressure, and V = full volume

(or p = total pressure, and V = partial volume)

$$\text{Mass of } CO_2 = \frac{0.1015 \times 10^2 \times 500 \times 10^{-6}}{0.189 \times 293}$$
$$= 9.162 \times 10^{-5} \text{ kg} = 0.09162 \text{ g} \quad \text{Ans.}$$

$$\text{Mass of } O_2 = \frac{0.0812 \times 10^2 \times 500 \times 10^{-6}}{0.26 \times 293}$$
$$= 5.33 \times 10^{-5} \text{ kg} = 0.0533 \text{ g} \quad \text{Ans.}$$

$$\text{Mass of } N_2 = \frac{0.8323 \times 10^2 \times 500 \times 10^{-6}}{0.297 \times 293}$$
$$= 4.782 \times 10^{-4} \text{ kg} = 0.4782 \text{ g} \quad \text{Ans.}$$

9.

$$pv = RT \quad \text{and} \quad R = \frac{R_0}{M}$$

$$\text{hence } pv = \frac{R_0}{M} \times T$$

$$\text{also } h = u + pv$$

$$\therefore h = u + R_0 \frac{T}{M}$$

$$\text{Hence } h_1 = u_1 + R_0 \frac{T_1}{M} \text{ and } h_2 = u_2 + R_0 \frac{T_2}{M}$$

$$u_1 - u_2 = 80$$

$$\therefore h_1 - h_2 = 80 + \frac{R_0}{M}(T_1 - T_2)$$

$$= 80 + \frac{8.314}{30}(200 - 100)$$

$$= 107.71 \text{ kJ/kg}$$

Steady flow energy equation:

$$h_1 + \frac{1}{2}c_1^2 + q = h_2 + \frac{1}{2}c_2^2 + w$$

As q and $w = 0$

$$h_1 - h_2 = \frac{1}{2}\left(c_2^2 - c_1^2\right)$$

$$\text{hence } 107.71 \times 10^3 = \frac{1}{2}\left(c_2^2 - 50^2\right)$$

$$c_2 = \sqrt{107.71 \times 10^3 \times 2 + 50^2}$$

$$c_2 = 466.82 \text{ m/s} \quad \text{Ans.}$$

10. 0.45 kg carbon monoxide $R = \dfrac{R_0}{M} = \dfrac{8.314}{28} = 0.2969$

0.23 kg oxygen $R = \dfrac{R_0}{M} = \dfrac{8.314}{32} = 0.2598$

0.77 kg nitrogen $R = \dfrac{R_0}{M} = \dfrac{8.314}{28} = 0.2969$

$$1.45 \, R = 0.45 \times 0.2969 + 0.23 \times 0.2598 + 0.77 \times 0.2969$$

$$1.45 \text{ kg mixture } R = 0.2911$$
$$pV = mRT$$
$$p = \frac{mRT}{V}$$
$$= \frac{1.45 \times 0.2911 \times 288}{0.4}$$
$$= 303.9 \text{ kN/m}^2$$

Total pressure = 3.039 bar Ans. (i)

$$pV = mRT$$
$$p = \frac{mRT}{V}$$
$$= \frac{0.77 \times 0.2911 \times 288}{0.4}$$
$$= 164.6 \text{ kN/m}^2$$

Partial pressure nitrogen = 1.646 bar Ans. (ii)

TEST EXAMPLES 6

1. For polytropic expansion:

$$p_1 V_1^n = p_2 V_2^n$$
$$2550 \times V_1^{1.3} = 210 \times 0.75^{1.3}$$
$$V_1^{1.3} = \frac{210 \times 0.75^{1.3}}{2550}$$
$$V_1 = 0.75 \times \sqrt[1.3]{\frac{210}{2550}}$$
$$= 0.1099 \text{ m}^3 \quad \text{Ans.}$$

2. Ratio of compression $= \dfrac{8.6}{1} = \dfrac{\text{Initial volume}}{\text{Final volume}} = \dfrac{V_2}{V_2}$

$$p_1 V_1^n = p_2 V_2^n$$

$$98 \times 8.6^{1.36} = 1828 \text{ kN/m}^2 \quad \text{Ans. (i)}$$

$$\frac{p_1 V_1}{T_1} = \frac{p_2 V_2}{T_2}$$

$$\frac{98 \times 8.6}{301} = \frac{1828 \times 1}{T_2}$$

$$= 380°C \quad \text{Ans. (ii)}$$

3. $$p_1 V_1^n = p_2 V_2^n$$

$$1750 \times 0.05^n = 122.5 \times 0.375^n$$

$$\frac{1750}{122.5} = \left\{\frac{0.375}{0.05}\right\}^n$$

$$n = 1.32 \quad \text{Ans.}$$

4.

$$\frac{T_1}{T_2} = \left\{\frac{p_1}{p_2}\right\}^{\frac{n-1}{n}}$$

$$\frac{n-1}{n} = \frac{1.35-1}{1.35} = \frac{7}{27}$$

$$\frac{293}{T_2} = \left\{\frac{101.3}{1420}\right\}^{\frac{7}{27}}$$

$$T_2 = 293 \times \left\{\frac{1420}{101.3}\right\}^{\frac{7}{27}} = 581.1K$$

$$581.1 - 273 = 308.1°C \quad \text{Ans.}$$

5.   For adiabatic expansion:

$$\frac{T_1}{T_2} = \left\{\frac{V_2}{V_1}\right\}^{\gamma-1}$$

$$\text{where } \gamma = \frac{c_p}{c_v} = \frac{1.005}{0.718} = 1.4$$

$$\gamma - 1 = 1.4 - 1 = 0.4$$

$$\frac{339}{275} = \left\{\frac{V_2}{0.014}\right\}^{0.4}$$

$$V_2 = 0.014 \times \sqrt[0.4]{\frac{339}{275}}$$

$$= 0.02362 \text{ m}^3 \quad \text{Ans.}$$

6.

$$\frac{T_1}{T_2} = \left\{\frac{p_1}{p_2}\right\}^{\frac{n-1}{n}}$$

$$\frac{305}{773} = \left\{\frac{117}{3655}\right\}^{\frac{n-1}{n}}$$

$$\text{or} \frac{773}{305} = \left\{\frac{3655}{117}\right\}^{\frac{n-1}{n}}$$

$$\log\left\{\frac{773}{305}\right\} = \log\left\{\frac{3655}{117}\right\} \times \left\{\frac{n-1}{n}\right\}$$

$$0.4039 = 1.4947 \times \frac{n-1}{n}$$

$$1.4947 = 1.0908n$$

$$n = \frac{1.4947}{1.0908} = 1.37$$

i.e. law of compression is:

$$pV^{1.37} = C \quad \text{Ans.}$$

7.

$$p_1V_1 = mRT_1$$

$$V_1 = \frac{1 \times 0.287 \times 473}{10 \times 100}$$

$$= 0.1142 \text{ m}^3$$

$$p_2V_2 = mRT_2$$

$$V_2 = \frac{1 \times 0.287 \times 398}{20 \times 100}$$

$$= 114.226 \text{ m}^3$$

$$\begin{array}{l} \text{Work transfer} \\ \text{(expansion)} \end{array} = \begin{array}{l} \text{Area under process straight} \\ \text{line of } pV \text{ diagram} \end{array}$$

$$= \text{Mean pressure} \times \text{Change of volume}$$

$$= 100 \times 15 \left(0.1142 - 0.0679\right)$$

$$= 69.45 \text{ kJ} \quad \text{Ans. (i)}$$

$$c_V = 1005 - 287 = 718 \text{ J/kg K}$$

$$\text{Change of internal energy} = mc_V \left(T_2 - T_1\right)$$

$$= 1 \times 0.718 \left(398 - 473\right)$$

$$= -53.85 \text{ kJ, i.e. decrease} \quad \text{Ans. (ii)}$$

$$\text{Heat transfer} = -53.85 + 69.45$$
$$= 15.6 \text{ kJ} \quad \text{Ans. (iii)}$$

$$\text{Change of enthalpy} = mc_p \left(T_2 - T_1 \right)$$
$$= 1 \times 1.005 \left(398 - 473 \right)$$
$$= -75 - 375 \text{ kJ, i.e. decrease} \quad \text{Ans. (iv)}$$

8.
$$P_1 V_1 = mRT_1$$

$$V_1 = \frac{1 \times 0.287 \times 486}{1 \times 100}$$
$$= 1.395 \text{ m}^3$$

$$p_2 V_2 = m_2 RT_2$$
$$m_2 = \frac{6 \times 100 \times 0.2}{0.287 \times 685}$$
$$= 0.6103 \text{ kg}$$

No heat transfer takes place (insulated).

No external work is done.

The process is at constant internal energy.

$$m_1 c_V T_1 + m_2 c_V T_2 = m_3 c_V T_3$$
$$1 \times 486 + 0.6103 \times 685 = 1.6103 \times T_3$$
$$T_3 = 561.4 \text{ K}$$

Final air temperature is 288.4°C   Ans. (i)

$$p_3 V_3 = m_3 RT_3$$
$$p_3 = \frac{1.6103 \times 0.287 \times 561.4}{1.595}$$
$$= 162.7 \text{ kN/m}^2$$

Final air pressure is 1.627 bar   Ans. (ii)

9.
$$pV = mRT$$

$$m = \frac{1 \times 100 \times 0.1}{0.287 \times 288}$$
$$= 0.1210 \text{ kg}$$

$$\frac{T_1}{T_2} = \frac{p_2}{p_1} \quad \text{as } V \text{ constant}$$

$$\frac{773}{288} = \frac{p_2}{1}$$

$$p_2 = 2.684 \text{ bar}$$

$$\frac{T_3}{T_2} = \left(\frac{p_3}{p_2}\right)^{\frac{n-1}{n}}$$

$$\frac{T_3}{773} = \left(\frac{1}{2.684}\right)^{\frac{0.5}{1.5}}$$

$$T_3 = 556.1 \text{ K}$$

Final temperature = 283.1°C   Ans. (i)

$$\text{Work done} = \frac{mR(T_2 - T_3)}{n-1}$$

$$= \frac{0.121 \times 0.287(773 - 556.1)}{0.5}$$

$$= 15.05 \text{ kJ}   \text{Ans. (ii)}$$

Heat supplied = $mc_v(T_2 - T_1)$ at constant volume

$$= 0.121 \times 0.718(773 - 288)$$

$$= 42.14 \text{ kJ}   \text{Ans. (iii)}$$

Heat transfer = $mA_v(T_3 - T_2)$

$$= 0.121 \times 0.718\,(556.1 - 773)$$

$$= -18.84 + 15.05$$

$$= -3.79 \text{ kJ}   \text{Ans. (iii)}$$

10.

$$T_1 = 783 \text{ K}$$

$$T_2 = 313 \text{ K}$$

$$\frac{p_1 V_1}{T_1} = \frac{p_2 V_2}{T_2}$$

$$\frac{36 \times 0.125}{783} = \frac{p_2 \times 1.5}{313}$$

$$p_2 = \frac{36 \times 0.125 \times 313}{783 \times 1.5}$$
$$= 1.199 \text{ bar or } 119.9 \text{ kN/m}^2 \quad \text{Ans.} \quad (i)$$

$$p_1 V_1^n = p_2 V_2^n \quad \text{or} \quad \frac{T_1}{T_2} = \left\{ \frac{V_2}{V_1} \right\}^{n-1}$$
$$36 \times 0.125^n = 1.199 \times 1.5^n$$
$$\frac{36}{1.199} = \left\{ \frac{1.5}{0.125} \right\}^n$$
$$n = 1.37 \quad \text{Ans.} \quad (ii)$$

$$p_1 V_1 = mRT_1$$
$$m = \frac{3600 \times 0.125}{0.258 \times 783} = 2.023 \text{ kg} \quad \text{Ans.} \quad (iii)$$

Increase in internal energy:

$$U_2 - U_1 = mc_v \left(T_2 - T_1 \right)$$
$$= 2.023 \times 0.71 \times (313 - 783)$$
$$= 2.023 \times 0.71 \times (-470)$$
$$= -675.1 \text{ kJ}$$

The minus sign indicates a *decrease* in internal energy.

Decrease in internal energy = 675.1 kJ   Ans. (iv)

$$\text{Work done} = \frac{p_1 V_1 - p_2 V_2}{n-1} \quad \text{or} \quad \frac{mR(T_1 - T_2)}{n-1}$$
$$= \frac{3600 \times 0.125 - 119.9 \times 1.5}{1.37 - 1}$$
$$= \frac{450 - 179.9}{0.37} = 730.2 \text{kg} \quad \text{Ans.} \quad (v)$$

Heat supplied = Increase in internal energy + External work done

---

$$= -675.1 + 730.2$$
$$= 55.1 \text{ kJ} \quad \text{Ans.} \quad (vi)$$

TEST EXAMPLES 7

1. Indicated mean effective pressure [kN/m²]

$$= \text{Mean height of diagram } [\text{mm}] \times \text{Spring scale } [\text{kN/m}^2\text{mm}]$$

$$= \frac{\text{Area of diagram}}{\text{Length of diagram}} \times \text{Spring scale}$$

$$= \frac{390}{70} \times 1.6 \times 10^2 = 891.6 \text{kN/m}^2$$

$\text{Ip} = p_m LAn \times 4 \ (\text{for 4 cylinders})$

$$= 891.6 \times 0.7854 \times 0.15^2 \times 0.2 \times \frac{5.5}{2} \times 4$$

$$= 34.66 \text{ kW} \quad \text{Ans.}$$

2. $\text{Ip} = P_m LAn \times 8 \ (\text{for 8 cylinders})$

$$= 1172 \times 0.7854 \times 0.75^2 \times 1.125 \times \frac{110}{60 \times 2} \times 8$$

$$= 4272 \text{ kW} \quad \text{Ans. (i)}$$

$$\text{bp} = \text{ip} \times \text{mech. efficiency}$$
$$= 4272 \times 0.86$$
$$= 3676 \text{ kW Ans.} \quad \text{(ii)}$$

3. $\text{Ip} = \dfrac{\text{bp}}{\text{mech. effic.}} = \dfrac{2250}{0.84} = 2680 \text{ kW}$

Let d = diameter of cylinders in metres
then $L = 1.25 \ d$ metres
$$\text{ip} = p_m LAn \times 6$$
$$2680 = 10 \times 10^2 \times 0.7854 \times d^2 \times 1.25 \ d \times 2 \times 6$$

$$d = \sqrt[3]{\frac{2680}{1000 \times 0.7854 \times 1.25 \times 2 \times 6}}$$

$$\left. \begin{array}{l} = 0.6103 \text{ m} = 610.3 \text{ mm} \\ L = 1.25 \times 610.3 = 763 \text{ mm} \end{array} \right\} \text{ Ans.}$$

4. Load on break = 480−84 = 369 N

Effective radius $= \dfrac{1}{2}(1220+24) = (1220+24) = 622$ mm $= 0.622$ m

Break power $= T_\omega$

$$= 396 \times 0.622 \times \frac{250 \times 2\pi}{60}$$

$$= 6449 \text{ W} = 60449 \text{ kW} \quad \text{Ans. (i)}$$

1 kWh = 3.6 MJ

Heat carried away by water in 1 h

$$= 0.9 \times 6449 \times 3.6 \times 10^3 \text{ kJ}$$

Heat received by water

$$= \text{mass} \times \text{spec. heat} \times \text{temp. rise}$$
$$Q\,[\text{kJ}] = m\,[\text{kg}] \times c_p\,[\text{kJ/kg K}] \times \theta\,[\text{K}]$$
$$= m \times 4.2 \times 18$$
$$m \times 4.2 \times 18 = 0.9 \times 6449 \times 3.6 \times 10^3$$
$$m = 2764 \text{ kg}$$

Mass of 1 l of fresh water = 1 kg

Quantity in litres = 276.4 l/h Ans. (ii)

5. Brake power [W] = Torque [Nm] × ω [rad/s]

When speed is 24.5 rev/s:

$$\text{bp} = T \times 2\pi \times 24.5 = 153.9\,T \text{ watts} = 0.1539\,T \text{ kW}$$

With all cylinders firing:

$$\text{bp} = 0.1539 \times 193.8 = 29.84 \text{ kW} \quad \text{Ans. (i)}$$
$$\text{no. 1 cyl. cut out, bp} = 0.1539 \times 130.8 = 20.13 \text{ kW}$$
$$\text{ip of no. 1 cyl.} = 29.84 - 20.13 \ = 9.71 \text{ kW}$$
$$\text{no. 2 cyl. cut out, bp} = 0.1539 \times 130.2 = 20.04 \text{ kW}$$
$$\text{ip of no. 2 cyl.} = 29.84 - 20.04 \ = 9.8 \text{ kW}$$
$$\text{no. 3 cyl. cut out, bp} = 0.1539 \times 129.9 = 20.0 \text{ kW}$$
$$\text{ip of no. 3 cyl.} = 29.84 - 20 \quad\ \ = 9.84 \text{ kW}$$
$$\text{no. 4 cyl. cut out, bp} = 0.1539 \times 131.1 = 20.18 \text{ kW}$$
$$\text{ip of no. 4 cyl.} = 29.84 - 20.18 \ = 9.66 \text{ kW}$$

Total ip of engine

$$= 9.71 + 9.8 + 9.84 + 9.66 = 39.01 \text{ kW} \quad \text{Ans. (ii)}$$

$$\text{Mech. effic.} = \frac{\text{Brake power}}{\text{Indicated power}}$$

$$= \frac{29.84}{39.01} = 0.7649 = 76.49\% \quad \text{Ans. (iii)}$$

6. $\text{Break thermal effic.} = \dfrac{3.6 \text{ [MJ/kWh]}}{\text{kg fuel/brake [kWh]} \times \text{cal. value [MJ/kg]}}$

$$= \frac{3.6}{0.255 \times 43.5} = 0.3245 \text{ or } 37.45\% \quad \text{Ans.} \quad \text{(i)}$$

$$\text{Indicated thermal efficiency} = \frac{\text{Break thermal efficiency}}{\text{Mechanical efficiency}}$$

$$= \frac{0.3245}{0.86} = 0.3773 \text{ or } 37.74\% \text{ Ans.} \quad \text{(ii)}$$

For each kg of fuel burned,

total mass of gases = 35 kg air + 1 kg fuel = 36 kg

Heat carried away in exhaust gases:

$$Q\,[\text{kJ}] = m\,[\text{kg}] \times c_p\,[\text{kJ/kg K}] \times \theta\,[\text{K}]$$
$$= 36 \times 1005 \times (393 - 26)$$
$$= 1.327 \times 10^4 \text{ kJ}$$
$$= 13.27 \text{ MJ/kg fuel}$$
$$\text{Heat supplied} = 43.5 \text{ MJ/kg fuel}$$

i.e. percentage heat in exhaust gases

$$= \frac{13.27}{43.5} \times 100 = 30.52\% \quad \text{Ans.} \quad \text{(iii)}$$

7. $\text{Ind. thermal effic.} = 100\% - (31.7 + 30.8)$

$$= 37.5\% \quad \text{Ans. (i)}$$

$$\text{Mech. effic.} = \frac{bp}{ip} = \frac{4060}{4960}$$
$$= 0.8185 \text{ or } 81.85\% \quad \text{Ans. (ii)}$$

Overall effic. = ind. thermal effic. × mech. effic.
$$= 0.375 \times 0.8185$$
$$= 0.307 \text{ or } 30.7\% \quad \text{Ans. (iii)}$$

Specific fuel consumption (indicated)

$$= \frac{27 \times 10^3}{24 \times 4960} = 0.2269 \text{ kg/ind. kWh} \quad \text{Ans. (iv)}$$

$$\text{Indicated thermal effic.} = \frac{3.6 [\text{MJ/kWH}]}{\text{kg fuel/ind. [kWh]} \times \text{cal. val. [MJ/kg]}}$$

$$0.375 = \frac{3.6}{0.2269 \times \text{cal. value}}$$

$$\text{Calorific value} = \frac{3.6}{0.375 \times 0.2269}$$

$$= 42.32 \text{ MJ/kg} \quad \text{Ans. (v)}$$

8. BEFORE ALTERATION:

$$V_1 = 87.5 + 12.5 = 100$$
$$V_2 = 12.5$$
$$p_1 V_1^{1.35} = p_2 V_2^{1.35}$$
$$0.97 \times 100^{1.35} = p_2 \times 12.5^{1.35}$$

$$P_2 = 0.97 \times \left\{ \frac{100}{12.5} \right\}^{1.35}$$

$$= 0.97 \times 8^{1.35} = 16.07 \text{ bar} \quad \text{Ans. (i)}$$

AFTER ALTERATION:

$$V_1 = 87.5 + 10 = 97.5$$
$$V_2 = 10$$
$$p_1 V_1^{1.35} = p_2 V_2^{1.35}$$
$$0.97 \times 97.51^{1.35} = p_2 \times 10^{1.35}$$

$$p_2 = 0.97 \times 9.75^{1.35} = 20.99 \text{ bar} \quad \text{Ans. (ii)}$$

9. Due to the 'hit and miss' governor cutting off the supply of gas, the number of power strokes per second is the number of explosions per second, which is $123 \div 60 = 2.05$

$$Ip = p_m LAn$$
$$= 3.93 \times 10^2 \times 0.7854 \times 0.18^2 \times 0.3 \times 2.05$$
$$= 6.152 \text{kW} \quad \text{Ans.} \,(\text{i})$$

$$\text{Mech. effic.} = \frac{\text{Brake power}}{\text{Indicated power}}$$
$$= \frac{4.33}{6.152} = 0.7039 = 70.39\% \quad \text{Ans.} \,(\text{ii})$$

Indicated thermal efficiency

$$= \frac{\text{Heat equivalent of work done in cyl. [KJ/h]}}{\text{Heat supplied [KJ/h]}}$$
$$= \frac{6.152 \times 3600}{3.1 \times 17.6 \times 10^3}$$
$$= 0.4059 = 40.59\% \quad \text{Ans.} \,(\text{iii})$$

Brake thermal effic. = ind. thermal effic. × mech. effic.
$$= 0.4059 \times 0.7039$$
$$= 0.2858 = 28.58\% \quad \text{Ans.} \,(\text{iv})$$

10. Mean piston speed $= 2\,Ln$

$$6 = 2 \times 1.5 \times n$$
$$n = 2 \text{ rev/s}$$
Brake power $= p_m LAn$
$$= 7 \times 100 \times 0.7854 \times 0.75^2 \times 1.5 \times 2$$
Brake power $= 927.8 \text{kW}$

$$\frac{\text{Power 1}}{\text{Swept volume 1}} = \frac{\text{Power 2}}{\text{Swept volume 2}}$$
$$\frac{370}{0.7854 \times D^2 \times 2.0D} = \frac{927.8}{0.7854 \times 0.75^2\,1.5}$$
$$D^3 = \frac{370 \times 0.5625 \times 1.5}{927.8 \times 2.0}$$
$$D = 0.552 \text{ m}$$
Cylinder bore $= 0.552 \text{ m} \quad \text{Ans.} \,(\text{i})$

Brake power $= p_m LAn$

$$370 = p_m \times 100 \times 0.7854 \times 0.552^2 \times 1.104 \times 2$$

$$p_m = 7.002 \text{ bar}$$

Brake mean effective pressure = 7.002 bar Ans. (ii)

TEST EXAMPLES 8

1.

$$p_1 V_1^{1.35} = p_2 V_2^{1.35}$$

$$1 \times 8.5^{1.35} = P_2 \times 1^{1.35}$$

$$P_2 = 8.5^{1.35} = 17.97 \text{ bar} \quad \text{Ans. (i)}$$

$$\frac{p_1 V_1}{T_1} = \frac{p_2 V_2}{T_2}$$

$$\frac{1 \times 8.5}{313} = \frac{17.97 \times 1}{T_2}$$

$$T_2 = \frac{313 \times 17.97}{8.5} = 661.8 \text{ K}$$

$$= 388.8°C \quad \text{Ans. (ii)}$$

$$\frac{T_3}{T_2} = \frac{p_3}{p_2}$$

$$T_3 = \frac{661.8 \times 31}{17.97} = 1141 \text{ K}$$

$$= 868°C \quad \text{Ans. (iii)}$$

2. $\dfrac{T_2}{T_1} = \left(\dfrac{V_1}{V_2}\right)^{\gamma-1}$ (Refer to Figure 8.1, p–V diagram)

$$T_2 = 323 \times 5^{0.4}$$

$$= 615\,\text{K}$$

Heat supplied $= c_v\left(T_3 - T_2\right)$

$$930 = 0.717\left(T_3 - 615\right)$$

$$T_3 = 1912 \text{ K}$$

Maximum temp. $= 1639°C$ Ans. (i)

$$\frac{T_3}{T_4} = \left(\frac{V_4}{V_3}\right)^{\gamma-1}$$

$$T_4 = \frac{1912}{5^{0.4}}$$

$$T_4 = 1004 \text{ K}$$

Work done = Heat supplied – Heat rejected

$$= c_v \left[\left(T_3 - T_2\right) - \left(T_4 - T_1\right)\right]$$

$$= 0.717 \left[(1912 - 615) - (1004 - 323)\right]$$

$$= 441.7 \text{ kJ/kg of working fluid} \quad \text{Ans.} \quad \text{(ii)}$$

$$\text{Ideal thermal efficiency of cycle} = \frac{441.7}{0.717 \times 1297} \times 100$$

$$= 47.5\% \quad \text{Ans. (iii)}$$

3.

$$\gamma = \frac{c_p}{c_v}$$

$$= \frac{1005}{718}$$

$$= 1.4$$

$$\frac{T_2}{T_1} = \left[\frac{V_1}{V_2}\right]^{\gamma-1}$$

$$T_2 = 330 \times 8^{0.4}$$

$$= 758 \text{ K}$$

Energy added $= c_v \left(T_3 - T_2\right)$

$$1250 = 0.718 \left(T_3 - 758\right)$$

$$T_3 = 2499 \text{ K}$$

Max. temperature = 2216°C   Ans. (i) (a)

$$P_1 V_1^{\gamma} = P_2 V_2^{\gamma}$$

$$P_2 = 8^{1.4} \times 1$$

$$= 18.37 \text{ bar}$$

$$\frac{P_2}{T_2} = \frac{P_3}{T_3}$$

$$P_3 = 2499 \times \frac{18.37}{758}$$

Max. pressure = 60.56 bar   Ans. (i)(b)

For pressure–volume diagram, refer to Figure 8.1   Ans.(ii) (a)

For temperature–entropy diagram, refer to Figure 11.5   Ans. (ii) (b)

4. Refer to Figure 8.3 (p–V diagram)

$$\gamma = \frac{c_p}{c_v}$$

$$= \frac{1005}{718}$$

$$= 1.4$$

$$P_2 = 1 \times (16)^{1.4}$$

Max. pressure = 48.5 bar   Ans. (i)

$$\frac{T_2}{T_1} = \left[\frac{V_1}{V_2}\right]^{\gamma-1}$$

$$T_2 = 330 \times 16^{0.4}$$

$$= 1000 \text{ K}$$

Energy added $= c_p \left(T_3 - T_2\right)$

$$1250 = 1.005 \left(T_3 - 1000\right)$$

$$T_3 = 2244 \text{ K}$$

Max. temperature = 1961°C Ans. (ii)

$$\frac{V_3}{V_2} = \frac{T_3}{T_2}$$

$$V_3 = 1 \times \frac{2244}{1000}$$

$$= 2.244$$

$$\frac{T_4}{T_3} = \left[\frac{V_3}{V_4}\right]^{\gamma-1}$$

$$T_4 = 2244 \times \left[\frac{2.244}{16}\right]^{0.4}$$

$$T_4 = 1023 \text{ K}$$

$$\frac{P_4}{P_1} = \frac{T_4}{T_1}$$

$$P_4 = 1 \times \frac{1023}{330}$$

$$= 3.1 \text{ bar}$$

$$\text{Energy removed} = c_v \left(T_4 - T_1 \right)$$

$$= 0.718 \left(1023 - 330 \right)$$

$$= 497.6 \text{ kJ/kg}$$

$$\text{Cycle efficiency} \left(\text{A.S.E.} \right) = \frac{1250 - 497.6}{1250} \times 100$$

$$= 60.19\% \quad \text{Ans. (iii)}$$

$$\text{m.e.p.} = \frac{p_2 (V_3 - V_2) + \dfrac{p_3 V_3 - p_4 V_4}{\gamma - 1} - \dfrac{p_2 V_2 - p_1 V_1}{\gamma - 1}}{V_1 - V_2}$$

$$= \frac{48.5(2.244 - 1) + \dfrac{48.5 \times 2.244 - 3.1 \times 16}{0.4} - \dfrac{48.5 \times 1 - 1 \times 16}{0.4}}{16 - 1}$$

$$= \frac{1}{15}(60.33 + 148.1 - 81.25)$$

$$= 8.479 \text{ bar} \quad \text{Ans. (iv)}$$

5. Referring to Figure 8.5, data given:

$$P_1 = 1 \, P_3 = P_4 = 41$$
$$V_1 = V_5 = 10.7 V_2 = V_3 = 1$$
$$T_1 = 305$$
$$T_4 = 1866$$
$$\gamma = \frac{c_p}{c_v} = \frac{1.005}{0.718} = 1.4$$

COMPRESSION PERIOD:

$$p_1 V_1^\gamma = p_2 V_2^\gamma \quad p_2 = \frac{1 \times 10.7^{1.4}}{1^{1.4}}$$

$$= 27.62 \text{ bar}$$

$$\frac{P_1 V_1}{T_1} = \frac{p_2 V_2}{T_2} \quad T_2 = \frac{305 \times 27.62 \times 1}{1 \times 10.7}$$

$$= 787.2 \text{ K}$$

$$= 514.2°C$$

PART COMBUSTION AT CONSTANT VOLUME:

Absolute temperature varies as absolute pressure,

$$\therefore T_3 = T_2 \times \frac{p_3}{p_2} = \frac{787.2 \times 41}{27.62} = 1169 K$$

$$= 896°C$$

PART COMBUSTION AT CONSTANT PRESSURE:

Volume varies as absolute temperature,

$$\therefore V_4 = V_3 \times \frac{T_4}{T_3} = \frac{1 \times 1866}{1169} = 1.596$$

EXPANSION PERIOD:

$$p_4 V_4^{1.4} = p_5 V_5^{1.4}$$

$$p_5 = \frac{41 \times 1.596^{1.4}}{10.7^{1.4}} = 2.858 \text{ bar}$$

Since $V_5 = V_1$, then $\frac{T_5}{T_1} = \frac{P_5}{P_1}$

$$T_5 = \frac{305 \times 2.858}{1} = 871.6 \text{ K}$$

$$= 598.6°C$$

Hence, pressures and temperatures required:

$$\left. \begin{array}{l} p_2 = 27.62 \text{ bar}, \theta_2 = 514.2°C \\ \theta_3 = 896°C \\ p_5 = 2.858 \text{ bar}, \theta_5 = 598.6°C \end{array} \right\} \text{ Ans. (i)}$$

Ideal thermal efficiency

$$= 1 - \frac{\text{Heat rejected}}{\text{Heat supplied}}$$

$$= 1 - \frac{c_V(T_5 - T_1)}{c_V(T_3 - T_2) + c_p(T_4 - T_3)}$$

$$= 1 - \frac{(T_5 - T_1)}{(T_3 - T_2) + \gamma(T_4 - T_3)}$$

$$= 1 - \frac{871.6 - 305}{(1169 - 787.2) + 1.4(1866 - 1169)}$$

$$= 1 - 0.4175$$

$$= 0.5825 = 58.25\% \quad \text{Ans. (ii)}$$

Note: Compression ratio and maximum pressures are low for this example compared with modern practice.

6. Cycle efficiency $= \dfrac{T_1 - T_2}{T_1} \times 100$

$$= \frac{1300 - 300}{1300} \times 100$$

$$= 76.9\% \quad \text{Ans. (i)}$$

Refer to Figure 11.4 (temperature–entropy) of Chapter 11. Also the description of practical constraints to the Carnot cycle – leading on to Figure 12.11 (temperature–entropy) for the vapour (steam) plant and the Rankine cycle (Chapter 12).

Difficulties with compressor for wet vapour   Ans. (ii)(a)

Substitution of feed pump to handle condensed liquid   Ans.   (ii)(b)

7.

$$\gamma = \frac{c_p}{c_V} = \frac{1000}{678} = 1.475$$

$$\frac{T_1}{T_2} = \left[\frac{p_1}{p_2} \right]^{\frac{\gamma - 1}{\gamma}}$$

$$T_2 = 333 \times 4.5^{\frac{0.475}{1.475}} = 540.5 \text{ K}$$

$$\frac{p_3}{T_3} = \frac{p_2}{T_2}$$

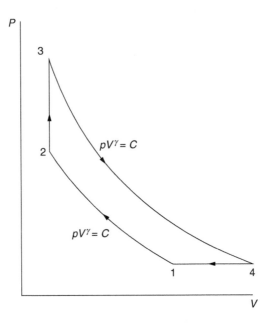

$$T_3 = \frac{1.35}{1} \times 540.5 = 729 \text{ K}$$

$$\frac{T_3}{T_4} = \left[\frac{p_3}{p_4}\right]^{\frac{\gamma-1}{\gamma}}, \quad p_3 = 4.5 \times 1.35 = 6.075$$

$$T_4 = 729 \left[\frac{1}{6.075}\right]^{\frac{0.475}{1.475}} = 407.8 \text{ K}$$

Thermal efficiency of the cycle

$$= 1 - \frac{c_p(T_4 - T_1)}{c_v(T_3 - T_2)} \times 100\%$$

$$= 1 - 1.47 \frac{(407.8 - 333)}{(729 - 540.5)} \times 100\%$$

$$= 41.5\% \text{ Ans.}$$

8.

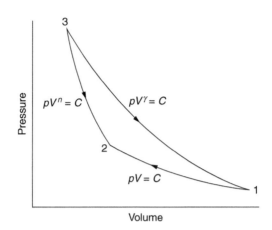

$$\frac{T_3}{T_1} = \left[\frac{p_3}{p_1}\right]^{\frac{\gamma-1}{\gamma}}$$

$$T_3 = 288 \times 4.5^{\frac{0.4}{1.4}} = 428 \text{ K}$$

$$\frac{T_3}{T_2} = \left[\frac{p_3}{p_2}\right]^{\frac{n-1}{n}} \quad \text{and} \quad T_1 = T_2$$

$$\therefore \frac{428}{288} = 2^{\frac{n-1}{n}}$$

$$\log 1.486 = \frac{n-1}{n} \times \log 2$$

$$0.172 \times n = (n-1) \times 0.3010$$
$$n = 2.33$$

From 1 to 2 work transfer $= -mRT_1 \ln r \quad r = p_2/p_1$
$$= -287 \times 288 \times \ln 2/10^3$$
$$= -57.29 \text{ kJ/kg} \quad \text{Ans. (i)(a)}$$

From 1 to 2 heat transfer $= -57.29$ kJ/kg   Ans. (ii)(a)

From 2 to 3   work transfer $= -\dfrac{mR(T_3 - T_2)}{n-1}$

$$= -\dfrac{287(428 - 288)}{10^3 (2.33 - 1)}$$

$$= -30.2 \text{ kJ/kg} \quad \text{Ans.} \quad (i)(b)$$

From 2 to 3   heat transfer $= \dfrac{\gamma - n}{\gamma - 1} \times (-30.2)$

$$= \left[\dfrac{1.4 - 2.33}{1.4 - 1} \right] \times (-30.2)$$

$$= 70.3 \text{ kJ/kg} \quad \text{Ans.} \quad (ii)(b)$$

From 3 to 1   work transfer $= \dfrac{mR(T_3 - T_1)}{\gamma - 1}$

$$= \dfrac{287(428 - 288)}{10^3 (1.4 - 1)}$$

$$= 100.5 \text{ kJ/kg} \quad \text{Ans.} \quad (i)(c)$$

From 3 to 1   heat transfer $= 0$ Ans. (ii) (c)

9.

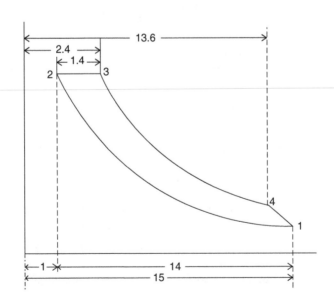

$$\text{Let } V_1 = 15, \text{ and } V_2 = 1$$
$$\text{then, stroke volume} = 15 - 1 = 14$$
$$\text{Fuel burning period} = \frac{1}{10} \times 14 = 1.4$$
$$\therefore V_3 = 1.4 + 1 = 24$$
$$V_4 = 1 + \frac{9}{10} \times 14 = 13.6$$
$$T_1 = 314 \text{ K}$$

COMPRESSION PERIOD:

$$\frac{T_2}{T_1} = \left\{ \frac{V_1}{V_2} \right\}^{n-1}$$
$$T_2 = 314 \times 15^{0.34} = 788.4 \text{ K}$$

∴ Temperature at end of compression

$$= 515.4°C \quad \text{Ans. (i)}$$

BURNING PERIOD:

$$\frac{T_3}{T_2} = \frac{V_3}{V_2}$$
$$T_3 = 788.4 \times 2.4 = 1892 \text{ K}$$

∴ Temperature at end of combustion

$$= 1619°C \quad \text{Ans. (ii)}$$

EXPANSION PERIOD:

$$\frac{T_4}{T_3} = \left\{ \frac{V_3}{V_4} \right\}^{n-1}$$
$$T_4 = 1892 \times \left\{ \frac{2.4}{13.6} \right\}^{0.34} = 1049 \text{ K}$$

∴ Temperature at end of expansion

$$= 776°C \quad \text{Ans. (iii)}$$

Note: This is not an ideal cycle.

10. Refer to Figure 8.3 (p–V diagram).

$$p_1 V_1^n = p_2 V_2^n$$
$$p_2 = 1 \times 12^{1.36}$$
$$= 29.38 \text{ bar}$$

$$\frac{T_2}{T_1} = \left[\frac{V_1}{V_2}\right]^{n-1}$$
$$T_2 = 353 \times 12^{0.36}$$
$$= 863.4\text{K}$$

$$\frac{T_3}{T_2} = \frac{V_3}{V_2}$$
$$V_3 = \frac{1923}{863.4}$$
$$p_3 V_3^n = p_4 V_4^n$$
$$p_4 = 29.38 \times \left[\frac{2.277}{12}\right]^{1.4}$$
$$= 2.780 \text{ bar}$$

$$\text{m.e.p} = \frac{p_2(V_3 - V_2) + \dfrac{p_3 V_3 - p_4 V_4}{n-1} - \dfrac{p_2 V_2 - p_1 V_1}{n-1}}{V_1 - V_2}$$

$$= \frac{29.38(2.227 - 1) + \dfrac{29.38 \times 2.227 - 2.78 \times 12}{0.4} - \dfrac{29.38 \times 1 - 1 \times 12}{0.36}}{12 - 1}$$

$$= \frac{1}{11}(36.05 + 80.18 - 48.28)$$
$$= 6.177 \text{ bar} \quad \text{Ans. (i)}$$

$$\text{Power} = p_m \, LAn$$
$$= \frac{6.177 \times 0.034 \times 200}{60}$$
$$= 70.0 \text{ kW} \quad \text{Ans. (ii)}$$

Note: This is not an ideal cycle.

TEST EXAMPLES 9

1. Clearance length $[mm] = \dfrac{\text{Clearance volume } [mm^3]}{\text{Area of cylinder } [mm^2]}$

$$= \frac{900 \times 10^3}{0.7854 \times 250^2} = 18.33 \text{ mm}$$

$$V_1 = 350 + 18.33 = 368.33 \text{ mm}$$

$$p_1 V_1^{1.25} = p_2 V_2^{1.25}$$

$$0.986 \times 368.3^{1.25} = 4.1 \times V_2^{1.25}$$

$$V_2 = 368.3 \times^{1.25} \sqrt{\frac{0.986}{4.1}}$$

$$= 117.8 \text{ mm}$$

Compression period $= V_1 - V_2$

$$= 368.3 - 117.8$$

$$= 250.5 \text{ mm} \quad \text{Ans.}$$

2. Refer to Figure 9.7 Ans. (i)

$$p_3 V_3^n = p_4 V_4^n$$

$$\left[\frac{p_3}{p_4}\right]^{\frac{1}{n}} = \frac{V_4}{V_3}$$

$$V_4 = \sqrt[1.3]{8} \times 0.08$$

$$= 0.3961 \text{ m}^3$$

Suction volume $= 1.08 - 0.3961$

$$= 0.6839 \text{ m}^3$$

Volumetric effic. $= \dfrac{0.6839}{1} \times 100$

$$= 68.39\% \quad \text{Ans. (ii)(a)}$$

Free air delivery $= 0.7854 \times 0.152^2 \times 0.105 \times 0.6839 \times 12$

$$= 0.01564 \text{ m}^3/\text{s} \quad \text{Ans. (ii)(b)}$$

3. Refer to Figure 9.7 Ans. (i)

$$pV = mRT$$
$$1 \times 100 \times V_1 = 0.0035 \times 0.287 \times 288$$
$$V_1 = 0.002893 \text{ m}^3$$

$$\text{Stroke volume} = V_1 - V_3$$
$$= 0.7854 \times 0.14^2 \times 018$$
$$= 0.002771 \text{ m}^3$$

$$\text{Clearance volume} = V_3$$
$$= 0.002893 - 0.002771$$
$$= 0.000122 \text{m}^3 \quad \text{Ans. (ii)(a)}$$

$$p_3 V_3^n = p_4 V_4^n$$
$$8.5 \times 0.000122^{1.32} = 1 \times V_4^{1.32}$$
$$V_4 = 0.0006166 \text{ m}^3$$
$$V_1 - V_4 = 0.002893 - 0.0006166$$
$$= 0.0022764 \text{ m}^3$$

$$\text{Volumetric effic.} = \frac{0.002276}{0.002771} \times 100$$
$$= 82.14\% \quad \text{Ans. (ii)(b)}$$

4. Refer to Figure 9.5 Ans. (i)

$$\text{Stroke volume} = 0.7854 \times 0.1^2 \times 0.12$$
$$= 0.0009424 \text{ m}^3$$

Let N be rev/min

$$\text{Suction volume/min} = 3 \times N \times 0.0009424$$
$$= \text{Free air delivery/min}$$
$$N = \frac{1.2}{3 \times 0.0009424}$$
$$\text{Operating speed} = 424.4 \text{ rev/min} \quad \text{Ans. (ii)(a)}$$

$$p_1 V_1^n = p_2 V_2^n$$
$$1 \times 0.0009424^{1.3} = 6.6 \times V_2^{1.3}$$
$$V_2 = 0.0002203 \text{ m}^3$$

$$\text{Work/cycle} = \frac{n}{n-1}(p_2 V_2 - p_1 V_1) \times 3$$
$$= \frac{3 \times 1.3 \times 10^2}{0.3} \times (6.6 \times 0.0002203 - 1 \times 0.0009424)$$
$$= 0.6651 \text{ kJ}$$

$$\text{Work/s} = 0.6651 \times \frac{424.4}{60}$$
$$= 4.704 \text{ kW}$$

$$\text{Input power} = \frac{4.704}{0.9}$$
$$= 5.227 \text{ kW} \quad \text{Ans. (ii)(b)}$$

5. Volume of free air (at 1.01 bar from the atmosphere) to make 22 m³ at 31.01 bar abs., at the same temperature

$$= 22 \times \frac{31.01}{1.01} \text{ m}^3$$

Volume of free air to make 22 m³ at 20.01 bar abs.

$$= 22 \times \frac{22.01}{1.01} \text{ m}^3$$

∴ Volume of free air to supply the difference

$$= \frac{22}{1.01}(31.01 - 20.01)$$
$$= \frac{22 \times 11}{1.01} = 239.6 \text{m}^3 \dots \quad \text{(i)}$$

Volume of free air dealt with by compressor per minute

$$= 0.7854 \left(0.35^2 - 0.075^2\right) \times 0.3 \times 0.92 \times 170$$
$$= 4.306 \text{ m}^3 / \text{min} \dots \quad \text{(ii)}$$

$$\therefore \text{Time} = \frac{239.6}{4.306} = 55.64 \text{ min Ans.}$$

6. Referring to Figure 9.7,

$$\text{Work/cycle} = \frac{n}{n-1}\left\{\left(p_2V_2 - p_1V_1\right) - \left(p_3V_3 - p_4V_4\right)\right\}$$

$$= \frac{n}{n-1}\left\{\left(p_2V_2 - p_1V_1\right) - \left(p_2V_3 - p_1V_4\right)\right\}$$

$$= \frac{n}{n-1}\left\{p_2\left(V_2 - V_3\right) - p_2\left(V_1 - V_4\right)\right\}$$

Piston swept volume $= (V_1 - V_3)$

$$= 0.7854 \times 0.2^2 \times 0.23 = 7.226 \times 10^3 \text{ m}^3$$

Working in litres of volume and bars of pressure:

$$V_3 = 364 \text{ cm}^3 = 0.364 \text{ l}$$
$$V_1 = \text{piston swept vol.} + \text{clearance vol.}$$
$$= 7.226 + 0.364 = 7.59$$

$$p_1V_1^n = p_2V_2^n$$
$$1 \times 7.59^{1.28} = 5 \times V_2^{1.28}$$
$$V_2 = \frac{7.59}{\sqrt[1.28]{5}} = 2.158$$

$$p_3V_3^n = p_4V_4^n$$
$$5 \times 0.364^{1.28} = 1 \times V_4^{1.28}$$
$$V_4 = 0.364 \times \sqrt[1.28]{5} = 1.28$$

Expressing pressure in kN/m² (1 bar = 10² kN/m²) and volumes in m³ (1 l = 10⁻³ m³) to obtain work in kJ:

$$\text{Work/cycle} = \frac{n}{n-1}\left\{p_2\left(V_2 - V_3\right) - p_1\left(V_1 - V_4\right)\right\}$$

$$= \frac{1.28}{0.28} \times 10^2 \times 10^{-3} \left\{5\left(2.158 - 0.363\right) - 1\left(7.59 - 1.28\right)\right\}$$

$$= 1.216 \text{ kJ}$$

$$\text{Indicated power}\left[\text{kW}\right] = 1.216\left[\text{kJ/cycle}\right] \times 2\left[\text{cycle/s}\right]$$

$$= 2.432\text{kW} \quad \text{Ans. (i)}$$

$$\text{Mean indicated press.}\left[\text{kN/m}^2\right] = \frac{\text{Area of diagram [kJ]}}{\text{Length of diagram [m}^3\text{]}}$$

$$= \frac{1.216}{7.226 \times 10^{-3}} = 168.3 \text{ kN/m}^2 = 1.683\text{bar} \quad \text{Ans. (ii)}$$

$$\text{Vol. effic.} = \frac{\text{Volume drawn inper stroke}}{\text{Piston swept volume}}$$

$$\frac{V_1 - V_4}{V_1 - V_3} = \frac{7.59 - 1.28}{7.226}$$

$$= 0.8732 \text{ or } 87.32\% \quad \text{Ans. (iii)}$$

7. Referring to Figure 9.7:

From pV/T = constant, volume rate of air induced at suction pressure and temperature

$$= \frac{5 \times 1.013 \times 297}{60 \times 0.98 \times 289} = 0.08853 \text{ m}^{3/S}$$

$$\text{Work/cycle} = \frac{n}{n-1}p_1\left(V_1 - V_4\right)\left[\left\{\frac{p_2}{p_1}\right\}^{\frac{n-1}{n}} - 1\right]$$

The above expression is the work per cycle when $(V_1 - V_4)$ is the volume drawn in per cycle. Similarly, if $(V_1 - V_4)$ is taken as the volume drawn in per second, the expression will give work per second, which is power, thus:

Power [kW] = work per second [kJ/s = kN m/s]

$$\frac{n}{n-1} = \frac{1.25}{0.25} = 5 \qquad \frac{n-1}{n} = \frac{1}{5}$$

$$V_1 - V_4 = 0.08853 \text{ m}^3/\text{s}$$
$$P_1 = 0.98 \times 10^2 \text{ kN/m}^2$$

$$\frac{P_2}{P_1} = \text{Pressure ratio} = 4.55$$

$$\text{Power} = \frac{n}{n-1} P_1 (V_1 - V_4) \left[\left\{ \frac{P_2}{P_1} \right\}^{\frac{n-1}{n}} - 1 \right]$$

$$= 5 \times 0.98 \times 10^2 \times 0.08853 (4.55^{\frac{1}{5}} - 1)$$
$$= 15.36 \text{kW} \quad \text{Ans. (i)}$$

Let stroke volume $(V_1 - V_3)$ on Figure 9.7 be represented by unity, then $V_3 = 0.05$ and $V_1 = 1.05$

$$\text{From } p_3 V_3^n = p_4 V_4^n$$
$$V_4 = 0.05 \times \sqrt[1.25]{4.55} = 0.1681$$

$$\text{Suction period} = V_1 - V_4$$
$$= 1.05 - 0.1681 = 0.8819$$

$$\text{Vol. effic.} = \frac{\text{Vol. drawn in per stroke}}{\text{Stroke volume}}$$

$$= \frac{0.8819}{1} = 0.8819 \quad \text{Ans. (ii)}$$

Actual volume of air drawn in per stroke [m³]

$$= \frac{0.08853 \text{ [m}^3/\text{s]}}{2 \times 8 \text{ [strokes/s]}}$$

$$\text{Piston swept vol.} = \frac{\text{Induced volume}}{\text{Volumetric efficiency}}$$

$$= \frac{0.08853}{2 \times 8 \times 0.8819}$$

Let d [m] = cyl. diameter, stroke = 1.2 d

$$\text{Piston swept vol.} = 0.7854\ d^2$$

$$d = \sqrt[3]{\frac{0.08853}{0.7854 \times 1.2 \times 2 \times 8 \times 0.8819}}$$

$$= 0.1881\ \text{m}$$

$$\left.\begin{array}{l} \text{Diameter of cylinder} = 188.1\ \text{mm} \\ \text{Stroke} = 1.2 \times 188.1 = 255.7\ \text{mm} \end{array}\right\} \text{Ans. (iii)}$$

8.

$$\dot{m} = \frac{pV}{RT}$$

$$\dot{m} = \frac{10^5 \times 17}{287 \times 306 \times 60}$$

$$\dot{m} = 0.323\ \text{kg/s}$$

$$\text{For minimum work}\ p_2 = \sqrt{p_3 p_1}$$

$$p_2 = \sqrt{16 \times 1}$$

$$P_2 = 4\ \text{bar}$$

$$\frac{T_1}{T_2} = \left[\frac{p_1}{p_2}\right]^{\frac{n-1}{n}}$$

$$\therefore T_2 = T_1 \left[\frac{p_1}{p_2}\right]^{\frac{n-1}{n}}$$

$$= 306(4)^{\frac{0.3}{1.3}}$$

$$T_2 = 421.4\ \text{K or}\ 148.4°\text{C}$$

$$\text{First stage power} = \frac{n}{n-1} mR(T_2 - T_1)$$

$$= \frac{1.3}{0.3} \times 0.323 \times 287(148.4 - 33) / 10^3$$

$$= 46.37\ \text{kW Ans. (i)}$$

Heat rejected to intercooler

$$= mc_p(T_2 - T_1)$$
$$= 0.323 \times 1005(148.4 - 33) \times 60/10^6$$
$$= 2025 \text{ MJ/min Ans. (ii)}$$

9. From Work/cycle $= \dfrac{n}{n-1}(p_3 V_3 - p_2 V_2) = $ constant, volume rate of air at suction pressure and temperature

$$= \frac{1.013 \times 0.6083 \times 300}{0.97 \times 288} = 0.6617 \text{ m}^3/\text{s}$$

Let stroke volume $V_1 - V_3$ in Figure 9.7 equal unity.

Then $V_3 = 0.06$, $V_1 = 1.06$

$$p_3 V_3^n = p_4 V_4^n$$

$$V_4 = 0.06 \times {}^{1.32}\!\!\sqrt{\frac{4.85}{0.97}} = 0.2031$$

Actual volume of air drawn in per stroke $= \dfrac{0.6617}{2 \times 5}\left[\dfrac{\text{m}^3/\text{s}}{\text{stroke/s}}\right]$

$$= 0.6617 \text{ m}^3$$

and $\dfrac{V_1 - V_4}{V_1 - V_3} = \dfrac{0.06617}{\text{Piston swept volume}}$

$\therefore$ Piston swept volume $= \dfrac{0.06617}{1.06 - 0.2031} = 0.07812 \text{m}^3$

also, Piston swept volume $= 0.7854 \, d^2 \times d$

$d = $ cylinder diameter and stroke

$$\therefore 0.07812 = 0.7854 d^3$$
$$d = 0.463 \text{ m Ans. (i)}$$

Isothermal efficiency

$$= \frac{\ln r}{\dfrac{n}{n-1}[r^{\frac{n-1}{n}} - 1]}$$

$$= \frac{\ln 5}{\dfrac{1.32}{0.32}\left[5^{\frac{0.32}{1.32}} - 1\right]}$$

$$= 0.818 \text{ or } 81.8\% \text{ Ans. (ii)}$$

10. Let r = stage pressure ratio for minimum work condition

Let s = number of stages of compression

$$\frac{p_2}{p_1} = \left(\frac{T_2}{T_1}\right)^{\frac{n}{n-1}}$$

$$r = \left(\frac{368}{308}\right)^{\frac{1.3}{0.3}}$$

$$r = 2.162$$

$$\text{now } p_2 = rP_1$$
$$7 = rp_2 = r^2p_1$$
$$p_4 = rp_3 = r^3p_1, \text{ etc.}$$
$$\therefore P_{s-1} = r^sp_1$$
$$\text{hence } 100 = 2.162^s \times 1$$
$$s = 5.97, \text{ say 6 stages Ans. (i)}$$

$$\text{Power input per stage} = \frac{n}{n-1}\ln R(T_2 - T_1)$$

$$= \frac{1.3}{0.3} \times 0.1 \times 0.287(368 - 308)$$

$$= 7.45 \text{ kW}$$

$$\text{Compressor power input} = 7.45 \times 6 \text{ stages}$$
$$= 44.7 \text{ kW Ans. (ii)}$$

TEST EXAMPLES 10

1. Tables page 2, water at 80°C, $h = 334.9$

… … 4, steam at 9 bar:

$$h = h_f + xh_{fg}$$
$$= 743 + 0.96 \times 2031 = 2693$$

$$\text{Heat energy transferred} = \text{Change in enthalpy}$$
$$= 2693 - 334.9$$
$$= 2358.1 \text{ kJ/kg}$$

2. Tables page 7, 30 bar 350°C, $h = 3117$

   ... 3, 0.06 bar dryness 0.88:

$$h = h_f + xh_{fg}$$
$$= 152 + 0.88 \times 2415 = 2277$$
$$\text{Enthalpy drop per kg} = 3117 - 2277$$
$$= 840 \text{ kJ/kg Ans. (i)}$$

At 0.5 kg of steam per second, total change in enthalpy in the steam through the turbine per second

$$= 0.5 \times 840 = 420 \text{ kJ/s}$$
$$\text{kilojoules per second} = \text{kilowatts}$$
$$\therefore \text{Power equivalent} = 420\text{kW Ans. (ii)}$$

3. Tables page 7, 20 bar 350°C, $h = 3138$, $v = 0.1386$

   ... 4, 20 bar 0.98 dry:

$$h = h_f + xh_{fg}$$
$$= 909 + 0.98 \times 1890 = 2761 \text{ kJ/kg}$$
$$v = xv_g$$
$$= 0.98 \times 0.09957$$
$$= 0.09757 \text{ m}^3/\text{kg}$$

Heat energy supplied to steam in superheaters

$$= 3138 - 2761 = 377 \text{ kJ/kg Ans. (i)}$$

% increase in specific volume

$$= \frac{0.1386 - 0.09757}{0.09757} \times 100$$
$$= 42.06\% \text{ Ans. (ii)}$$

4. Tables page 4, 4 bar, $v_g = 0.4623$

   8 bar, $v_g = 0.2403$

For 8 bar, dryness 0.94,

$$v = xv_g = 0.94 \times 0.2403 = 0.2258 \text{ m}^3 \text{ kg}$$

$$p_1 v_1^n = P_2 v_2^n$$
$$8 \times 0.2258^{1.12} = 4 \times v_2^{1.12}$$
$$v_2 = 0.2258 \times 1.12\sqrt{2} = 0.4194 \text{ m}^3 / \text{kg}$$

As 0.4194 m³/kg is the specific volume of wet steam at 4 bar, let its dryness fraction be x:

$$v = xv_g$$
$$0.4194 = x \times 0.4623$$
$$x = 0.9072 \quad \text{Ans.}$$

5. Tables page 4, steam 2.4 bar, $h_g = 2715$

    ... 2, water 42°C, $h = 175.8$
    ... 3, water 99.6°C, $h = 417$

See Figure 10.2. Consider one kg of steam from boiler, let x kg be tapped off low pressure turbine to heater, then $(1 - x)$ kg passes through condenser and as water to hotwell.

Enthalpy of heating steam and water entering heater = Enthalpy of water leaving heater

---

$$x \times 2715 + (1 - x) \times 175.8 = 1 \times 417$$
$$2715x + 175.8 - 175.8x = 417$$
$$2539.2x = 241.2$$
$$x = 0.095$$
$$\therefore \% \text{ of steam tapped off} = 9.5\% \quad \text{Ans.}$$

6. Tables page 4, steam 16 bar, $h_f = 859, h_{fg} = 1935$
    steam 8 bar, $h_f = 721, h_{fg} = 2048$

$$\text{Enthalpy after throttling} = \text{Enthalpy before}$$
$$721 + x \times 2048 = 859 + 0.98 \times 1935$$
$$x \times 2048 = 2034$$
$$x = 0.9931 \quad \text{Ans.}$$

7.
$$t_s = 165°C$$

$$\left.\begin{array}{l} p = 7 \text{ bar} \\ v_{g1} = 0.2728 \text{ m}^3/\text{kg} \end{array}\right\} \text{tables, page 4}$$

$$V_1 = 0.75 \times 0.2728 \times 0.2$$
$$= 0.0409 \text{ m}^3$$

$$V_2 = 0.0818 \text{ m}^3$$

$$v_{g2} = \frac{0.0818}{0.2}$$
$$= 0.4090 \text{ m}^3/\text{kg}$$
$$t = 355°C$$

i.e. using tables, page 7, interpolating at 7 bar with v_g.0.4090

Final temperature = 355°C   Ans. (i)

$$\text{Work done} = p(V_2 - V_1)$$
$$= 7 \times 100 \ (0.0818 - 0.0409)$$
$$= 28.64 \text{ kJ} \text{ Ans. (ii)}$$

$$\text{Heat energy transfer} = H_1 - H_2$$
$$= 0.2 \left[h_g - \left(h_f + x h_{fg} \right) \right]$$
$$= 0.2 \ [3175 - (697 + 0.75 \times 2067)]$$
$$= 185.5 \text{ kJ} \text{ Ans. (iii)}$$

8. Tables page 4, 15 bar, $h_f = 845$, $h_{fg} = 1947$

1.1 bar, $t_s = 102.3$, $h_g = 2680$

Dryness fraction by separator,

$$x_1 = \frac{m_2}{m_2 + m_1} = \frac{10}{10.55} = 0.9479$$

Dryness fraction by throttling calorimeter:

$$\text{Enthalpy before throttling} = \text{Enthalpy after}$$
$$845 + x_2 \times 1947 = 2680 + 2(111 - 102.3)$$
$$x_2 \times 1947 = 1852.4$$
$$x_2 = 0.9513$$

Dryness fraction of steam:

$$x = x_1 \times x_2$$
$$= 0.9479 \times 0.9513$$
$$= 0.9018 \ \text{Ans.}$$

9. *1st Case:* absolute press. $= 1.9 + 1 = 2.9$ bar

Tables page 4, when temp, of steam is 130°C, $p = 2.7$ bar, $v_g = 0.6686$ m³/kg

$$\text{Mass of steam} = \frac{4.25}{0.6686} = 6.356 \ \text{kg Ans.} \quad \text{(i)(a)}$$

Air pressure = total press. − steam press.
$$= 2.9 - 2.7 = 0.2 \ \text{bar} = 20 \ \text{kN/m}^2$$

$$pV = mRT$$

$$\therefore m = \frac{20 \times 4.25}{0.287 \times 403}$$
$$= 0.7348 \ \text{kg Ans.} \quad \text{(ii)(a)}$$

2nd Case: absolute press. $= 6.25 + 1 = 7.25$ bar

Tables page 4, when temp, of steam is 165°C,

$$p = 7 \ \text{bar}, \ v_g = 0.2728 \ \text{m}^3\text{/kg}$$

$$\text{Mass of steam} = \frac{4.25}{0.2728} = 15.58 \ \text{kg Ans.} \quad \text{(i)(b)}$$

Air pressure = Total press. − Steam press.
$$= 7.25 - 7 = 0.25 \ \text{bar} = 25 \ \text{kN/m}^2$$

$$pV = mRT$$

$$\therefore m = \frac{25 \times 4.25}{0.287 \times 438}$$
$$= 0.845 \ \text{kg Ans.} \quad \text{(ii)(b)}$$

10. $p_{wv} = 0.02337$ bar

Using tables, page 2, t_s 20°C

$$P_A = 1 - 0.02337$$
$$= 0.97663 \text{ bar}$$

$$\frac{p_N}{p_O} = \frac{V_N}{V_O} \text{ using partial volumes}$$

$$\frac{p_N}{0.97663 - p_N} = \frac{0.79}{0.21}$$

$$P_N = 3.674 - 3.762 p_N$$
$$P_N = 0.7715 \text{ bar}$$
$$P_O = 0.97663 - 0.7715$$
$$= 0.2051 \text{ bar}$$

$$\text{Partial pressures} = \left\{ \begin{array}{l} 0.2051 \text{ bar, oxygen} \\ 0.7715 \text{ bar, nitrogen} \\ 0.02337 \text{ bar, water vapour} \end{array} \right\} \quad \text{Ans.} \quad \text{(i)}$$

$$p_A^v = R_A T$$
$$v = \frac{0.287 \times 293}{100 \times 0.97663}$$
$$= 0.861 \text{ m}^3 / \text{kg of dry air}$$
$$v_g = 57.84 \text{ m}^3 / \text{kg}$$

using tables, page 2, t_s 20°C

$$\text{kg of water vapour} = \frac{0.861}{57.84}$$
$$= 0.01488 \text{ per kg of dry air}$$
$$\% \text{ Absolute humidity} = 1.488 \quad \text{Ans.} \quad \text{(ii)}$$

TEST EXAMPLES 11

1. Tables page 4, 17 bar, $s_f = 2.372$, $s_{fg} = 4.028$

$$s = 2.372 + 0.95 \times 4.028$$
$$= 6.198 \text{ kJ/kg K} \quad \text{Ans.}$$

2. Tables page 4, 195°C (14 bar), $s_f = 2.284$, $s_{fg} = 4.185$

    $$s = 2.284 + 0.9 \times 4.185$$
    $$= 6.05 \text{kJ/kg} \ \text{Ans.}$$

3. Tables page 4, 5.5 bar, $s_g = 6.790$

    Tables page 3, 0.2 bar, $s_f = 0.832$, $s_{fg} = 7.075$

    Entropy after expansion = entropy before

    $$0.832 + x \times 7.075 = 6.79$$
    $$x \times 7.075 = 5.958$$
    $$x = 0.8422 \ \text{Ans.}$$

4. Tables page 7. Note that 17 bar is not in the table specifically, therefore the value needs to be calculated by finding the 1 bar increase/decrease between 15 and 20 bar:

    15 bar 350°C, $s = 7.102$

    20 bar 350°C, $s = 6.957$

    For 5 bar increase $s = 0.145$ decrease

    $$\text{For 2 bar increases} = s = \frac{2}{5} \times 0.145 = 0.058 \text{ decrease}$$

    ∴ 17 bar 350°C, $s = 7.102 - 0.058 = 7.044$

    Page 4, 1.7 bar, $s_f = 1.475$, $s_{fg} = 5.707$

    Entropy after expansion = entropy before
    $$1.475 + x \times 5.707 = 7.044$$
    $$x \times 5.707 = 5.569$$
    $$x = 0.9759 \ \text{Ans.}$$

5. Tables page 4, 22 bar, $h_g = 2801$, $s_g = 6.305$

    $$7 \text{ bar}, h_g = 2764$$
    $$1.4 \text{ bar}, s_f = 1.411, s_{fg} = 5.835$$

    THROTTLING PROCESS:

    Enthalpy after = enthalpy before

    i.e. Enthalpy at 7 bar = 2801

    Throttled steam is therefore superheated.

Tables page 7, interpolating:

$$\text{Enthalpy of superheat} = 2801 - 2764 = 37$$
$$\text{At 7 bar, } h = 2846 \text{ for } 200°C$$
$$h = 2764 \text{ for } 165°C \text{ (sat. temp.)}$$
$$\text{increase } h = 82 \text{ for } 35°C \text{ increase}$$
$$\text{difference in temp, for } h = 37, = \frac{37}{82} \times 35 = 15.8°C$$

Degree of superheat at 7 bar = 15.8°C Ans. (i)

Entropy of steam at 7 bar with 15.8°C of superheat:

$$7 \text{ bar } 200°C \ \ s = 6.888$$
$$165°C \ s = 6.709$$
$$\text{for increase } 35°C \ s = 0.179 \text{ increase}$$
$$\ldots \ldots \ 15.8°C \ s = \frac{15.8}{35} \times 0.179 = 0.0808$$

∴ Entropy at 7 bar, 15.8°C superheat

$$= 6.709 + 0.0808 = 6.7898$$
$$\text{Increase in entropy} = 6.7898 - 6.305$$
$$= 0.4848 \text{ kJ/kg K} \ \ \text{Ans. (ii)}$$

ISENTROPIC EXPANSION from 7 to 1.4 bar:

Entropy after = Entropy before

$$1.411 + x \times 5.835 = 6.7898$$
$$x \times 5.835 = 5.3788$$
$$x = 0.9217 \ \ \text{Ans. (iii)}$$

TEST EXAMPLES 12

1. Tables page 4, 8 bar, $h_g = 2769$
$$5 \text{ bar}, h_f = 640 \ h_{fg} = 2109$$
$$v_g = 0.3748$$

$$\text{Enthalpy drop} = 2769 - (640 + 0.97 \times 2109)$$
$$= 83 \text{ kJ/kg}$$
$$\text{Velocity} = \sqrt{2 \times 83 \times 10^3} = 407.40 \text{ m/s} \quad \text{Ans.} \quad \text{(i)}$$

Spec. vol. of steam at exit $= 0.97 \times 0.3748 \text{ m}^3/\text{kg}$

$$\text{Mass flow} \left[\text{kg/s}\right] = \frac{\text{Volume flow} \left[\text{m}^3/\text{s}\right]}{\text{Spec. vol.} \left[\text{m}^3/\text{kg}\right]}$$
$$= \frac{\text{Area} \left[\text{m}^2\right] \times \text{velocity} \left[\text{m/s}\right]}{\text{Spec. vol.} \left[\text{m}^3/\text{kg}\right]}$$
$$= \frac{14.5 \times 10^{-4} \times 407.4}{0.97 \times 0.3748}$$
$$= 1.625 \text{ kg/s} \quad \text{Ans.} \quad \text{(ii)}$$

2. Referring to Figure 12.7:

$$\text{Angle between } u \text{ and } v_{r1} = 180° - \beta_1$$
$$= 180° - 33° = 147°$$

$$\text{Angle between } v_1 \text{ and } v_{r1} = \beta_1 - \alpha_1$$
$$= 33° - 20° = 13°$$

By sine rule:

$$\frac{u}{\sin 13°} = \frac{v_1}{\sin 147°}$$
$$u = \frac{450 \times \sin 13°}{\sin 147°}$$

$$\text{Rotationals peed [rev/s]} = \frac{\text{Linear velocity [m/s]}}{\text{Circumference [m]}}$$
$$= \frac{185.9}{\pi \times 0.66} = 89.64 \text{ rev/s} \quad \text{Ans. (ii)}$$

3. Referring to Figure 12.10:

$$v_{a1} = v_1 \sin \alpha_1 = 243 \times \sin 23° = 94.95 \text{ m/s}$$
$$v_{w1} = v_1 \cos \alpha_1 = 243 \times \cos 23° = 223.7 \text{ m/s}$$
$$x = v_{w1} - u = 223.7 - 159 = 64.7 \text{ m/s}$$

$$\tan \beta_1 = \frac{v_{a1}}{x} = \frac{94.95}{64.7} = 1.468$$

Blade inlet angle = 55° 44′ Ans. (i)

Since the combined vector diagram of inlet and exit velocities is symmetrical, $vw_2 = x,$

$$V_w = V_{wl} + V_{w2} = 223.7 + 64.7 = 288.4 \text{ m/s}$$
$$\text{Force on blades} = mv_w$$
$$= 0.9 \times 288.4 = 259.5 \text{ N} \quad \text{Ans. (ii)}$$

$$\text{Power } [W] = \text{Force } [N] \times \text{Velocity } [m/s]$$
$$= 259.5 \times 159$$
$$= 4.127 \times 10^4 \text{ W} = 41.27 \text{ kW} \quad \text{Ans. (iii)}$$

4.

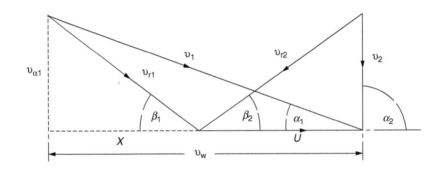

$$u = \pi \times 0.6 \times 100$$
$$= 188.5 \text{ m/s}$$
$$V_{r2} = \frac{188.5}{\cos 35°}$$
$$= 2301 \text{ m/s}$$

Let $v_{r1} = v_{r2}$ (see assumption following)
$$V_{a1} = 230.1 \times \sin 35°$$
$$= 132 \text{ m/s}$$
$$V_w = 2 \times 188.5$$
$$= 377 \text{ m/s}$$
$$\tan \alpha_1 = \frac{132}{377}$$
$$\alpha_1 = 19.36°$$
Nozzle angle = 19.36° Ans. (i)(a)

$$V_1 = \frac{377}{\cos \alpha_1}$$

$$= 399.6\,\text{m/s} \quad \text{Ans.} \quad \text{(i)(b)}$$

$$\text{Power} = m v_w u$$

$$= 1 \times 377 \times 188.5 \times 10^{-3}$$

$$= 71.06\,\text{kW} \quad \text{Ans.} \quad \text{(i)(c)}$$

Assumed – no frictional losses across the blades   Ans. (ii)

5. From h–s chart, $h_1 = 2795$ kJ/kg, $s_1 = 6.27$ kJ/kg K

$h_2 = 2795$ kJ/kg

$h_3 = 2455$ kJ/kg

$h_4 = 2870$ kJ/kg

$h_5 = 2615$ kJ/kg, $s_5 = 7.63$ kJ/kg K

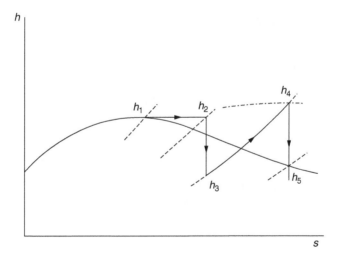

Changes in enthalpy during each stage, final – initial

Stage 1 $h_2 - h_1 = 0$

Stage 2 $h_3 - h_2 = -340$ kJ/kg

Stage 3 $h_4 - h_3 = 415$ kJ/kg

Stage 4 $h_5 - h_4 = -255$ kJ/kg   Ans.

Overall changes in entropy $= s_5 - s_1 = 1.36$ kJ/kg K   Ans.

Condition of steam at end of final expansion is *dry saturated.*   Ans.

6.
$$p_T = p_1 \left[\frac{2}{\gamma+1} \right]^{\frac{\gamma}{\gamma-1}}$$

$$= 7 \left[\frac{2}{1.67+1} \right]^{\frac{1.67}{1.67-1}}$$

$$= 3.407 \text{ bar}$$

$$T_T = T_1 \left[\frac{p_T}{p_1} \right]^{\frac{\gamma-1}{\gamma}}$$

$$T_T = 423 \left[\frac{3.407}{7} \right]^{\frac{1.67-1}{1.67}}$$

$$= 316.9 \text{ K or } 43.87°C$$

Enthalpy drop to throat $= c_p[T_1 - T_T]$

i.e., $h = 833.9[423 - 316.87]$

$h = 88501.8$ J/kg

$$\text{Velocity of gas at throat} = \sqrt{2h}$$
$$v_T = \sqrt{2 \times 88501.8}$$
$$v_T = 420.7 \text{m/s Ans.}$$

$$P_T V_T = mRT_T$$
$$\dot{V}_T = \frac{\dot{m}RT_T}{p_T}$$
$$= \frac{0.25 \times 2078.5 \times 316.9}{3.407 \times 10^5}$$
$$= 0.4833 \text{ m}^3/\text{s}$$

$$\text{Throat area} = \frac{\dot{V}_T}{v_T} = \frac{0.4833}{420.7}$$
$$= 0.001149 \text{ m}^2 \text{ or } 11.49 \text{cm}^2 \text{ Ans.}$$

7. Velocity at nozzle exit V_1

$$= \sqrt{2 \times 312.5 \times 10^3 \times 0.9}$$
$$= 750 \text{ m/s}$$

Refer Figure 12.7: $v_{a1} = v_1 \sin \alpha_1 = 750 \times \sin 20° = 256.5 \text{ m/s}$
$v_{w1} = v_1 \cos \alpha_1 = 750 \times \cos 20° = 704.9 \text{ m/s}$

$$x = \frac{v_{a1}}{\tan \beta_1} = \frac{265.6}{\tan 35°} = 366.4 \text{ m/s}$$

$$\text{Blade velocity} = 8 = v_{w1} - x$$
$$= 704.9 - 366.4$$
$$= 388.5 \text{ m/s} \quad \text{Ans. (i)}$$

Absolute velocity of exit steam is in the direction of the turbine axis, therefore $\alpha_2 = 90°$

$$\tan \beta_2 = \frac{v_2}{u_1} = \frac{204}{338.5} = 0.6027$$

Exit angle of blades = 31°5′ Ans. (ii)

$$v_{r1} = \frac{v_{a1}}{\sin \beta_1} = \frac{256.5}{\sin 35°} = 447.2 \text{ m/s}$$

$$v_{r2} = \frac{v_2}{\sin \beta_2} = \frac{204}{\sin 35° 5'} = 395.2 \text{m/s}$$

Loss of kinetic energy of steam across the blades

$$= \frac{1}{2}\dot{m}\,(v_{r1}^2 - v_{r2}^2)$$
$$= \frac{1}{2} \times 1 \times (447.2^2 - 395.2^2)$$
$$= 21900 \text{ J} = 21.9 \text{ kJ/kg steam} \quad \text{Ans. (iii)}$$

$$\text{Axial thrust} = m(v_{a1} - v_{a2})$$
$$= 1 \times (256.5 - 204)$$
$$= 52.5 \text{ N/kg of steam} \quad \text{Ans. (iv)}$$

Power $= mv_w u$

Since the steam leaves the turbine axially, that is, at 90° to the blade movement, there is no velocity of whirl at exit, $v_{w2} = 0$, that is, $v_w = V_{w1}$

$$\text{Power} = 1 \times 704.9 \times 338.5$$
$$= 2.386 \times 10^5 \text{ W}$$
$$= 238.6 \text{ kW/kg of steam Ans. (v)}$$

$$\text{Diagram efficiency} = \frac{\text{Work done on blades}}{\text{Work supplied}}$$
$$= \frac{\dot{m}v_w u[J/S]}{\frac{1}{2}\dot{m}v_1^2[J/2]} = \frac{2uv_w}{v_1^2}$$
$$= \frac{2 \times 338.5 \times 704.9}{750^2}$$
$$= 0.8484 \text{ or } 84.84\% \text{ Ans. (vi)}$$

8. Tables page 7, 15 bar 250°C, $h = 2925$, $s = 6.711$

    Page 3, 016 bar, $h_f = 232$, $s_f = 0.772$
    $h_{fg} = 2369$, $s_{fg} = 7.213$
    Entropy after expansion = Entropy before

$$0.772 + x \times 7.213 = 6.711$$
$$x \times 7.213 = 5.939$$
$$\therefore \text{Dryness fraction } x = 0.8235 \text{ Ans. (i)}$$

At 0.16 bar, $h = h_f + xh_{fg}$
$$= 232 + 0.8235 \times 2369 = 2183$$

$$\text{Rankine efficiency} = \frac{h_1 - h_2}{h_1 - h_{j2}}$$
$$= \frac{2925 - 2183}{2925 - 232} = \frac{742}{2693}$$
$$= 0.2755 \text{ or } 27.55\% \text{ Ans. (ii)}$$

9. Tables page 4, 14 bar, $h_g = 2790$ $v_g = 0.1408$

10 bar, $h_f = 763$, $h_{fg} = 2015$, $v_g = 0.1944$

$$p_1 v_1^{1.135} = p_2 v_2^{1.135}$$
$$14 \times 0.1408^{1.135} = 10 \times v_2^{1.135}$$
$$v_2 = 0.1408 \times \sqrt[1.35]{1.4} = 0.1893 \ m^3/kg$$

Spec. vol. of dry steam at 10 bar is 0.1944 m³/kg, therefore:

$$\text{Dryness after expansion} = \frac{0.1893}{0.1944} = 0.974 \quad \text{Ans. (i)}$$

$$\text{Spec. enthalpy drop} = 2790 - (763 + 0.974 \times 2015)$$
$$= 64 \ kJ/kg \quad \text{Ans. (ii)}$$

$$\text{Velocity} = \sqrt{2 \times 64 \times 10^3} = 357.8 \ m/s \quad \text{Ans. (iii)}$$

$$\text{Volume flow [m}^3/\text{s]} = \text{Area [m}^2] \times \text{Velocity [m/s]}$$

For mass flow of 1 kg/s:

$$\text{Area [m}^2] = \frac{0.1893}{357.8} = 5.293 \times 10^{-4} \ m^2$$
$$5.293 \times 10^{-4} \times 10^6 = 529.3 \ mm^2 \quad \text{Ans. (iv)}$$

10. Referring to Figure 12.15:

$$\gamma = \frac{C_p}{C_v} = \frac{1.005}{0.718} = 1.4$$
$$r_p^{(\gamma-1)/\gamma} = 5.7^{0.4/1.4} = 1.644$$
$$\frac{T_2}{T_1} = \left\{ \frac{p_2}{p_1} \right\}^{\frac{\gamma-1}{\gamma}}$$
$$T_2 = 294 \times 1.644 = 483.3 \ K$$

Temperature at end of compression = 210.3°C   Ans. (i)

$$\frac{T_4}{T_3} = \left\{ \frac{p_4}{p_3} \right\}^{\frac{\gamma-1}{\gamma}}$$

$$T_4 = \frac{953}{1.644} = 579.7 \text{ K}$$

Alternatively, $\dfrac{T_4}{T_3} = \dfrac{T_1}{T_2}$ because pressure ratios are equal

$$\therefore T_4 = \frac{953 \times 294}{483.3} = 579.7 \text{K}$$

Temperature at end of expansion = 306.7°C   Ans. (ii)

Heat energy supplied per kg

$$= m \times c_p \times (T_3 - T_2)$$
$$= 1 \times 1.005 \times (953 - 483.3)$$
$$= 472.1 \text{ kJ/kg Ans. (iii)}$$

Increase in internal energy per kg from inlet to exhaust

$$= m \times c_v \times (T_4 - T_1)$$
$$= 1 \times 0.718 \times (579.7 - 294)$$
$$= 205.1 \text{ kJ / kg Ans. (iv)}$$

Ideal thermal effic. $= 1 - \dfrac{1}{r_p^{(\gamma-1)/\gamma}}$

$$= 1 - \frac{1}{1.644} = 0.3918 \text{ Ans. (v)}$$

Alternatively,

Thermal effic. $= 1 - \dfrac{T_4 - T_4}{T_3 - T_2}$

$$= 1 - \frac{579.9 - 294}{953 - 483.3} = 0.3919$$

TEST EXAMPLES 13

1. Tables page 4, water 130°C, $h = 546$

    ... 7, steam 30 bar 375°C, by interpolation,

    h at 30 bar 400°C = 3231

    h at 30 bar 350°C = 3117

    difference for 50°C = 114

    difference for 25°C = 57

    i.e. h at 30 bar 375°C = 3117 + 57= 3174

    Heat energy transferred to steam per hour

    $$= 30000 \times (3174 - 546) = 30000 \times 2628 \text{ kJ/h}$$

    $$\text{Hourly fuel consumption} = \frac{53 \times 10^3}{24} = 2209 \text{ kg/h}$$

    Heat energy supplied by fuel per hour

    $$= 2209 \times 42 \times 10^3 \text{ kJ/h}$$
    $$\text{Efficiency} = \frac{30000 \times 2628}{2209 \times 42 \times 10^3}$$
    $$= 0.85 \text{ or } 85\% \text{ Ans. (i)}$$

    Heat energy supplied to plant by fuel [kJ/s = kW]

    $$= \frac{2209 \times 42 \times 10^3}{3600}$$

    Energy converted into engine power

    $$= 0.13 \times \text{Energy supplied}$$

    $$= \frac{0.13 \times 2209 \times 42 \times 10^3}{3600} = 3349 \text{ kW Ans. (ii)}$$

Tables page 2, h_{fg} at 100°C = 2256.7

Evaporative capacity from and at 100°C

$$= \frac{30000 \times 2628}{2256.7} = 34920 \text{ kg/h Ans. (iii)}$$

$$= \frac{0.13 \times 2209 \times 42 \times 10^3}{3600} = 3349 \text{ kW Ans. (ii)}$$

Equivalent evaporation, per kg fuel, from and at 100°C

$$= \frac{34920}{2209} = 15.81 \text{ kg steam/kgfuel Ans. (iv)}$$

2. Solids in initially + Solids put in = Solids in finally

Water in × Initial × Amount × Feed	= Water in × Final
bolier ppm of feed ppm	bolier ppm

$$3.5 \times 40 + 0.875 \times 24 \times \text{Feed ppm} = 3.5 \times 2500$$
$$21 \times \text{Feed ppm} = 8610$$
$$\text{Feed ppm} = 410 \quad \text{Ans}$$

3. Mass of solids put in = Mass of solids blown out + Mass of solids in evaporated output

Mass of feed × feed ppm = Mass of blow out × Blow out ppm + Mass evaporated × evaporated ppm

Let x be mass flow per day of sea water feed

then $(x - 10)$ is mass flow per day of brine discharge

$$x \times 31250 = (x - 10) \times 78125 + 10 \times 250$$
$$x = 16.613$$
$$(x - 10) = 6.613$$

Mass flow of sea water feed = 16.613 tonne/day   Ans. (i)

Mass flow of brine discharge = 6.613 tonne/day   Ans. (ii)

4. Per kg of fuel:

$$\text{Available hydrogen} = H_2 - \frac{O_2}{8}$$

$$= 33.7 \times 0.85 + 144 \times 0.1275 \text{ kg}$$

$$= 28.64 + 18.36 = 47 \text{ MJ/kg Ans. (i)}$$

$$\text{Cal. value} = 33.7 \, C + 144 \left(H_2 - \frac{O_2}{8} \right)$$

$$= 33.7 \times 0.85 + 144 \times 0.1275$$

$$= 28.64 + 18.36 = 47 \text{ MJ/kg Ans. (i)}$$

$$\text{Stoichiometric air} = \frac{100}{23} \left\{ 2\tfrac{2}{3} C + 8 \left(H_2 - \frac{O_2}{8} \right) \right\}$$

$$= \frac{100}{23} \left\{ 2\tfrac{2}{3} \times 0.85 + 8 \times 0.1275 \right\}$$

$$\text{Actual air} = 1.5 \times 14.29 = 21.44 \text{ kg air/kg fuel Ans. (ii)}$$

Products of combustion per kg of fuel burned

$$= 21.44 + 1 \text{ kg fuel} = 22.44 \text{ kg}$$

Heat energy carried away

$$= \text{Mass} \times \text{Spec. heat} \times \text{Temp. rise}$$

$$= 22.44 \times 1.005 \times (553 - 304) = 5614 \text{ kJ/kg fuel}$$

as a percentage of the heat energy supplied

$$= \frac{5614}{47 \times 10^3} \times 100 = 11.95\% \text{ Ans. (iii)}$$

5. $\text{Available hydrogen} = H_2 - \dfrac{O_2}{8}$

$$= 0.13 - \frac{0.02}{8} = 0.1275 \text{ kg}$$

$$\text{Cal. value} = 33.7 \times 0.84 + 144 \times 0.1275$$
$$= 28.31 + 18.36 = 46.67 \text{ MJ/kg} \quad \text{Ans. (i)}$$

$$\text{Stoichiometric air} = \frac{100}{23} \times \text{oxygen required}$$
$$= \frac{100}{23} \left\{ 2\tfrac{2}{3} \times 0.84 + 8 \times 0.1275 \right\}$$
$$= \frac{100}{23} \times 3.26 = 14.18 \text{ kgair/kg fuel} \quad \text{Ans. (ii)}$$

Per kg fuel burned:

Mass of gases in the products from

22 kg air + (1 kg fuel – 0.01 kg incombustibles)

$$= 22.99 \text{ kg}$$

Mass of oxygen in 22 kg of air

$$= 0.23 \times 22 = 5.06 \text{ kg}$$

Surplus oxygen = 5.06 – 3.26 = 1.8 kg

Mass of nitrogen in 22 kg of air

$$= 0.77 \times 22 = 16.94 \text{ kg}$$

CO_2 formed = 32/3 × 0.84 = 3.08 kg

H_2O formed = 9 × 0.13 = 1.17 kg

% composition of gases by mass   Ans. (iii):

$$CO_2 = \frac{3.08}{22.99} \times 100 = 13.4\%$$

$$H_2O = \frac{1.17}{22.99} \times 100 = 5.09\%$$

$$O_2 = \frac{1.8}{22.99} \times 100 = 7.83\%$$

$$N_2 = \frac{16.94}{22.99} \times 100 = 73.68\%$$

6. Per kg of fuel burned:

$$\text{Stoichiometric air} = \frac{100}{23}\left\{2\tfrac{2}{3}C + 8\left(H_2 - \frac{O_2}{8}\right)\right\}$$

$$= \frac{100}{23}\left\{2\tfrac{2}{3} \times 0.85 + 8 \times 0.1275\right\}$$

$$= \frac{100}{23} \times 3.216 = 13.98 \text{ kg air/kg fuel}$$

$$CO_2 = 32/3 \times 0.855 = 3.135 \text{ kg}$$

When air supply is stoichiometric:

Mass of products of combustion

$$= 13.98 \text{ kg air} + (1 \text{ kg fuel} - 0.01 \text{ kg impurities})$$
$$= 13.98 + 0.99 = 14.97 \text{ kg}$$
$$\% \ CO_2 = \frac{3.135}{14.97} \times 100 = 20.94\% \quad \text{Ans.} \quad \text{(i)}$$

When air supply is 25% excess:

Mass of products of combustion

$$= 1.25 \times 13.98 \text{ kg air} + 0.99 = 18.47 \text{ kg}$$
$$\% \ CO_2 = \frac{3.135}{14.97} \times 100 = 16.97\% \quad \text{Ans.} \quad \text{(ii)}$$

When air supply is 50% excess:

Mass of products of combustion

$$= 1.5 \times 13.98 + 0.99 = 21.96 \text{ kg}$$
$$\% \ CO_2 = \frac{3.135}{21.96} \times 100 = 14.28\% \quad \text{Ans.} \quad \text{(iii)}$$

When air supply is 75% excess:

Mass of products of combustion

$$= 1.75 \times 13.98 + 0.99 = 25.45 \text{ kg}$$
$$\% \ CO_2 = \frac{3.135}{21.45} \times 100 = 12.31\% \quad \text{Ans.} \quad \text{(iv)}$$

$$\text{Stoichiometric air} = \frac{100}{23}\left[2\tfrac{2}{3} \times 0.84 + 0.14 \times 8\right]$$

7.
$$= 14.6 \text{ kg/kg of fuel}$$

$$\text{Actual air supplied} = 14.6 \times 1.2$$
$$= 17.52 \text{ kg/kg of fuel}$$

i.e. $17.52 \times 0.23 = 4.03$ kg of O_2
$17.52 - 4.03 = 13.49$ kg of N_2
CO_2, $3.667 \times 0.84 = 3.08$ kg/kg of fuel

i.e. $2.667 \times 0.84 = 2.24 \; O_2$
$H_2O, 0.14 \times 9 = 1.26$ kg/kg of fuel

i.e. $0.14 \times 8 = 1.12 \; O_2$,
$3.36 \; O_2$
$O_2, 403 - 3.36 = 0.67$ kg/kg of fuel Ans.

DFG	m	M	N		N%
CO_2	308	44	$3.08 \div 44 = 0.07$		12.22
O_2	0.67	32	$0.67 \div 32 = 0.021$		3.65
N_2	13.49	28	$13.49 \div 28 = 0.482$		84.13
			Total = 0.573		100
H_2O	1.26	18	$1.26 \div 18 = 0.07$		

Total mass of gases (wet and dry) $= 3.08 + 0.67 + 13.49 + 1.26$
$$= 18.5 \text{ kg/kg of fuel}$$
Alternatively $17.52 + 0.98 = 18.5$ kg/kg of fuel
Mass flow rate of gases $= m = 100 \times 18.5 = 1850$ kg/h

From $pV = mRT$, $R = \dfrac{R_0}{M}$

$$pV = \frac{m}{M} R_0 T$$

$$\frac{m}{M} = 0.573 + 0.07 = 0.643$$

$$\therefore V = \frac{0.643 \times 8.3143 \times 523}{1 \times 10^2}$$
$$= 27.95 \text{ m}^3/\text{kg of fuel}$$
$$V = 27.95 \times 100$$
$$= 2795 \text{ m}^3/\text{h Ans.}$$

8.

$$\text{Stoichiometric air} = \frac{100}{23}\left(2\tfrac{2}{3}C + 8H + S\right)$$
$$= 4.348\left(2.667 \times 0.86 + 8 \times 0.12 + 0.02\right)$$
$$= 14.23 \text{ kg/kg fuel}$$
$$\text{Actual air} = 20.00 \text{ kg/kg fuel}$$
$$\text{Excess air} = 5.77 \text{ kg/kg fuel}$$

Mass products of combustion per kg fuel:

$$CO_2 = 32/3 \times 0.86 = 3.154 \text{ kg}$$
$$H_2O = 9 \times 0.12 = 1.08$$
$$SO_2 = 2 \times 0.02 = 0.004$$
$$\text{Excess } O_2 = 0.23 \times 5.77 = 1.329$$
$$\text{All } N_2 = 0.77 \times 20 = 15.40$$
$$\text{Total} = 21.003 \text{ kg}$$

% mass analysis of the wet flue gases:

$$CO_2 = \frac{315.4}{21} = 15.01$$

$$H_2O = \frac{108}{21} = 5.14 \text{ Ans. (i)}$$

$$SO_2 = \frac{4}{21} = 0.19$$

$$O_2 = \frac{132.9}{21} = 6.32$$

$$N_2 = \frac{1540}{21} = 73.33$$

% volume analysis of the wet flue gases:

DFG	m%	M	N	N%
CO_2	15.01	44	0.341	9.89
H_2O	5.14	18	0.286	8.30
SO_2	0.19	64	0.003	0.09
O_2	6.32	32	0.198	5.73
N_2	73.33	28	2.619	76.00
			3.447	

Ans. (ii)

9. Mass of 1 mol of the fuel $= 12 \times 6 + 1 \times 6$

$$= 78 \text{ kg}$$

$$H_2 \text{ fraction by mass } = \frac{6}{78} = 0.0769$$

$$C \text{ fraction by mass } = \frac{72}{78} = \frac{0.9231}{1.0000}$$

$$\text{Stoichiometric air} = \frac{100}{23}(2.667 \times 0.9231 + 0.0769 \times 8)$$

$$= 13.38 \text{ kg/kg fuel Ans. (i)}$$

Mass of gas $= 14.38$ kg/kg fuel

$$CO_2 = 3.667 \times 0.9231 = 3.385$$
$$H_2O = 0.0769 \times 9 = 0.692$$
$$N_2 = 0.77 \times 13.38 = \frac{10.303}{14.380}$$

$$CO_2 = \frac{3.385}{14.38} \times 100 = 23.54\%$$

$$H_2O = \frac{0.692}{14.38} \times 100 = 4.81\% \text{ Ans. (ii)}$$

$$N_2 = \frac{10.303}{14.38} \times 100 = 71.65\%$$

$$\text{DFG} = 14.38 - 0.692 = 13.688 \text{ kg/kg fuel}$$

$$CO_2 = \frac{3.385}{13.688} \times 100 = 24.73\%$$

$$N_2 = \frac{10.303}{13.688} \times 100 = 75.27\%$$

Ans. (iii)

10.

Let x be C mass

DFG	$m\%$	M	N	$N\%$
CO_2	24.73	44	0.562	17.29
N_2	75.27	28	2.688	82.71
Total			3.250	

then $(1 - x)$ is H_2 mass

DFG	N	M	m	$m\%$
CO_2	10.8	44	475.2	15.83
CO	0.8	28	22.4	0.75
O_2	7.2	32	230.4	7.68
N_2	81.2	28	2273.6	75.75
Total			3001.6	

$$\text{Relative gas mass} = 3001.6$$
$$\text{Relative C mass} = 12(10.8 + 0.8)$$
$$= 139.2$$

$$\text{Dry flue gas mass} = \frac{3001.6 \times x}{139.2}$$
$$= 21.56x \text{ kg}$$

Water vapour $= 9 - 9x$ kg

Total gases $= 9 - 12.56x$ kg

Mass of air supplied $= 8 - 12.56$ kg/kg fuel (1)

$$\text{Mass of } N_2 \text{ supplied} = \frac{2273 \times x}{139.2}$$
$$= 16.33 \text{ kg}$$

$$\text{Mass of air supplied} = \frac{16.33 \times 100}{77}$$
$$= 21.21x \text{ kg/kg fuel } (2)$$

From (1) and (2):

$$8 - 12.56x := 21.21x$$
$$x = 0.925$$
$$(1 - E) = 0.075$$

Mass percentage of $C = 92.5\%$ in the fuel

Mass percentage of $H_2 = 7.5\%$ in the fuel Ans.

TEST EXAMPLES 14

Refer to Figure 14.2 for all solutions.

1. Specific enthalpy gain of refrigerant through evaporator

$$= h_1 - h_4 = 320 - 135 = 185 \text{ kJ/kg}$$

(h_4 being equal to h_3 because there is no change of enthalpy in the throttling process through the expansion valve)

$$\text{Refrig. effect } [\text{kJ/h}] = \text{Mass flow } [\text{kg/h}] \times (h_1 - h_4)[\text{kJ/kg}]$$
$$= 5 \times 601 \times 85$$
$$= 5.55 \times 10^4 \text{ kJ/h or } 55.5 \text{ MJ/h Ans.}$$

2. From tables page 15, R134a,

$$571.62 \text{ kPa } h_f = 227.45$$
$$132.72 \text{ kPa } t_s = -20°C \, h_f = 173.67 h_g = 386.44$$

$h_{fg} = h_g - h_f = 386.44 - 173.67 = 212.77$

Since the saturation temperature at 132.72 kPa is −20°C and the refrigerant at this pressure leaves the evaporator at −10°C, it is superheated by 10°. h at 132kPa superheated 10° = 394.63

Throttling between condenser exit and evaporator inlet:

Enthalpy after (h_4) = enthalpy before (h_3)

$$173.67 + (x_4 \times 212.77) = 227.45$$
$$x_4 \times 212.77 = 53.78$$
$$x_4 = 0.2528 \quad \text{Ans. (i)}$$

$$\text{Refrig. effect per kg} = h_1 - h_4 = h_1 - h_3$$
$$= 394.63 - 227.45$$
$$= 167.18 \text{ kJ/kg} \quad \text{Ans. (ii)}$$

3. From tables page 12, NH_3:

$$8.57 \text{ bar } h_{f.} = 275.1 h_g = 1462.6$$
$$1.902 \text{ bar } h_f = 89.8 h_g = 1420.0 v_g = 0.6237$$
$$h_{fg} = 1420 - 89.8 = 1330.2$$

Specific enthalpy drop through condenser

$$= h_2 - h_3 = 1462.6 - 275.1 = 1187.5 \text{ kJ/kg}$$

Heat rejected in condenser (by 2 kg)

$$= 2 \times 1187.5 = 2375 \text{ kJ/min} \quad \text{Ans. (i)}$$

Specific enthalpy gain in evaporator

$$h_1 - h_4 = h_1 - h3$$
$$= (89.8 + 0.96 \times 1330.2) - 275.1$$
$$= 1366.8 - 275.1 = 1091.7 \text{ kJ/kg}$$

Refrigerating effect

$$= 2 \times 1091.7 = 2183.4 \text{ kJ/min} \quad \text{Ans. (ii)}$$

Specific volume of refrigerant leaving evaporator and entering compressor = 0.96 x 0.6237 m³/kg

Volume taken into compressor per minute

$$= 2 \times 0.96 \times 0.6237 = 1,198 \text{ m}^3/\text{min Ans. (iii)}$$

4. Since there is a change only in the dryness fraction of the refrigerant through the evaporator, the enthalpy of saturated liquid (h_f) and of evaporation (h_{fg}) being the same for h_2 as for h_4 then:

Specific enthalpy gain of the CO_2 through evaporator

$$= h_1 - h_4$$
$$= \left(h_f + 0.92h_{fg}\right) - \left(h_f + 0 - 28h_{fg}\right)$$
$$= (0.92 - 0.28) \times 290.7$$
$$= 0.64 \times 290.7 = 186.1 \text{ kJ/kg} \quad \text{Ans. (i)}$$

Heat to be extracted from water to make 1 kg of ice

$$= 4.2 \times 14 + 335 + 2.04 \times 5$$
$$= 58.8 + 335 + 10.2 = 404 \text{ kJ/kg}$$

Let m [kg] = mass of ice made per second

when 0.5 kg/s = mass flow of CO_2

Heat transfer:

$$\text{from water} = \text{to } CO_2$$
$$m \times 404 = 0.5 \times 186.1$$
$$m = \frac{0.5 \times 186.1}{404} = 0.2303 \text{ kg/s}$$

Mass of ice in tonnes per 24 h

$$= \frac{0.2303 \times 3600 \times 24}{10^3} = 19.9 \text{ tonne/day} \quad \text{Ans. (ii)}$$

5. Quantity of heat extracted per kg of water

$$= 4.2 \times 18 + 335 + 2.04 \times 7$$
$$= 75.6 + 335 + 14.28 = 424.88 \text{ kJ/kg}$$

Heat energy extracted per second (refrigerating effect)

$$= \frac{2.5 \times 10^3}{24 \times 3600} \times 424.88 \text{ kJ/s}$$

Energy supplied per second, kJ/s = kW = 2.25

$$\text{Coeff. of performance} = \frac{\text{Heat energy extracted}}{\text{Heat energy supplied}}$$
$$= \frac{2.5 \times 10^3 \times 424.88}{24 \times 3600 \times 2.25}$$
$$= 5.464 \quad \text{Ans. (i)}$$

If ice at 0°C was made from water at 0°C, heat extracted would be equal to enthalpy of fusion only = 335 kJ/kg

i.e. Capacity from and at 0°C

$$= \frac{2.5 \times 424.88}{335} = 3.171 \text{ tonne/day Ans. (ii)}$$

$$= 3.171 \text{ tonne/day} \quad \text{Ans. (ii)}$$

6. At 1097.7 kPa the saturated temperature = 20°C therefore at 50°C the refrigerant is superheated by 30°. This means that at the compressor discharge the value $h_2 = 384.3$

At 611.1 kPa the saturated temperature = –15°C therefore at 0°C the refrigerant is superheated by 15°. This means that at the compressor suction (evaporator exit) the value $h_1 = 365.3$

h_3 – condenser outlet = h_f at 1097.7 kPa = 229.9

h_4 – evaporator inlet = h_3 = 229.9

$$\text{the Coeff. of performance} = \frac{\text{Refrigerating effect } [\text{kJ/kg}]}{\text{Work transfer } [\text{kJ/kg}]}$$

$$= \frac{h_1 - h_4}{h_2 - h_1}$$

$$= \frac{135.4}{19.0}$$

$$= 7.13 \quad \text{Ans.}$$

7. Tables page 12, NH_3:

$$14.7 \text{ bar, sat. temp.} = 38°C$$
$$\therefore \text{Vapour is superheated } (63 - 38) = 25°$$

By interpolation:

$$14.7 \text{ bar } 50°C \text{ supht}, h = 1620.1$$
$$14.7 \text{ bar no supht}, h = 1472.6$$
$$\text{increase for } 50° = 147.5$$
$$\text{increase for } 25° = 1/2 \times 147.5 = 73.75$$

$\therefore$ 14.7 bar 25° supht, $h = 1472.6 + 73.75 = 1546.35 = h_2$

14.7 bar 50°C supht, $s = 5.340$

14.7 bar no supht, $s = 4.898$

increase for 50° $= 0.442$

increase for 25° $= 1/2 \times 0.442 = 0.221$

$\therefore$ 14.7 bar 25° supht, $s = 4.898 + 0.221 = 5.119 = s_2$

2.077 bar, $h_f = 98.8$, $h_g = 1422.7$

$h_{fg} = 1422.7 - 98.8 = 1323.9$

$s_f = 0.404 s_g = 5.593$

$s_{fg} = 5.593 - 0.404 = 5.189$

Isentropic compression in compressor:

$$s_1 = s_2$$
$$0.404 + x_1 \times 5.189 = 5.119$$
$$X_1 = 0.9086$$
$$h_1 = 98.8 + 0.9086 \times 1323.9 = 1301.8$$
$$h_4 = h_3 = h_f \text{ at 14.7 bar} = 362.1$$

Refrig. effect [kJ/s] $= (h_1 - h_4)$ [kJ/kg] $\times$ mass flow [kg/s]

$$= (1301.8 - 362.1) \times 0.15$$
$$= 140.9 \text{ kJ/s Ans. (i)}$$

Work transfer [kJ/s] $= (h_2 - h_1)$ [kJ/kg] $\times$ mass flow [kg/s]

$$= (1546.35 - 1301.8) \times 0.15$$
$$= 36.68 \text{ kJ/s Ans. (ii)}$$

$$\text{Coeff. of performance} = \frac{\text{Refrigerating effect}}{\text{Work transfer}}$$

$$= \frac{140.9}{36.68} = 3.842 \text{ Ans. (iii)}$$

8. Thermodynamic data is as follows;

At 84.35 kPa $h_1 = h_g = 380.27$ kJ/kg

$s_1 = s_g = 1.7512$ kJ/kg K

$v_1 = v_g = 0.2258$ m³/kg

Therefore, the mass flow of the refrigerant $= \dfrac{\text{flow rate}}{\text{kinematic viscosity (v)}}$

$$\text{Mass flow} = \dfrac{0.15}{0.2258} = 0.664 \text{ kg/s}$$

For isentropic compression, entropy remains constant and therefore $s_1 = s_2 = 1.7512$ kJ/kg K. However, the specific entropy at 488.33 kPa should be 1.7194 kJ/kg K. The degree of superheat can be calculated by finding the incremental increase between the saturated value of sg at 488.33 kPa and the value of s at 10 degrees of superheat at 488.33 kPa.

For 10 degrees of superheat the rise in s = 1.7524 – 1.7194 = 0.0318 however the actual rise in entropy (s) is only to 1.7512. This means that instead of there being 10 degrees of superheat there is only 1.7524 – 1.7194 = t × 0.318 = 9.64°C.

Using the same maths, the actual value of enthalpy at the end of compression can be found as;

$$h_2 = 406.93 + \dfrac{9.64}{10}(416.57 - 406.93)$$
$$h_2 = 416.22 \text{ kj/kg}$$
$$\text{At 488.33 kPa } h_4 = h_3 = h_f = 220.46 \text{ kj/kg}$$

$$\text{the coefficient of performance} = \dfrac{\text{refrigerating effect [kJ / kg]}}{\text{work transfer [kJ / kg]}}$$

$$= \dfrac{h_1 - h_4}{h_2 - h_1}$$

$$= \dfrac{380.27 - 220.46}{416.22 - 380.27}$$

$$= 4.45 \quad \text{Ans (i)}$$

$$\text{Power input} = m\,[h_2 - h_1]$$

$$= 0.664 \times [416.22 - 380.27]$$

$$= 23.87 \text{ kW} \quad \text{Ans. (ii)}$$

9. At 40°C specific enthalpy

$$\text{Leaving compressor} = h_2$$
$$h_2 = h_g = 1473.3 \text{ kJ/kg}$$
$$s_2 = 4.877 \text{ kJ/kg K}$$

With isentropic compression $s_1 = s_2 = 4.877$ kJ/kg K

$$\therefore 4.877 = 0.44 + x(5.563 - 0.44)$$
$$x = 0.8661 \text{ dry}$$
$$\text{at} -16°C \text{ i.e. } h_1 = h_f + 0.8661(h_g - h_f)$$
$$h_1 = 107.9 + 0.8661 \ (1425.3 - 107.9)$$
$$= 1248.9 \text{ kJ/kg}$$

If the ammonia leaves the heat exchanger as saturated liquid then at 40°C $h_3 = h_f = 371.9$ kJ/kg

$$\text{Work done in compressor} = h_2 - h_1$$
$$= 1473.3 - 1248.9$$
$$= 224.4 \text{ kJ/kg}$$

$$\text{Heat transfer in condenser} = h_2 - h_1$$
$$= 1473.3 - 371.9$$
$$= 1101.4 \text{ kJ/kg}$$

$$\text{c.o.p.} = \frac{h_2 - h_3}{h_2 - h_1}$$
$$= \frac{1101.4}{224.4} = 4.908 \text{ Ans.}$$

$$\text{Mass of air changed } m_1 = \frac{pV}{RT}$$
$$= \frac{1.013 \times 1200}{287 \times 293} \times 10^5$$
$$= 1445.6 \text{ kg/h}$$

$$m_1 \times c_p \times (T_2 - T_1) = (h_2 - h_3) \times m$$
$$1445.6 \times \frac{1005}{10^3}(293 - 286) = 1101.4 \times m$$
$$m = 9.233 \text{ kg/h Ans.}$$

10. Using tables on page 15 of Rogers and Mayhew:

$$\text{At } 25°C \; h_f = 234.52 \; h_g = 412.23$$
$$\text{At } -15°C \; h_f = 180.16 \; h_g = 389.49 \text{ therefore } h_{fg} = 209.33$$

$$\text{Refrigerating effect} = h_1 - h_4$$
$$= 389.49 - 234.52$$
$$= 180.16 \text{ kJ/kg}$$

Cooling load = refrigerating effect × mass flow

No undercooling and therefore $h_3 = h_4$

Therefore, the cooling load= refrigerating effect × mass flow (m)

$$70.5 = 180.16 \times m$$
$$M = 0.39 \text{kg/s} \quad \text{Ans. (i)}$$

$h_2 = 422.41$, which is 665.25 kPa and 35°C, which is 10 degrees of superheat

Therefore, the work done by the compressor = $h_2 - h_1$

$$= 422.41 - 389.49$$
$$= 32.92 \text{ kW}$$

Compressor work done = 32.92×0.39
$$= 12.88 \text{ kW} \qquad \text{Ans. (ii)}$$

$$h_4 = h_f + xh_{fg}$$
$$\text{Therefore:} \quad 234.52 = 180.16 + 209.33x$$
$$54.36 = 209.33x$$

X = 0.2597, Which is the dryness fraction after the expansion valve Ans. (iii)

SELECTION OF EXAMINATION QUESTIONS – MANAGEMENT LEVEL

1. The efficiency of an auxiliary boiler is 70% when dry saturated steam at 8 bar is generated from feed water at 43°C. If the calorific value of the fuel is 42.5 MJ/kg and it is burned at the rate of 6 tonne/day, calculate (i) the hourly steam production, (ii) the equivalent evaporation per kg of fuel from and at 100°C.

2. An electric motor was tested by coupling it to a dynamometer and the brake load was 112 N at 40 rev/s. A steady flow of water passed through the brake and was raised in temperature from 15°C to 48°C. If the power absorbed by this brake is given by $Wn/330$ where W is the brake load in newtons and n is the rotational speed in rev/s, and assuming 98% of the heat generated at the brake is carried away by the cooling water, find the quantity of water passing through the dynamometer in litres per minute.

3. (i) A refrigerated hold of 580 m^2 surface area is lined with a 100 mm thick layer of insulation of thermal conductivity 017 W/m K. The interior and exterior surface temperatures of the insulation are 0°C and 17°C, respectively. Calculate the heat flow rate from the hold.

   (ii) The exterior surface temperature is increased to 50°C. Calculate the additional thickness of insulating material of thermal conductivity 0.07 W/m K required to be placed on the inside surfaces to keep the heat flow rate at the same level.

4. The bore of an IC engine is exactly 300 mm at 20°C. The diameters of the piston at 20°C are 298 mm at the crown and 299 mm at mid-depth of the body. Under working conditions the mean temperatures are: piston crown 250°C, piston body

100°C, cylinder 180°C. Take the coefficients of linear expansion of the piston and cylinder materials as 1.2×10^{-5} and $1.1 \times 10^{-5}/°C$, respectively, and calculate the diametrical clearances at the crown and mid-depth of the piston under working conditions.

5. An indicator card taken off one cylinder of a six-cylinder, two-stroke, single-acting engine has an area of 2850 mm² and length 75 mm when running at 1.75 rev/s. A height of 1 mm on the card represents 0.4 bar. If the cylinder bore is 550 mm and stroke 850 mm, calculate the indicated power of the engine assuming equal powers are developed in all cylinders.

6. A vessel of volume 0.65 m³ contains air at 27.6 bar and 18°C. Calculate the final pressure after 3.5 kg of air is added if the final temperature is 20.5°C. Take R for air $= 0.287$ kJ/kg K.

7. 1 kg of steam at 7.0 bar, 0.95 dry, is expanded according to the law $pV^{1.3} = $ constant, until the pressure is 3.5 bar. Calculate the final dryness fraction of the steam.

8. A turbine plant consisting of HP and LP units is supplied with steam at 15 bar, 300°C. The steam is expanded in the HP and leaves at 2.5 bar 0.97 dry. At this point some steam is bled off to the feed heater and the remaining steam passes to the LP where it is expanded to 0.14 bar 0.84 dry. If the same quantity of work transfer takes place in each unit, calculate the amount of steam bled off expressed as a percentage of the steam supplied.

9. The mass analysis of a fuel is 87% carbon, 11% hydrogen and 2% oxygen. Calculate the volume of air in cubic metres at 1.0 bar and 25°C required for stoichiometric combustion per kg of fuel. Take the values: R for air $= 0.287$ kJ/kg K. Mass analysis of air $= 23\%$ oxygen, 77% nitrogen. Atomic weights, hydrogen 1, carbon 12, oxygen 16.

10. Gas at pressure 0.95 bar, volume 0.2 m³ and temperature 17°C is compressed until the pressure is 2.75 bar and volume 0.085 m³. Calculate the compression index and the final temperature.

11. A ship's cold room has dimensions of 9 m × 4 m × 2.5 m and is lined on the inside with 15 mm thick boarding. The deck, deckhead and bulkheads are of 10 mm thick steel and a 70 mm thick layer of cork insulation is sandwiched between the steel plate and the boarding. The thermal conductivities of steel, cork and boarding are 45 W/m K, 0.06 W/m K and 0.11 W/m K, respectively. The surface heat transfer coefficients at the exposed board inner and steel outer surfaces are 1.62 W/m² K and 13 W/m² K, respectively. Calculate the cooling load required to maintain the cold room at −6°C when the ambient temperature is 27°C.

12. A boiler working at 16 bar produces 9000 kg of steam per hour from feed water at 95°C, the dryness fraction of the steam being 0.98. If the boiler efficiency is 87% and the calorific value of the fuel 42 MJ/kg, calculate the daily fuel consumption.

13. The scavenge ports of a two-stroke diesel engine are just covered when the piston is 800 mm from the top of its stroke. The contents of the cylinder at this instant are at a pressure of 1.21 bar and temperature 40°C. The piston diameter is 700 mm and the clearance is equivalent to 70 mm. Find the mass of air taken in per cycle if the scavenge efficiency is 0.95. R for air = 0.287 kJ/kg K.

14. The diameter of an air compressor cylinder is 130 mm, the stroke is 180 mm and the clearance volume is 73 cm³. The pressure in the cylinder at the beginning of the stroke is 1.0 bar and the pressure during delivery is constant at 4.6 bar. Taking the law of compression as $pV^{1.2}$ = constant, calculate the distance moved by the piston during the delivery period and express this as a fraction of the stroke.

15. In a single-stage impulse turbine the steam leaves the nozzles at a velocity of 500 m/s at 18° to the plane of rotation of the blades, and the linear velocity of the blades is 230 m/s. Neglecting friction across the blades and assuming the steam leaves the blade wheel in an axial direction, calculate (i) the inlet angle of the blades so that the steam enters without shock and (ii) the outlet blade angle.

16. The clearance volume of a reciprocating compressor of 100 mm bore and 100 mm stroke is 5% of its swept volume. At the end of the suction stroke the air in the cylinder is at 1 bar 25°C.
    (i) Show the cycle on a pressure volume diagram.
    (ii) Calculate the mass of air in the cylinder at the beginning of the delivery stroke.
    (iii) Explain why the compressor takes in less than the swept volume of air during each suction stroke.

    Note: for air R = 287 J/kg K.

17. In an NH_3 refrigerating plant the ammonia leaves the condenser as a saturated liquid at 10.34 bar. The evaporator pressure is 2.265 bar and the refrigerant leaves the evaporator as a vapour 0.95 dry. If the circulation of the refrigerant through the plant is 4 kg/min, calculate (i) the dryness fraction at inlet to the evaporator, (ii) the refrigerating effect per minute and (iii) the volume of refrigerant taken into the compressor per minute.

18. (i) Write down the combustion equations for the complete combustion of carbon to carbon dioxide, hydrogen to water and sulphur to sulphur dioxide.
    (ii) A fuel oil of mass analysis 86.3% carbon, 12.8% hydrogen and 0.9% sulphur is burned with 25% excess air. The flue gases are at 1.5 bar 370°C. Calculate the mass and volume of flue gases per kg of fuel burned.

Note: Air contains 23% oxygen by mass;

relative atomic masses: hydrogen 1, carbon 12, oxygen 16, sulphur 32;

for flue gases R = 276 J/kg K.

19. Water at 100°C flows through a steel pipe of 150 mm inside diameter and 160 mm outside diameter. The surface heat transfer coefficients at the inside and outside surfaces are 240 W/m² K and 12 W/m² K, respectively. The air surrounding the pipe is at 15°C. The thermal resistance of the steel may be neglected and the diameter of the pipe is large compared with its wall thickness. Calculate the rate of heat loss from the water per metre length of pipe.

20. A volume of 0.8 m³ of steam at 17 bar 0.95 dry is passed through a reducing valve and throttled to 6 bar. Calculate the dryness fraction and the volume after throttling.

21. At the beginning of a voyage a boiler contains 6 tonne of water having 120 ppm dissolved solids. The feed rate is 1250 kg/h and after 24 h the boiler water contains 1080 ppm dissolved solids. Calculate the average dissolved solids in the feed water.

22. 1.5 m³ of wet steam at 2.8 bar are blown into 36 kg of water at 16°C and the resulting temperature of the mixture is 55°C. Calculate the dryness fraction of the steam.

23. A single stage impulse turbine has a mean blade diameter of 600 mm and a blade velocity of 120 m/s. The nozzle angle is 18° and the enthalpy drop across the nozzles is 465 kJ/kg.

    Determine:
    (i)   the turbine rotational speed;
    (ii) the blade inlet angle;
    (iii) the axial component of the steam at the blade inlet.

24. The air in a ship's saloon is maintained at 19°C and is changed twice every hour from the outside atmosphere which is at 7°C. The saloon is 27 m × 15 m × 3 m high. Calculate the kilowatt loading to heat this air, taking the saloon to be at atmospheric pressure = 1.013 bar, R for air = 0.287 kJ/kg K, c_p = 1.005 U/kg K.

25. A six-cylinder, single-acting, four-stroke oil engine of 200 mm stroke and 225 mm bore runs at 5 rev/s when the mean effective pressure is 17 bar. If the mechanical efficiency is 85%, calculate the indicated and brake powers.

26. A single-cylinder, single-acting compressor of 200 mm stroke and 100 mm bore runs at 5 rev/s and takes in air at 1 bar, 17°C. It is used to charge a 3 m³ capacity receiver from 1 bar, 25°C to 7 bar, 25°C. The compressor has negligible piston clearance.
    (i) Sketch the cycle on a pressure volume diagram.
    (ii) Calculate the time required to charge the receiver.

27. An ammonia refrigerator has a cooling load of 3.5 kW. The ammonia leaves the condenser as a liquid at 24°C, is throttled to 2.68 bar and leaves the evaporator as a dry saturated vapour.

    (i) Sketch component and pressure enthalpy diagrams.

    (ii) Calculate the mass flow rate of the ammonia.

28. A boiler working at 15 bar generates 7000 kg of steam per hour. The steam leaves the boiler steam drum dry and saturated and then passes through the superheater tubes at constant pressure. The flue gases enter the nests of superheaters at 822°C and leave at 690°C. The fuel consumption is 750 kg/h and 24 kg of air are supplied per kg of fuel burned. Find (i) the temperature of the superheated steam and (ii) the mass of injection water to the de-superheater, at 21°C, required to de-superheat each kg of steam. Take the specific heat of the flue gases as 1.007 kJ/kg K.

29. A gas initially at 12 bar, 216°C and volume 9900 cm³ is expanded in a cylinder. The volumetric expansion is 6 and the index of expansion is 1.33. Calculate the final volume, pressure and temperature.

30. 1 kg of steam initially occupying a volume of 008 m³ at 40 bar is expanded to 2 bar according to the law $pV^{1.3}$ = constant.

    Calculate:

    (i) the work transfer;

    (ii) the heat transfer.

31. Steam leaves the nozzles and enters the blade wheel of a single-stage impulse turbine at a velocity of 840 m/s and at an angle of 20° to the plane of rotation. The blade velocity is 350 m/s and the exit angle of the blades is 25° 12′. Due to friction, the steam loses 20% of its relative velocity across the blades. Calculate (i) the blade inlet angle, (ii) the magnitude and direction of the absolute velocity of the steam at exit.

32. The mass composition of a fuel oil is 84.6% carbon, 11.4% hydrogen, 0.4% sulphur, 2.4% oxygen and 1.2% impurities. Calculate the calorific value of the fuel and the theoretical mass of air required for stoichiometric combustion of 1 kg of fuel. Take the values:

	Calorific atomic value MJ/kg	Atomic weight
Hydrogen	144	1
Carbon	33.7	12
Sulphur	9.3	32
Oxygen	–	16

Mass analysis of air = 23% oxygen, 77% nitrogen.

33. In a Freon-12 refrigerating plant, the refrigerant leaves the condenser with a specific enthalpy of 50 kJ/kg. The pressure in the evaporator is 1.826 bar and the refrigerant leaves the evaporator at this pressure and at a temperature of 0°C. Calculate (i) the dryness fraction of the freon at inlet to the evaporator, and (ii) the refrigerating effect per minute if the flow rate of the refrigerant is 0.4 kg/s.

34. A single-stage, double-acting air compressor deals with 18.2 m³ of air per minute measured at conditions of 1.01325 bar, 15°C. The condition at the beginning of compression is 0.965 bar, 27°C and the discharge pressure is 4.82 bar. The compression is according to the law $pV^{1.32}$ = constant. If the mechanical efficiency of the compressor is 0.9, calculate the input power required to drive the compressor.

35. The mean area of indicator cards taken from a cylinder of a double-acting, two-stroke engine is 346 mm² and the length is 75 mm. The spring used in the indicator deflects 1 mm under a force of 60 N and the movement of the stylus is six times that of the indicator piston. The diameter of the indicator piston is 7 mm. Calculate (i) the mean effective pressure. If the diameter of the engine cylinder is 600 mm, stroke 900 mm and rotational speed 2.1 rev/s, calculate (ii) the indicated power per cylinder.

36. At the entrance of a nozzle of circular cross-section, the velocity of the steam is 457 m/s and the specific volume is 0.2765 m³/kg. At exit, the velocity is 1524 m/s and the specific volume 7.404 m³/kg. If the mass flow of steam through the nozzle is 0.315 kg/s, calculate the entrance and exit diameters in millimetres.

37. A quantity of air of volume 0.2 m³ at 1.1 bar and 15°C is heated at constant pressure until its temperature is 150°C, and then compressed to 7.15 bar according to the law $pV^{1.32}$ = constant. Calculate (i) the amount of heat energy transferred to the air at constant pressure and (ii) the temperature at the end of compression. For air, $R = 0.287$ kJ/kg K, $c_p = 1.005$ kJ/kg K.

38. A cold store wall 6 m long and 3 m high is constructed of 120 mm thick brick with an inside layer of 80 mm thick cork insulation. The thermal conductivity of the brick is 1.15 W/m K, while that of cork is 0.043 W/m K. The inner and outer wall surface temperatures are –4°C and 21°C, respectively. Calculate:
(i)   the amount of heat flow through the wall in 24 h;
(ii) the interface temperature between the cork and the brick.

39. The clearance volume of a reciprocating compressor of 381 mm stroke is 7% of the swept volume. The piston travels 267 mm from bottom dead centre to the point at which the delivery valves open and raises the air pressure from 1.013 bar to 4 bar. The compression process is polytropic.
(i)   Sketch the cycle on a pV diagram.
(ii) Calculate the value of the polytropic index n.

40. The mass analysis of a hydrocarbon fuel A is 88.5% carbon and 11.5% hydrogen. Another hydrocarbon fuel B requires 6% more air than fuel A for stoichiometric combustion. Calculate the mass analysis of fuel B, taking the following values: atomic weights, carbon 12, hydrogen 1, oxygen 16, mass content of oxygen in air = 23%.

41 The stroke of an internal combustion engine is 90 mm and the clearance volume is equivalent to a linear clearance of 15 mm. If the clearance is reduced by 2.5 mm, find the pressure at the end of compression before and after the alteration, taking the initial pressure in each case as 1.0 bar and the index of compression as 1.33.

42. The kinetic energy of the steam jet leaving the nozzles of a single-stage impulse turbine is equivalent to 250 kJ/kg. The entrance and exit angles of the blades are equal at 35°, and the steam leaves the blade wheel in an axial direction. Neglecting friction across the blades and assuming shockless flow, calculate (i) the nozzle angle and (ii) the blade velocity.

43. A four-cylinder, single-acting, two-stroke engine develops 600 kW indicated power when the mean effective pressure is 12.56 bar and the speed is 4.5 rev/s.
    (i) If the stroke is 25% greater than the cylinder diameter, calculate the diameter of the cylinders and the stroke to the nearest millimetre.
    (ii) When burning fuel of calorific value 42 MJ/kg the fuel consumption is 0.225 kg/kWh (indicated), find the indicated thermal efficiency.

44. The refrigerating effect of a plant using ammonia as the refrigerant is 800 kJ/min. At the exit points of the components the conditions of the refrigerant are:
    Evaporator, dry saturated vapour at 1.902 bar
    Compressor, vapour at 7.529 bar and 66°C
    Condenser, saturated liquid at 7.529 bar

    Calculate (i) the mass flow of the refrigerant through the plant, in kg/min, (ii) the heat rejected in the condenser, in kJ/min, and (iii) the output power of the compressor in kW.

45. Steam enters a de-superheater at 30 bar and 400°C and leaves at the same pressure as dry saturated steam. The temperature of the water injected into the de-superheater is 38°C. Calculate (i) the mass of injection water used per kg of steam de-superheated and (ii) the percentage change in volume from that occupied by 1 kg of superheated steam to the volume occupied by the dry saturated steam resulting from the mixture of 1 kg of superheated steam and its injected water.

46. A gas is compressed in a cylinder from 1 bar and 35°C at the beginning of the stroke to 37 bar at the end of the stroke. If the clearance volume is 850 cm³ and the compression index 1.32, calculate the stroke volume and the temperature at the end of compression.

47. When 1 kg of a fuel containing only carbon and hydrogen is burned in air, 15.6 kg of exhaust containing only nitrogen, carbon dioxide and water vapour is produced.

Determine the fuel mass analysis.

Air contains 23% oxygen by mass.

Atomic mass relationships: hydrogen 1, carbon 12, oxygen 16

48. (i) A ship's ice-making machine produces 27.2 kg/h of ice at 0°C from water at 14.4°C with a coefficient of performance of 7.51. Calculate the power input to the machine.

(ii) Using the concept of a reversed heat engine, explain why refrigerators require some power input.

Note: Specific heat capacity of water is 4.186 kJ/kg K.

Specific enthalpy of fusion of ice is 332.6 kJ/kg.

49. The following data relate to a single-stage impulse turbine:

Blade:	mean diameter	1.32 m
	inlet angle	34°
	outlet angle	30°
Steam:	inlet axial velocity component	550 m/s
	relative outlet velocity	850 m/s
	flow rate	0.0833 kg/s
Turbine:	power developed	67 kW

Determine the turbine rotational speed.

50. Water is kept at a temperature of 66°C in a closed tank 1 m × 0.75 m × 1.6 m. The heat loss from the water must not exceed 200 W when the ambient temperature is 18°C. The tank is to be lagged with insulating material of thermal conductivity 0.048 W/m K. The exterior surface heat transfer coefficient is 1 W/m² K. The thermal resistances of the tank walls and lid and the heat loss through the base are negligible.

Calculate:

(i) the thickness of lagging required;

(ii) the additional thickness of lagging required to reduce the heat loss to 100 W.

SOLUTIONS TO EXAMINATION QUESTIONS – MANAGEMENT LEVEL

1. From steam tables:

   Steam 8 bar, h_g = 2769 kJ/kg

   Water 43°C, by interpolation,

   $$h \text{ at } 44°C = 184.2$$
   $$h \text{ at } 42°C = 175.8$$
   $$\text{Increase for } 2° = 8.4$$
   $$\text{Increase for } 1° = 4.2$$
   $$\therefore h \text{ at } 43°C = 175.8 + 4.2 = 180 \text{ kJ/kg}$$

   $$\text{Fuel consumption} = \frac{6 \times 10^3}{24} = 250 \text{ kg/h}$$

   $$\text{Boiler efficiency} = \frac{\text{heat energy transferred to steam } [\text{kJ/h}]}{\text{heat energy supplied by fuel } [\text{kJ/h}]}$$

   $$0.7 = \frac{\text{mass of steam } [\text{kg/h}] \times (2769 - 180)}{250 \times 42.5 \times 10^3}$$

   $$\text{Mass of steam/}h = \frac{0.7 \times 250 \times 42.5 \times 10^3}{2589}$$

   $$= 2873 \text{ kg/h Ans. (i)}$$

   From tables, h_{fg} at 100°C = 2256.7 kJ/kg

Equivalent evaporation from and at 100°C, per kg of fuel,

$$= \frac{2873 \times (2769 - 180)}{250 \times 2256.7}$$

$$= 13.18 \text{ kg steam/kg fuel Ans. (ii)}$$

2. Brake power $= \dfrac{Wn}{330} = \dfrac{112 \times 40}{330} = 13.58 \text{ kW}$

Energy at brake = 13.58 kJ/s

Energy carried away by water

$$= 0.98 \times 13.58 = 13.3 \text{ kJ/s}$$

From steam tables:

$$\text{Water } 48°C, h = 200.9$$
$$15°C, h = 62.9$$

Enthalpy gain of cooling water

$$= 200.9 - 62.9 = 138 \text{ kJ/kg}$$

Heat energy transferred to water [kJ/s]

$$= \text{Mass flow}[\text{kg/s}] \times \text{spec. enthalpy gain } [\text{kJ/kg}]$$

$$\therefore \text{Mass flow} = \frac{13.3}{138} = 0.09638 \text{kg/s}$$

$$0.09638 \times 60 = 5.783 \text{ kg/min}$$

$$= 5.783 \text{ l/min Ans.}$$

$$Q = \frac{kAt(T_1 - T_2)}{S}$$

$$= \frac{0.17 \times 580 \times 1(0 - 17)}{0.1}$$

$$= -16762 \text{ W i.e. into the hold}$$

$$= -16.762 \text{kW Ans. (i)}$$

For the two thicknesses:

$$T_1 - T_3 = \frac{Q}{At}\left\{ \frac{S_1}{k_1} + \frac{S_2}{k_2} \right\}$$

$$0 - 50 = \frac{-16762}{580 \times 1}\left\{ \frac{S_1}{0.07} + \frac{0.1}{0.17} \right\}$$

$$S_1 = 0.07 \left\{ \frac{50 \times 580 \times 1}{16762} - \frac{0.1}{0.17} \right\}$$
$$= 0.07993 \text{ m}$$
$$= 79.93 \text{ mm} \quad \text{Ans. (ii)}$$

4. Linear (diametrical) expansion $= \alpha \times d \times (\theta_2 - \theta_1)$

Increase in cylinder diameter

$$= 1.1 \times 10^{-5} \times 300 \times (180 - 20) = 0.5279 \text{ mm}$$
$$\text{Working diameter} = 300 + 0.5279 = 300.5279 \text{ mm}$$

Increase in piston crown diameter

$$= 1.2 \times 1^{-5} \times 298 \times (250 - 20) = 0.8224 \text{ mm}$$
$$\text{Working diameter} = 298 + 0.8224 = 298.8224 \text{ mm}$$

Increase in piston body diameter

$$1.2 \times 10^{-5} \times 299 \times (100 - 20) = 0.2871 \text{ mm}$$
$$\text{Working diameter} = 299 + 0.2871 = 299.2871 \text{ mm}$$

Diametrical clearances at working temperatures:

$$\left. \begin{array}{l} \text{Piston crown} = 300.5279 - 298.8224 = 1.7055 \text{ mm} \\ \text{Piston body} \ = 300.5279 - 299.2871 = 1.2408 \text{ mm} \end{array} \right\} \text{Ans.}$$

5. Mean height of card $= \dfrac{2850}{75} = 38 \text{ mm}$

$$\text{Mean effective pressure} = 38 \times 0.4 \text{ bar}$$
$$38 \times 0.4 \times 10^2 = 1520 \text{ kN/m}^2$$

Indicated power of 6 cylinders

$$= p_m LAN \times 6$$
$$= 1520 \times 0.7854 \times 0.55^2 \times 0.85 \times 1.75 \times 6$$
$$= 3224 \quad \text{Ans.}$$

6. $pV = mRT$

Initial mass of air:

$$m = \frac{pV}{RT} = \frac{27.6 \times 10^2 \times 0.65}{0.287 \times 291} = 21.48 \text{ kg}$$

$$\text{Final mass} = 21.48 + 3.5 = 24.98 \text{ kg}$$

$$\text{Final pressure}, p = \frac{mRT}{V}$$

$$= \frac{24.98 \times 0.287 \times 293.5}{0.65}$$

$$= 3238 \text{ kN/m}^2 = 32.38 \text{ bar} \quad \text{Ans.}$$

7. From steam tables:

$$7 \text{ bar}, v_g = 0.2728 \quad 3.5 \text{ bar}, v_g = 0.5241$$

Volume of 1 kg of steam at 7 bar, 0.95 dry

$$= 0.95 \times 0.2728 = 0.2592 \text{ m}^3$$

$$P_1 v_1^{1.3} = P_2 v_2^{1.3}$$

$$v_2 = 0.2592 \times \frac{\sqrt[1.3]{7}}{3.5}$$

$$= 0.4417 \text{ m}^3$$

Specific volume of dry sat. steam at 3.5 bar is 0.5241 m³/kg therefore dryness fraction of expanded steam

$$= \frac{0.4417}{0.5241} = 0.8428 \quad \text{Ans.}$$

8. From steam tables:

$$15 \text{ bar } 300°C, \quad h = 3039$$

$$2.5 \text{ bar}, \quad h_f = 535 \quad h_{fg} = 2182$$

$$0.14 \text{ bar}, \quad h_f = 220 \quad h_{fg} = 2376$$

Specific enthalpy at 2.5 bar, 0.97 dry

$$= 535 + 0.97 \times 2182 = 2651 \text{ kJ/kg}$$

Specific enthalpy at 0.14 bar, 0.84 dry

$$= 220 + 0.84 \times 2376 = 2215 \text{ kJ/kg}$$

Specific enthalpy drop through HP

$$= 3039 - 2651 = 388 \text{ kJ/kg}$$

Specific enthalpy drop through LP

$$= 2651 - 2215 = 436 \text{ U/kg}$$

Total enthalpy drop through each unit is to be the same, let 1 kg of steam be supplied and x kg bled off, then 1 kg passes through HP and $(1 - x)$ kg passes through LP.

$$1 \times 388 = (1 - x) \times 436$$
$$436x = 436 - 388$$
$$x = \frac{48}{436} = 0.1101$$

Expressed as a percentage

Amount bled off $= 11.01\%$ Ans.

9. Mass of air required per kg of fuel

$$= \frac{100}{23} \times \text{oxygen required}$$

$$= \frac{100}{23}\left\{2\tfrac{2}{3}C + 8\left(H_2 - \frac{O_2}{8}\right)\right\}$$

$$= \frac{100}{23}\left\{2\tfrac{2}{3}C + 8H_2 - O_2\right\}$$

$$= \frac{100}{23}\left\{2\tfrac{2}{3} \times 0.87 + 8 \times 0.11 - 0.02\right\}$$

$$= \frac{100}{23} \times 3.18 = 13.82 \text{ kg air/kg fuel}$$

$$pV = mRT$$

$$V = \frac{13.82 \times 0.287 \times 298}{1 \times 10^2}$$

$$= 11.82 \text{ m}^3 \text{ air/kg fuel}   \text{Ans.}$$

10. $p_1 V_1 n = p_2 V_2 n$

$$0.95 \times 0.2^n = 2.75 \times 0.085^n$$

$$\frac{p_1 V_1}{T_1} = \frac{p_2 V_2}{T_2}$$

$$T_2 = \frac{290 \times 2.75 \times 0.085}{0.95 \times 0.2} = 356.8 \text{ K}$$

Final temp. $= 83.8°C$ Ans. (ii)

11. Area of top and bottom $= 9 \times 4 \times 2$

$$\text{Area of sides} = 9 \times 2.5 \times 2$$

$$\text{Area of ends} = 4 \times 2.5 \times 2$$

$\therefore$ Area exposed to heat source $= 72 + 45 + 20 = 137 \text{ m}^2$

Heat passing across inner film:

$$Q = h_i \, At \left(T_i - T_1 \right) \quad \therefore T_i - T_1 = \frac{Q}{h_i AT}$$

Temperature rise across the three thicknesses:

$$T_1 - T_0 = \frac{Q}{At}\left\{ \frac{S_1}{k_1} + \frac{S_2}{k_2} + \frac{S_3}{k_3} \right\}$$

Heat passing across outer film:

$$Q = h_o \, At \left(T_4 - T_0 \right) \quad \therefore T_4 - T_0 = \frac{Q}{h_o At}$$

$$T_i - T_4 = \frac{Q}{At}\left\{ \frac{1}{h_i} + \frac{S_1}{k_1} + \frac{S_2}{k_2} + \frac{S_3}{k_3} + \frac{1}{h_o} \right\}$$

$$(-6 - 27) = \frac{Q}{137 \times 1}\left\{ \frac{1}{1.62} + \frac{0.015}{0.11} + \frac{0.07}{0.06} + \frac{0.01}{45} + \frac{1}{13} \right\}$$

$$-33 \times 137 = Q(0.6172 + 0.1364 + 1.167 + 0.0002 + 0.077)$$

$$Q = -2263.3 \text{ W, i.e. into the room}$$

$$= 2.2633 \text{ kW cooling load \quad Ans.}$$

Note : $\dfrac{1}{U}$ equals the term above in brackets, i.e. $U = 0.5006$

$$Q = UAt\left(T_i - T_4 \right)$$

12. From steam tables:

$$\text{Steam 16 bar,} \quad h_f = 859 \quad h_{fg} = 1935$$
$$\text{Water } 95°C, \quad h = 398$$
$$\text{Boiler steam} \quad h_1 = 859 + 0.98 \times 1935 = 2755$$
$$\text{Feed water} \quad h_w = 398$$

Heat energy transferred to water to make steam

$$= h_1 - h_w = 2755 - 398 = 2357 \text{ kJ/kg}$$
$$= 2357 \times 9000 \text{ kJ/h}$$

Heat energy released by m kg of fuel per hour

$$= 42 \times 10^3 \times m \text{ kJ/h}$$

$$\text{Boiler efficiency} = \frac{\text{Heat energy transferred to steam}}{\text{Heat energy supplied by fuel}}$$

$$0.87 = \frac{2357 \times 9000}{42 \times 10^3 \times m}$$

$$m = \frac{2357 \times 9000}{0.87 \times 42 \times 10^3} = 580.6 \text{ kg/h}$$

Tonnes of fuel used per day

$$= 580.6 \times 24 \times 10^3 = 13.93 \text{ tonne/day} \quad \text{Ans.}$$

13. When scavenge ports are just closed, distance from cylinder cover to piston = $800 + 70 = 870$ mm

$$\text{Volume enclosed} = 0.7854 \times 0.7^2 \times 0.87 \text{ m}^3$$
$$\text{Volume of air taken in (scavenge effic. being 0.95)}$$
$$= 0.95 \times 0.7854 \times 0.7^2 \times 0.87$$
$$= 0.3181 \text{ m}^3$$
$$pV = mRT$$
$$m = \frac{pV}{RT} = \frac{1.21 \times 10^2 \times 0.3181}{0.287 \times 313}$$
$$= 0.4285 \text{ kg} \quad \text{Ans.}$$

14. $\text{Linear clearance} \left[\text{mm}\right] = \frac{\text{Volumetric clearance} \left[\text{mm}^3\right]}{\text{Area of cylinder} \left[\text{mm}^2\right]}$

$$= \frac{73 \times 10^3}{0.7854 \times 130^2} = 5.5 \text{ mm}$$

Representing volumes by linear dimensions:

$$V_1 = 180 + 5.5 = 185.5 \text{ mm}$$
$$P_1V_1^{1.2} = P_2V_2^{1.2}$$
$$V_2 = \frac{185.5}{\sqrt[1.2]{4.6}} = 52 \text{ mm}$$

Movement of piston during delivery period

$$= 52 - 5.5 = 46.5 \text{ mm} \quad \text{Ans. (i)}$$

as a fraction of the stroke

$$= \frac{46.5}{180} = 0.2583 \text{ Ans. (ii)}$$

15.

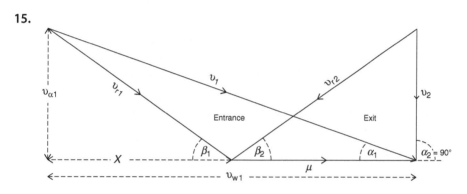

$$v_{a1} = v_1 \sin-1 = 500 \times \sin 18° = 154.5 \text{ m/s}$$
$$v_{w1} - = V_1 \cos-1 = 500 \times \cos 18° = 475.5 \text{ m/s}$$
$$x = v_{wl} - 8 = 475.5 - 230 = 245.5$$
$$\tan \beta_1 = \frac{v_{a1}}{x} = \frac{154.5}{245.5} = 0.6292$$
$$\therefore \text{ Inlet blade angle} = 32° \ 11' \quad \text{Ans. (i)}$$

Neglecting friction across the blades;

$$V_{r2} = V_{r1} = \frac{V_{a1}}{\sin +1} = \frac{154.5}{\sin 32°11'} = 290 \, m/s$$

$$\cos \beta_2 = \frac{u}{V_{r2}} = \frac{230}{290} = 0.7930$$

$$\therefore \text{ Outlet blade angle } = 37° \, 32' \text{ Ans. (ii)}$$

16. Refer to Figure 9.3 Ans. (i):

$$\text{Stroke volume}(s) = 0.7854 \times 0.1^2 \times 01$$
$$= 0.0007854 \, m^3$$
$$\text{Clearance volume } (c) = 0.0007854 \times 0.05$$
$$= 000003927 \, m^3$$
$$V_1 = s + A, \text{ i.e. } 0.0008247$$
$$pV = mRT$$
$$m = \frac{1 \times 100 \times 0.0008247}{0.287 \times 298}$$
$$= 00009642 \text{ Ans. (ii)(i)}$$

The air trapped in the clearance space re-expands and reduces the intake volume on each suction stroke, Ans. (iii).

17. From tables for NH_3,

$$10.34 \, bar, \quad h_f = 303.7$$
$$2.265 \, bar, \quad h_f = 107.9 \quad h_g = 1425.3 \quad v_g = 0.5296$$
$$2.265 \, bar, \quad h_f = 107.9 \quad h_g = 1425.3 \quad v_g = 0.5296$$

Refer to Figure 14.2:

Throttling process through expansion valve:

$$\text{Enthalpy after } (h_4) = \text{enthalpy before } (h_3)$$
$$107.9 + x_4 \times 1317.4 = 303.7$$
$$x_4 \times 1317.4 = 195.8$$
$$x_4 = 0.1486 \text{ Ans. (i)}$$

Enthalpy gain of refrigerant through evaporator

$$= h_1 - h_4$$
$$= (107.9 + 0.95 \times 1317.4) - (107.9 + 0.1486 \times 1317.4)$$
$$= (0.95 - 0.1486) \times 1317.4$$
$$= 1055 \text{ kJ/kg}$$

Refrigerating effect per minute

$$= 4 \times 1055 = 4220 \text{ kJ/min} \quad \text{Ans. (ii)}$$

Specific volume of refrigerant entering compressor

$$= 0.95 \times 0.5296 \text{ m}^3\text{/kg}$$

Total volume entering compressor per minute

$$= 4 \times 0.95 \times 0.5296 = 2013 \text{ m}^3\text{/min} \quad \text{Ans. (iii)}$$

18.
$$\left.\begin{array}{l} C + O_2 = CO_2 \\ 2H_2 + O_2 = 2H_2O \\ S + O_2 = SO_2 \end{array}\right\} \text{ Ans. (i)}$$

$$\text{Stoichiometric air} = \frac{100}{23} \times \text{oxygen required}$$
$$= \frac{100}{23}\left(2\tfrac{2}{3}C + 8H_2 + S\right) \text{kg air/kg fuel}$$
$$= \frac{100}{23}(2.667 \times 0.863 + 8 \times 0.128 + 0.009)$$
$$= 14.5 \text{ kg}$$

Actual air is $14.5 \times 1.25 = 18.125$

Mass of the gases/kg fuel burned $= 19.125$ kg Ans. (ii)(a)

$$pV = mRT$$
$$V = \frac{19.125 \times 0.276 \times 643}{1.5 \times 100}$$

Volume of flue gases/kg fuel burned

$$= 22.63 \text{ m}^3 \text{ Ans. (ii)(b)}$$

19. Heat passing across water film:

$$Q = h_i At (T_i - T) \quad \therefore T_i - T = \frac{Q}{h_i At}$$

Heat passing across air film:

$$Q = h_o At (T - T_o) \quad \therefore T - T_o = \frac{Q}{h_o At}$$

$$T_i - T_o = \frac{Q}{At} \left\{ \frac{1}{h_i} + \frac{1}{h_o} \right\}$$

$$100 - 85 = \frac{Q}{\pi \times 0.155 \times 1} \left\{ \frac{1}{240} + \frac{1}{12} \right\}$$

$$Q = \frac{85 \times \pi \times 0.155 \times 1 \times 240}{21}$$

$$= 473.1 \, W \quad \text{Ans.}$$

Note: Area A of large diameter pipe, of small thickness, is mean circumference (per unit length).

20. From steam tables:

$$17 \, bar, \quad h_f = 872 \quad h_{fg} = 1923 \quad v_g = 0.1167$$

$$6 \, bar, \quad h_f = 670 \quad h_{fg} = 2087 \quad v_g = 0.3156$$

Throttling process:

Enthalpy after = Enthalpy before

$$670 + x \times 2087 = 872 + 0.95 \times 1923$$
$$x \times 2087 = 872 + 1827 - 670$$
$$\pi : \times 2087 = 2029$$
$$x = 0.9721 \quad \text{Ans. (i)}$$

Spec. vol. of steam before throttling

$$= 0.95 \times 0.1167 \, m^3/kg$$

Mass of steam passed through

$$= \frac{0.8}{0.95 \times 0.1167} \, kg$$

Spec. vol. of steam after throttling

$$= 0.9721 \times 0.3156 \text{ m}^3/\text{kg}$$

$$\text{Final volume of steam} = \frac{0.8 \times 0.9721 \times 0.3156}{0.95 \times 0.1167}$$

$$= 2.214 \text{m}^3 \text{ Ans. (ii)}$$

21. Total feed in 24 h

$$= 1250 \times 10^{-3} \times 24 = 30 \text{ tonne}$$

Solids in boiler initially		+	Solids put in		=	Solids in finally		
water in	×	Initial	+	Amount	×	Feed	=	Water in	×	Final
boiler		ppm		of feed		ppm		boiler		ppm

$$6 \times 120 + 30 \times \text{feed ppm} = 6 \times 1080$$

$$30 \times \text{feed ppm} = 6480 - 720$$

$$\text{feed ppm} = \frac{5760}{30}$$

$$= 192 \text{ ppm } \text{ Ans.}$$

22. From steam tables:

$$\text{Steam 2.8 bar, } h_f = 551 \text{ } h_{fg} = 2171 \text{ } v_g = 0.6462$$

$$\text{Water } 16°\text{C}, h = 67.1$$

$$\text{Water } 55°\text{C}, h = 230.2$$

$$\text{Let } x = \text{dryness fraction}$$

$$m = \text{mass of steam } [\text{kg}]$$

Total enthalpy of steam water before mixing = Total enthalpy of water after mixing

$$m(551 + x \times 2171) + (36 \times 67 - 1) = (m + 36) \times 230.2$$

$$551m + 2171mx + 2415 = 230.2m + 8287$$

$$320.8m + 2171mx = 5872$$

$$320.8 + 2171x = \frac{5872}{m} \quad \dots \dots \dots \dots \dots \text{ (i)}$$

Also, spec. vol. of steam $= x \times 0.6462 \text{ m}^3/\text{kg}$

∴ mass of 1.5 m³ of wet steam

$$m = \frac{1.5}{x \times 0.6462}\ kg \quad \dots\ \dots\ \dots \quad (ii)$$

Substituting for m from (ii) into (i):

$$320.8 + 2171x = \frac{5872 \times 0.6462x}{1.5}$$
$$320.8 + 2171x = 2528x$$
$$320.8 = 357x$$
$$x = 0.8984 \ \text{Ans.}$$

23.

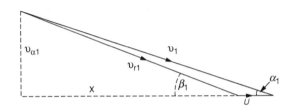

$$u = \pi dn$$
$$120 = \pi \times 0.6 \times n$$
$$n = 63.65 \ \text{rev/s} = 3819.0 \ \text{rev/min} \ \ \text{Ans. (i)}$$
$$v_1 = \sqrt{2 \times 10^3 \times \text{Spec. enthalpy drop}}$$
$$= \sqrt{2 \times 10^3 \times 465} = 974 \ \text{m/s}$$
$$x = v_1 \cos \alpha_1 - u$$
$$= 974 \cos 18° - 120$$
$$= 806.4$$
$$v_{a1} = 974 \sin 18°$$
$$= 301 \ \text{m/s}$$
$$\tan \beta_1 = \frac{301}{974}$$
$$\beta_1 = 17.2°$$
Blade inlet angle = $17.2°$ Ans. (ii)

Axial component of steam at blade inlet = 301 m/s Ans. (iii)

24. Volume of air to be heated every hour

$$= 27 \times 15 \times 3 \times 2 = 2430 \ \text{m}^3$$

Mass of air to be heated every hour:

$$pV = mRT$$

$$m = \frac{pV}{RT}$$

$$= \frac{1.013 \times 10^2 \times 2430}{0.287 \times 291} = 2937 \text{ kg/h}$$

Heat energy supplied = Mass × Spec. heat × Temp. rise

Energy supplied per second

$$= \frac{2937}{3600} \times 1.005 \times (19 - 7)$$

$$= 9837 \text{ kJ/s}$$

Kilowatt loading = 9.837 kW Ans.

25. For a four-stroke, single-acting engine:

$$n = \text{rev/s} \div 2 = 2.5$$

For six cylinders:

$$ip = p_m LАn \times 6$$

$$= 17 \times 10^2 \times 0.7854 \times 0.225^2 \times 0.2 \times 2.5x \times 6$$

$$= 202.8 \text{kW} \text{Ans. (i)}$$

$$bp = ip \times \text{mech. efficiency}$$

$$= 202.8 \times 0.85$$

$$= 172.36 \text{ kW Ans.(ii)}$$

26. Refer to Figure 9.4a   Ans. (i):

$$\text{Volume of receiver} = 3 \text{ m}^3$$

$$\text{Volume of free air} = 3 \times \frac{7}{1} = 21 \text{m}^3$$

(at the same temperature)

$$\text{Volume to compress} = 21 - 3 = 18 \text{ m}^3 \text{ at } 25°C$$

$$\frac{V_1}{T_1} = \frac{V_2}{T_2}$$

$$\frac{V_1}{290} = \frac{18}{298}$$

$$V_1 = 17.52 \text{ m}^3$$

$$\text{Volume to compress} = 17.52 \text{ m}^3 \text{ at } 17°C$$

$$\text{Suction volume/s} = 0.7854 \times 0.1^2 \times 0.2 \times 5$$

$$= 0.007854 \text{ m}^3 /s$$

$$\text{Time to charge receiver} = \frac{17.52}{0.007854 \times 60}$$

$$= 37.18 \text{ min} \quad \text{Ans. (ii)}$$

27. Refer to Figures 14.2 and 14.3 Ans. (i):

$$h_3 = h_4 = 294.1 \text{ kJ/kg}$$

i.e. from tables, page 12, at 24°C

$$h_4 = h_f + x_4 \left(h_g - h_f \right)$$
$$294.1 = 126.2 + x_4 \left(1430.5 - 126.2 \right)$$

i.e. from tables, page 12, at 2.68 bar

$$x_4 = 0.1287$$
$$\text{Heat extracted} = h_1 - h_4$$
$$= h_g - x_4 h_{fg}$$
$$= 1430.5 - 0.1287 \left(1430.5 - 126.2 \right)$$
$$= 1262.2 \text{ kJ/kg}$$
$$3.5 = m \times 1262.6$$
$$m = 0.00308 \text{ kg/s}$$

$$\text{Mass flows rate of ammonia} = 0.00308 \text{ kg/s} \quad \text{Ans. (ii)}$$

28. From steam tables: 15 bar, $h_g = 2792$

Mass of flue gases per kg of fuel

$$= 1 \text{ kg fuel} + 24 \text{ kg air} = 25 \text{ kg}$$

Heat energy transferred from flue gases per hour

$$= \text{Mass} \times \text{Spec. heat} \times \text{Temp. change}$$
$$= 25 \times 750 \times 1.007 \times (822 - 690)$$
$$= 2.492 \times 10^6 \text{ kJ/h}$$

Let h = spec. enthalpy of the superheated steam,

Heat energy transferred to steam per hour to superheat it

$$= 7000 \times (h - 2792)$$

Heat gained by steam = Heat lost by gases
$$= 7000 \times (h - 2792) = 2.492 \times 10^6$$
$$h - 2792 = \frac{2.492 \times 10^6}{7000}$$
$$h = 356 + 2792 = 3148 \text{ kJ/kg}$$

From superheated steam tables: h for 15 bar 350°C reads 3148

∴ Temperature of steam = 350°C Ans. (i)

From steam tables: water 21°C, h = 88.0

Enthalpy of 1 kg sup. steam and m kg water entering de-superheater	=	Enthalpy of $(1 + m)$ kg of dry sat. steam leaving de-superheater

$$(1 \times 3148) + m \times 88 = (1 + m) \times 2792$$
$$3148 + 88m = 792 + 2792m$$
$$356 = 2704m$$
$$m = 0.4316 \text{ kg Ans. (ii)}$$

29. Ratio of expansion $= \dfrac{\text{Final volume}}{\text{Initial volume}}$

∴ Final volume $= 6 \times 9900 = 59400 \text{ cm}^3$

or 59.4 l or 0.0594 m³ Ans. (i)
$$p_1 V_1^{1.33} = p_2 V_2^{1.33}$$

where V_1 and V_2 may be represented by 1 and 6, respectively.

$$12 \times 1^{1.33} = p_2 \times 6^{1.33}$$

$$p_2 = \frac{12}{6^{1.33}} = 1.108 \text{ bar} \quad \text{Ans. (ii)}$$

30.

$$\frac{p_1 V_1}{T_1} = \frac{p_2 V_2}{T_2}$$

$$\frac{12 \times 1}{489} = \frac{1.108 \times 6}{T_2}$$

$$T_2 = \frac{489 \times 1.1018 \times 6}{12} = 270.9 \text{ K}$$

$$= -2.1^\circ\text{C} \quad \text{Ans. (iii)}$$

$$P_1 v_1^{\,n} = P_2 v_2^{\,n}$$

$$40 \times 0.08^{1.3} = 2 \times v_2^{\,1.3}$$

$$v_2 = 0.8013$$

$$0.8856 x = 0 - 8013$$

$$x = 0.9048$$

$$\text{Work transfer} = \frac{p_1 v_1 - p_2 v_2}{n-1}$$

$$= \frac{40 \times 10^2 \times 0.08 - 2 \times 10^2 \times 0.8013}{0.3}$$

$$= 532.5 \text{ kJ} \quad \text{Ans. (a)}$$

$$U_1 + q_{in} = u_2 + w_{out}$$

At 40 bar, 0.08 m³/kg, steam is 450°C (tables)

$$3010 + q_{in} = (505 + 0 - 9048 \times 2025) + 532.5$$

$$q_{in} = -3010 + 2337.2 + 532.5$$

Heat transfer $= -140 - 3$ kJ i.e. loss Ans. (b)

Note: This is non-flow work, u not h is used.

$$u_1 = h - p v_g$$

$$= 3330 - 40 \times 10^2 \times 0.08$$

$$= 2337.2 \text{ (as above)}$$

Similarly for u_2, by calculation

31. Referring to Figure 12.7:

$$V_{a1} = V_1 \sin \alpha_1 = 840 \times \sin 20° = 287.4 \text{ m/s}$$
$$V_{wi} = V_1 \cos \alpha_1 = 840 \times \cos 20° = 789.4 \text{ m/s}$$
$$x = V_{w1} - u = 789.4 - 350 = 439.4 \text{ m/s}$$
$$\tan \beta_1 = \frac{V_{a1}}{x} = \frac{287.4}{439.4} = 0.6539$$

i.e. Blade inlet angle $= 33°11'$ Ans. (i)

$$V_{r1} = \frac{V_{a1}}{\sin \beta_1} = \frac{287.4}{\sin 33°11'} = 525 \text{m/s}$$
$$V_{r2} = 0.8 \times 525 = 420 \text{ m/s}$$
$$V_{a2} = V_{r2} \times \sin \beta_2 = 420 \times \sin 25° \ 12' = 178.8 \text{ m/s}$$
$$V_{w2} + u = V_{r2} \times \cos \beta_2 = 420 \times \cos 25° \ 12' = 380 \text{ m/s}$$
$$V_{w2} = 80 - 350 = 30 \text{ m/s}$$
$$\tan \varphi = \frac{V_{w2}}{V_{a2}} = \frac{30}{178.8} = 0.1678$$
$$\varphi = 9°32'$$
$$V_2 = \frac{V_{a2}}{\cos \varphi} = \frac{178.8}{\cos 9°32'} = 181.3 \text{ m/s}$$

Absolute velocity of steam at exit:

$$\left. \begin{array}{l} \text{Magnitude} = 181.3 \text{m/s} \\ \text{Direction to axis} = 9°32' \\ \text{or direction to plane of wheel} = 80° \ 28' \end{array} \right\} \text{ Ans. (ii)}$$

32. Available hydrogen $= H_2 - \dfrac{O_2}{8}$

$$= 0.114 - \frac{0.024}{8} = 0.111 \text{ kg}$$

$$\text{c.v.} = 33.7C + 144\left[H_2 - \frac{O_2}{8} \right] + 9.3S$$
$$= 33.7 \times 0.846 + 144 \times 0.111 + 9.3 \times 0.004$$
$$= 28.51 + 15.99 + 0.0372$$
$$= 44.5372 \text{ MJ/kg Ans. (i)}$$

$$\text{Air required} = \frac{100}{23} \times \text{Oxygen required}$$

$$= \frac{100}{23} \left\{ 2\frac{2}{3}\, C + 8 \left[H_2 - \frac{O_2}{8} \right] + S \right\}$$

$$= \frac{100}{23} \left\{ 2\frac{2}{3} \times 0.846 + 8 \times 0.111 + 0.004 \right\}$$

$$= \frac{100}{23} \times 3.148$$

$$= 13.69 \text{ kg air/kg fuel} \quad \text{Ans. (ii)}$$

33. From tables, Freon-12, 1.826 bar:

$$h_f = 22.33 \quad h_g = 180.97$$
$$h_{fg} = 180.97 - 22.33 = 158.64$$

Sat. temp. at 1.826 bar = −15°C, therefore freon at evaporator outlet at 0°C is superheated by 15°.

1.826 bar 15° superheat, $h = 190.15$

Referring to Figure 14.2:

Throttling effect through expansion valve between condenser outlet and evaporator inlet:

$$\text{Enthalpy after throttling} \left(h_4 \right) = \text{Enthalpy before} \left(h_3 \right)$$
$$22.33 + x_4 \times 158.64 = 50$$
$$x_4 \times 158.64 = 27.67$$
$$x_4 = 0.1744 \quad \text{Ans. (i)}$$

$$\text{Refrig. effect/kg} = h_1 - h_4 \,(\text{note } h_4 = h_3)$$
$$= 190.15 - 50$$
$$= 140 - 15 \text{ kJ/kg}$$

$$\text{Refrigerating effect per minute}$$
$$= 0.4 \times 60 \times 140.15$$
$$= 3364 \text{ kJ/min} \quad \text{Ans. (ii)}$$

34.

$$\frac{p_1 V_1}{T_1} = \frac{p_2 V_2}{T_2}$$

$$\frac{1.01325 \times 18.2}{288} = \frac{0.965 \times V_2}{300}$$

$$V_2 = 19.91 \text{ m}^3/\text{min}$$

$$p_2 V_2{}^n = p_3 V_3{}^n$$

$$0.965 \times 19.91^{1.32} = 4.82 \times V_3{}^{1.32}$$

$$V_3 = 5.891 \text{ m}^3/\text{min}$$

$$\text{Work/cycle} = \frac{n}{n-1}(p_3 V_3 - p_2 V_2)$$

$$\text{Work/s} = \frac{1.32 \times 10^2}{0.32}\left(\frac{482 \times 5.891}{60 \times 2} - \frac{0.965 \times 19.91}{60 \times 2}\right)$$

$$= 31.56 \text{ kW}$$

$$\text{Input power} = \frac{31.56}{0.9}$$

$$= 35.06 \text{ kW} \quad \text{Ans.}$$

35. For 1 mm movement of the stylus (1 mm on height of card) indicator piston deflection = 1/6 mm and this would be under a force of $60 \div 6 = 10$ N in the indicator cylinder.

$$\text{Force [N]} = \text{Pressure [N/m}^2\text{]} \times \text{Area [m}^2\text{]}$$

$$10 = \text{Pressure [N/m}^2\text{]} \times 0.7854 \times 7^2 x \times 10^{-4}$$

∴ pressure scale on card per mm of height

$$= \frac{10}{0.7854 \times 7^2 \times 10^{-6}}$$

$$= 2.598 \times 10^5 \text{ N/m}^2 = 2.598 \text{ bar}$$

$$\text{Mean height of diagram [mm]} = \frac{\text{Area [mm}^2\text{]}}{\text{Length [mm]}}$$

$$= \frac{346}{75} \text{ mm}$$

$$\therefore \text{Mean effective press.} = \frac{346}{75} \times 2.598$$

$$= 11.99 \text{ bar} \quad \text{Ans. (i)}$$

$$\text{I.p.} = p_m LAn$$

where $n = 2 \times$ rev/s for a double-acting two-stroke

$$\text{I.p.} = 11.99 \times 10^2 \times 0.7854 \times 0.6^2 \times 0.9 \times 2.1 \times 2$$
$$= 1281\,\text{kW} \ \ \text{Ans. (ii)}$$

36.
$$\text{Cross-sect. area} \left[\text{m}^2\right] = \frac{\text{Volume flow}\left[\text{m}^3/\text{s}\right]}{\text{Velocity}\left[\text{m/s}\right]}$$

$$= \frac{\text{Mass flow}\left[\text{kg/s}\right] \times \text{Spec. vol.}\left[\text{m}^3/\text{kg}\right]}{\text{velocity}\left[\text{m/s}\right]}$$

$$\text{Diameter} = \frac{\sqrt{\text{Area}}}{0.7854}$$

$$\text{Entrance diameter} = \frac{\sqrt{0.315 \times 0.2765}}{457 \times 0.7854}$$
$$= 0.01558\,\text{m} = 15.58\,\text{mm} \ \ \text{Ans. (ii)}$$

37. $pV = mRT$

$$m = \frac{pV}{RT} = \frac{1.1 \times 10^2 \times 0.2}{0.287 \times 288} = 0.2662\,\text{kg}$$

Heat energy supplied [kJ]

$$= \text{Mass}\left[\text{kg}\right] \times \text{Spec. heat}\left[\text{kJ/kgK}\right] \times \text{Temp. rise}\left[\text{K}\right]$$
$$= 0.2662 \times 1.005 \times (423 - 288)$$
$$= 36.1\,\text{kJ} \ \ \text{Ans. (i)}$$

$$P_1 V_1^{1.32} = P_2 V_2^{1.32}$$
$$V_2 = 0.2 \times \frac{\sqrt[1.32]{1.1}}{7.15}$$
$$= 0\ 04844\,\text{m}^3$$
$$\frac{p_1 V_1}{T_1} = \frac{p_2 V_2}{T_2}$$
$$\frac{1.1 \times 0.2}{288} = \frac{7.15 \times 0.04844}{T_2}$$
$$T_2 = \frac{288 \times 7.15 \times 0.04844}{1.1 \times 0.2} = 453.4\,\text{K}$$
$$= 180.4\text{°C} \ \ \text{Ans. (ii)}$$

38. For the two thicknesses:

$$T_1 - T_3 = \frac{Q}{At}\left\{\frac{S_1}{k_1} + \frac{S_2}{k_2}\right\}$$

$$-4 - 21 = \frac{Q}{6\times3\times1}\left\{\frac{0.12}{1.15} + \frac{0.08}{0.043}\right\}$$

$$Q = \frac{-25\times18}{0.1043+1.860}$$

$$= -229.09 \text{ W, i.e. into the store}$$

$$Q = 229.09\times3600\times24 \text{ J/day}$$

$$= 19.793 \text{ MJ/day Ans. (i)}$$

$$T_1 - T_2 = \frac{-2229.09}{6\times3\times1}\left\{\frac{0.08}{0.043}\right\}$$

$$-4 - T_2 = -23.67$$

$$T_2 = 19.67°C \text{ Ans. (ii)}$$

39. Refer to Figure 9.2 Ans. (i):

$$\text{Clearance volume} = 007\times381$$

$$= 26.67 \text{ mm}$$

$$V_1 = 381+26.67$$

$$= 407.67 \text{ mm}$$

$$V_2 = 407.67-267$$

$$= 140.67 \text{ mm}$$

$$P_1V_1^n = P_2V_2^n$$

$$1.013\times407.67^n = 4\times0.1407^n$$

$$2.898^n = 3.949$$

$$n = 1.291 \text{ Ans. (ii)}$$

40. Stoichiometric air required per kg of fuel A

$$= \frac{100}{23}\{2\tfrac{2}{3}\times0.885 + 8\times0.115\}$$

$$= \frac{100}{23}\times3.28 \text{ kg}$$

Stoichiometric air required per kg of fuel B

$$= \frac{100}{23}\{2\tfrac{2}{3}C + 8H_2\}$$

and this is 6% more than for fuel A, therefore,

$$= \frac{100}{23}\left\{2\tfrac{2}{3}C + 8H_2\right\} = 1.06 \times \frac{100}{23} \times 3.28$$

Cancelling $\dfrac{100}{23}$ and multiplying throughout by $\tfrac{3}{8}$:

$$C + 3H_2 = 1.304 \text{ (i)}$$

Also, fractional analysis of fuel B:

$$C + H_2 = 1$$
$$\text{i.e. } C = 1 - H_2 \tag{ii}$$

Substituting value of C from (ii) into (i):

$$C + 3H_2 = 1.304$$
$$1 - H_2 + 3H_2 = 1.304$$
$$2H_2 = 0.304$$
$$H_2 = 0.152$$
$$\text{and, } C = -0.152 = 0.848$$

Mass analysis of fuel B = 84.8% carbon, 15.2% hydrogen Ans.

41. Before alteration:

$$V_1 = 90 + 15 = 150$$
$$V_2 = 15$$
$$P_1 V_1^{1.33} = P_2 V_2^{1.33}$$
$$1 \times 105^{1.33} = p_2 \times 15^{1.33}$$
$$p_2 = \left\{\frac{100}{15}\right\}^{1.33} = 13.3 \text{ bar Ans. (i)}$$

42.

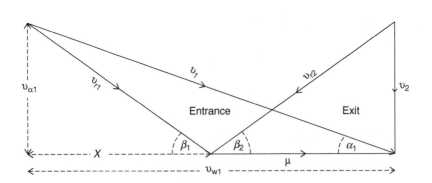

$$\text{Kinetic energy} = \tfrac{1}{2}\,mv^2$$

Kinetic energy being in joules when m is the mass in kg and v is the velocity in m/s, hence,

$$250 \times 10^3 = \tfrac{1}{2} \times 1 \times v^2$$
$$v = \sqrt{2 \times 250 \times 10^3} = 707.1 \text{m/s}$$

With no friction, $v_{r2} = v_{r1}$

and, since $\beta_2 = \beta_1$

then $x = 8 = \tfrac{1}{2} v_{w1}$

$$\frac{V_{a1}}{x} = \tan \beta_1 \quad \therefore \quad \frac{v_1 \sin \alpha_1}{\tfrac{1}{2} v_1 \cos \alpha_1} = \tan 35°$$

V_1 cancels, $\sin \alpha_{1+} \cos \alpha_1 = \tan \alpha_1$, therefore

$$2 \tan \alpha_1 = \tan 35°$$

$$\tan \alpha_1 = \frac{0.7002}{2} = 0.3501$$

Nozzle angle $\alpha_1 = 19° \ 18'$ Ans. (i)

$$\text{Blade velocity } u = \tfrac{1}{2} v_{w1} = \tfrac{1}{2} v_1 \cos \alpha_1$$
$$= \tfrac{1}{2} \times 7071 \times \cos 19° \ 18'$$
$$= 333.7 \text{m/s} \quad \text{Ans. (ii)}$$

43. Let d = diameter of cylinder

$$\text{then } 1.25d = \text{stroke}$$

For 4 cylinders:

$$\text{I.p.} = p_m LAn \times 4$$
$$600 = 12.56 \times 10^2 \times 0.7854 d^2 \times 1.25 d \times 4.5 \times 4$$
$$d = \sqrt[3]{\frac{600}{1256 \times 0.7854 \times 1.25 \times 4.5 \times 4}}$$
$$= 0.3\text{m} = 300 \text{ mm}$$

$$\left.\begin{array}{l} \text{Cylinder diameter} = 300 \text{ mm} \\ \text{Stroke} = 1.25 \times 300 = 375 \text{ mm} \end{array}\right\} \text{Ans. (i)}$$

Indicated thermal efficiency

$$= \frac{3.6 \, [\text{MJ/kWh}]}{\text{kg fuel/ind.} \, [\text{kWh}] \times \text{c.v.} \, [\text{MJ/kg}]}$$

$$= \frac{3.6}{0.225 \times 42}$$

$$= 0.381 \quad \text{or} \quad 38.1\% \quad \text{Ans. (ii)}$$

44. From NH_3 tables:

$$1.902 \text{ bar, } h_g = 1420$$

$$7.529 \text{ bar, } h_f = 256, \text{ sat. temp.} = 16°C$$

$\therefore$ at 66°C vapour is superheated 50°

$$7.529 \text{ bar } 50° \text{ superheat, } h = 1591.7$$

Referring to Figure 14.2:

Enthalpy gain per kg through evaporator

$$= h_1 - h_4 \, (\text{note } h_4 = h_3)$$

$$= 1420 - 256 = 1164 \text{ kJ/kg}$$

Refrigerating effect $[\text{kJ/min}]$

$$= \text{mass flow} \, [\text{kg/min}] \times \text{enthalpy gain} \, [\text{kJ/kg}]$$

$$\therefore m = \frac{800}{1164} = 0.6873 \text{ kg/min Ans.}$$

Enthalpy drop per kg through condenser

$$= h_2 - h_3$$

$$= 1591.7 - 256 = 1335.7 \text{ kJ/kg}$$

Heat rejected in condenser

$$= 0.6873 \times 1335.7 = 918.1 \text{ kJ/min} \quad \text{Ans. (ii)}$$

Enthalpy gain per kg in compressor

$$= h_2 - h_1$$

$$= 1591.7 - 1420 = 171.7 \text{ kJ/kg}$$

Energy given to refrigerant in compressor

$$= 0.6873 \times 171.7 \text{ kJ/min}$$

Power [kW = kJ/s]

$$\frac{0.6873 \times 171.7}{60}$$

$$= 1.967 \text{ kW Ans. (iii)}$$

45. From saturated steam tables:

Steam 30 bar, $h_g = 2803$ $v_g = 0.06665$

Water 38°C, $h = 159 - 1$

From superheated steam tables:

Steam 30 bar 400°C, $h = 3231$ $v = 0.0993$

Let m [kg] = mass of injection water per kg of superheated steam.

Mixing in de-superheater:

Enthalpy before mixing = Enthalpy after

$$1 \times 3231 + m \times 159 - 1 = (1 + m) \times 2803$$

$$3231 + 159 - 1m = 2803 + 2803m$$

$$428 = 2643.9m$$

$$m = 0.1618 \text{ kg Ans. (i)}$$

Volume of 1 kg superheated steam $= 0.0993 \text{ m}^3$

Volume of $(1 + m)$ kg of dry saturated steam

$$= 1.1618 \times 0.06665 = 0.07743 \text{ m}^3$$

Percentage change in volume

$$= \frac{0.0993 - 0.07743}{0.0993} \times 100$$

$$= 22.03\% \text{ Ans. (ii)}$$

46. Let V_1 = volume of gas in cylinder at beginning of stroke, this is stroke volume + clearance volume.

V_2 = volume of gas in cylinder at end of stroke, this is the clearance volume.

$$p_1 V_1^{1.32} = p_2 V_2^{1.32}$$

$$1 \times V_1^{1.32} = 37 \times 850^{1.32}$$

$$V_1 = 850 \times \sqrt[1.32]{37} = 13100 \text{ cm}^3$$

Stroke volume $= 13100 - 850 = 12250 \text{ cm}^3$

$$12.25 \text{ l or } 0.01225 \text{ m}^3 \text{ Ans. (i)}$$

$$\frac{p_1 V_1}{T_1} = \frac{p_2 V_2}{T_2}$$

$$\frac{1 \times 13100}{308} = \frac{37 \times 850}{T_2}$$

$$T_2 = \frac{308 \times 37 \times 850}{13100} = 739.4 \text{ K}$$

$$= 466.4°C \text{ Ans.}$$

47.

Let x be kg of C/kg fuel

$1 - x$ is kg of H_2/kg fuel

Oxygen required for $C = 2.667x$

Oxygen required for $H_2 = 8(1 - x)$

$$\text{Stoichiometric air} = \frac{100}{23}(2 - 667x + 8 - 8x)$$

Exhaust gas includes 1 kg of fuel burned.

$$15.6 - 1 = \frac{100}{23}(2.667x + 8.8x)$$

$$3.358 - 8 = -5.333x$$

$$x = 0.8704$$

$$1 - x = 0.1296$$

$$\left.\begin{array}{l} \text{Mass of carbon} = 0.8704 \text{kg} \\ \text{Mass of hydrogen} = 0.1296 \text{kg} \end{array}\right\} \text{ Ans.}$$

48.　　　　Heat extracted $= 27.2 (4.186 \times 144 + 332.6)$ kJ/h

$$= \frac{10686.3 \times 10^3}{3600} \text{ J/s}$$

$$= 2968.4$$

$$\text{c.o.p.} = \frac{\text{heat extracted by refrigerant}}{\text{work done on refrigerant}}$$

$$\text{Work done on refrigerant} = \frac{2968.4}{7.51}$$

$$= 394 \text{ W}$$

Power input to machine $= 395$ W Ans. (i)

For a heat engine, heat is taken in and heat is given out (at a lower temperature) and overall there is a net work output. Theoretically the cycle can be operated in reverse (reversibility), that is, heat taken in at a lower temperature and given out at a high temperature but it does require a net work input (heat pump or refrigerator). Ans. (ii)

49.

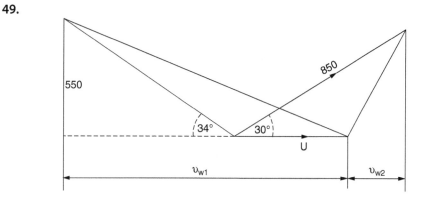

$$V_{w1} + V_{w2} = \frac{550}{\tan 34°} + 850 \cos 30°$$

$$= 8154 + 736.1$$

$$V_w = 1551.5 \text{ m/s}$$

$$\text{Power} = \text{force} \times \text{velocity}$$

$$= mv_w u$$

$$1000 \times 67 = 0.0833 \times 1551.5 \times \pi \times 1.32 \times n$$

$$n = 124.9 \text{ rev/s}$$

Turbine rotational speed $= 124.9$ rev/s Ans.

50.

$$\text{Area of top} = 1 \times 0.75$$
$$\text{Area of sides} = 1 \times 1.6 \times 2$$
$$\text{Area of ends} = 1 - 6 \times 0.5 \times 2$$
$$\text{Area exposed to heat source} = 0.75 + 3.2 + 2.4$$
$$= 6.35 \text{ m}^2$$

Temperature fall across the lagging:

$$T_i - T = \frac{QS}{Atk}$$

Heat passing across air film:

$$Q = h_o At(T - T_o) \therefore T - T_o = \frac{Q}{h_o At}$$

$$T_i - T_o = \frac{Q}{AT}\left\{\frac{S}{0.048} + \frac{1}{1}\right\}$$

$$66 - 18 = \frac{200}{6.35 \times 1}\left\{\frac{S}{0.048} + 1\right\}$$

$$S = 0.048\left\{\frac{48 \times 6.38}{200} - 1\right\}$$

$$= 0.02515\,m$$

$$= 25.15\,mm\ \ Ans.\ (i)$$

$$S' = 0.048\left\{\frac{48 \times 6.35}{100} - 1\right\}$$

$$= 0.0983\,m$$

$$= 98.3\,mm$$

$$S' - S = 73.15\,mm\ \ Ans.\ (ii)$$

SELECTION OF EXAMINATION QUESTIONS – HIGHER LEVEL

1. A simple open-cycle gas turbine plant operates at a pressure ratio of 5:1 and at an air/fuel ratio of 50:1. The power output of the plant is 5 MW with maximum and minimum temperatures in the cycle of 925 K and 300 K, respectively. The isentropic efficiencies of the compressor and turbine are 0.82 and 0.86, respectively. Calculate:
   (i)   the temperature of the exhaust gases leaving the turbine;
   (ii) the mass of fuel used per second.

   Note: For air: $c_p = 1005$ J/kg K, $\dfrac{c_p}{c_v} = 1.4$.

   For the turbine gases: $c_p = 1150$ J/kg K, $\dfrac{c^p}{c_v} = 1.3$.

2. A vessel of volume 15 m³ contains air and dry saturated steam at a total pressure of 0.09 bar and temperature 39°C. Taking R for air = 0.287 kJ/kg K, calculate the masses of (i) steam and (ii) air in the vessel, and (iii) give the mass ratio of air to steam.

3. An air compressor is to compress 0.4 kg/s of air from 100 kN/m² and 20°C to 1500 kN/m² and 40°C. The air inlet velocity is negligible and the exit velocity is 100 m/s. The cooling water for the compressor has a mass flow rate of 0.1 kg/s and an inlet temperature of 20°C. Power input to the compressor is 13 kW. Calculate the cooling water outlet temperature.

   Note: cp air = 1 005 kJ/kg K, c_p water = 418 kJ/kg K.

4. The mass analysis of the fuel burned in a boiler is 87% carbon, 11% hydrogen and 2% oxygen, and the fuel is burned at the rate of 1.8 tonne/h. Calculate the mass flow rate [kg/s] of each of the constituents of the flue gases if combustion is stoichiometric. Express the composition of the flue gases as percentages by mass.

Atomic weights: hydrogen 1, carbon 12, nitrogen 14, oxygen 16. Mass composition of air: 23% oxygen, 77% nitrogen.

5. Dry saturated steam enters a convergent–divergent nozzle at 9 bar, and the pressures at the throat and exit are 5 bar and 0.14 bar, respectively. The specific enthalpy drop of the steam from entrance to throat is 107 kJ/kg, and from entrance to exit it is 633 kJ/kg. Assuming 8% of the enthalpy drop is lost to friction in the divergent part of the nozzle, calculate the areas in mm² of the nozzle at the throat and exit to pass 23 kg of steam per minute.

6. In an R134a refrigerating plant the compressor takes the refrigerant in at 84.35 kPa and discharges it at 770 kPa and 40°C. At condenser outlet the refrigerant is saturated liquid at 770 kPa. If compression is isentropic and the flow of the refrigerant is 15 kg/min, calculate the refrigerating effect and the coefficient of performance.

7. (i) A quantity of a perfect gas is compressed at constant temperature from initial pressure p_1 to final pressure p_2. Show that the area under the pV curve for this process can be written as:

$$RT \ln \left(\frac{p_1}{p_2} \right)$$

Note: Area under p–V curve is given by $\int p dV$.

(ii) An isothermal compression from 1 bar to 8 bar takes place on a perfect gas. The initial volume of the gas is 0.25 m³. Calculate the heat energy transfer during the process.

8. The compressor of an open-cycle gas turbine unit receives air at 1 bar, 18°C and delivers it at 4 bar, 200°C to the combustion chamber where the temperature is raised at constant pressure to 650°C. The products of combustion pass through the turbine, which has an isentropic efficiency of 0.85, and exhaust at 1 bar. The power required by the compressor is provided by the turbine. Calculate:

(i) the isentropic efficiency of the compressor;

(ii) the net work output of the turbine per kilogram of air.

Note: For air $c_p = 1\ 005$ kJ/kg K, $\gamma = 1.4$

For combustion gases $c_p = 1.15$ kJ/kg K, $\gamma = 1.333$

9. In a steam turbine plant steam expands in the high pressure turbine from 50 bar, 500°C to 6 bar with an isentropic efficiency of 0.9. The steam is then reheated at

constant pressure to 500°C before entering the low pressure turbine. In the low pressure turbine steam is expanded to 0.05 bar with an isentropic efficiency of 0.85. Neglecting feed pump work,

(i) sketch the expansion and reheat processes on a temperature–entropy diagram;

(ii) calculate using chart and tables the thermal efficiency of the plant.

10. The walls of a cold room consist of an outer layer of wood of thickness 30 mm and thermal conductivity 0.18 W/m K, and a cork lining of thickness 70 mm and thermal conductivity 0.05 W/m K. If the surface heat transfer coefficient from and to each exposed surface is 10 W/m² K and the heat flow through the wall is 24 W/m², calculate (i) the temperature differences across the thicknesses of the wood and cork, (ii) the total temperature difference between the outside atmosphere and inside of room and (iii) the temperature of the room when the external ambient temperature is 20°C.

11. A surface-type feed heater is supplied with steam at 2 bar and 0.95 dry. The temperature of the feed water entering and leaving the heater are 55°C and 105°C, respectively. The mass flow rate of feed water through the heater is 20 000 kg/h and the overall heat transfer coefficient is 4540 W/m² K. Calculate, stating any assumptions made:

(i) the mass flow rate of heating steam in kg/hour;

(ii) the effective heating surface area of the feed heater.

For the feed water: Specific heat capacity = 418 kJ/kg K.

Note: Logarithmic mean temperature difference

$$\theta_m = \frac{\theta_1 - \theta_2}{\ln\left(\dfrac{\theta_1}{\theta_2}\right)}$$

where θ_1 = temperature difference between hot and cold fluid in inlet,

θ_2 = temperature difference between hot and cold fluid at outlet.

12. In a test on a single-cylinder, two-stroke, diesel engine, the mean effective pressure was 8.9 bar, running speed 2.3 rev/s, brake load 8 kN, brake radius 1.25 m, specific fuel consumption 0.251 kg/kWh (brake). The diameter of the cylinder is 360 mm, stroke 780 mm, calorific value of the fuel 41.5 MJ/kg. Calculate (i) the indicated power, (ii) brake power, (iii) indicated thermal efficiency and (iv) the total heat energy loss per second.

13. The volumetric analysis of a mixture of gases shows it to contain 80% hydrogen and 20% oxygen. A vessel holds 0.7 m³ of this mixture at 38°C and 3.5 bar. Calculate the masses of hydrogen and oxygen in the vessel.

Universal Gas Constant = 8·3143 kJ/mol K.

Atomic mass relationships: hydrogen = 1, oxygen =16.

14. In a single-stage impulse turbine the steam enters the nozzle at 7 bar, 300°C and is discharged at 1.2 bar dry saturated, directed at 20° to the plane of rotation. The blade velocity is 40% of the steam jet velocity and the relative velocity of the steam at exit is 80% of the relative velocity at entrance. The outlet angle of the blades is 35° and the steam flow 0.5 kg/s. Calculate (i) the axial thrust and (ii) the power developed.

15. Air expands isentropically through a convergent nozzle from 6 bar 260°C, to 4 bar. The velocity of the air at nozzle inlet is 900 m/s and the nozzle cross-sectional area at exit is 0.025 m². Calculate:
    (i)   the air velocity at nozzle exit;
    (ii) the mass flow rate of air;
    (iii) the nozzle cross-sectional area at inlet.

For air: c_p = 1005 J/kg K, $\dfrac{c_p}{c_v}$ = 1.4.

16. The pressure, volume and temperature of a gas mixture sample in a closed vessel is 101 bar, 500 cm³ and 20°C, respectively, and is composed of 14% carbon dioxide and 86% nitrogen by volume. Taking R for CO_2 = 0.189 kJ/kg K, calculate (i) the partial pressure of each gas and (ii) the mass of carbon dioxide in the sample.

17. A perfect gas is compressed, polytropically from 1 bar, 22°C, 0.037 m³ to 35 bar, 420°C. Determine:
    (i)   the index of compression;
    (ii) the work done;
    (iii) the change of internal energy;
    (iv) the heat transfer.

For the gas c_v = 718 J/kg K, R = 282 J/kg K.

18. In a Freon-12 refrigeration plant the refrigerant leaves the condenser as a liquid at 25°C. The refrigerant leaves the evaporator as dry saturated vapour at −15°C and leaves the compressor at 6.516 bar and 40°C. The cooling load is 73.3 kW. Calculate:
    (i)   the mass flow rate of refrigerant;
    (ii) the compressor power;
    (iii) the coefficient of performance of the plant.

19. In a compression–ignition engine working on the ideal dual-combustion cycle, the volumetric compression ratio is 12.5:1.

The cycle consists of (a) adiabatic compression from 1.013 bar, 35°C, (b) heat received at constant volume to a maximum pressure of 40 bar, (c) heat received at constant pressure to a maximum temperature of 1425°C, (d) adiabatic expansion to the initial volume, (e) heat rejected at constant volume. Make a sketch of the pV diagram and calculate the mean effective pressure. Take $\gamma = 1.4$ for air and products of combustion.

20. A rigid vessel contains a mixture of superheated steam and air at a total pressure of 0.02 bar and at 30°C. The steam/air mass ratio is 20:1. Assume the superheated steam has the properties of a perfect gas. Calculate:

(i) the partial pressures of steam and air;

(ii) the density of the mixture.

Note: For air $R = 287$ J/kg K. For superheated steam $R = 462$ J/kg K.

21. The mass analysis of a fuel burned in a boiler is 85.5% carbon, 13.5% hydrogen and 1% oxygen, and the air supply is 25% in excess of the minimum required for stoichiometric combustion. Calculate (i) the percentage mass analysis of the wet flue gases, (ii) the percentage volumetric analysis of the dry flue gases. Take mass composition of air = 23% oxygen, 77% nitrogen. Atomic weights: hydrogen 1, carbon 12, nitrogen 14, oxygen 16.

22. The wall of a cold room is composed of two materials, an inner material of thickness 100 mm, having a thermal conductivity of 0.115 W/m K and a layer of cork of thermal conductivity 0.06 W/m K. The external air temperature is 24°C and the room air temperature is –23°C. The surface heat transfer coefficient of exposed surfaces is 12 W/m² K. The heat transfer through the wall is 30 W/m². Calculate:

(i) the temperature of the exposed surfaces;

(ii) the temperature of the interface;

(iii) the thickness of the cork.

23. In a refrigerating plant using freon as the refrigerating agent, the refrigerant leaves the condenser as saturated liquid at 15°C, leaves the evaporator and enters the compressor at 1.509 bar and –5°C, and is delivered from the compressor into the condenser at 4.914 bar, 45°C. Calculate the coefficient of performance.

24. The gravimetric analysis of a liquid fuel is: carbon 82%, hydrogen 18%. Assume stoichiometric combustion. Calculate:

(i) the air/fuel ratio by mass;

(ii) the volumetric analysis of the wet products of combustion.

Air contains 23% oxygen by mass.

Atomic mass relationships: oxygen = 16, nitrogen = 14, carbon = 12, hydrogen = 1.

25. Steam expands in a turbine from 25 bar 320°C to 0.04 bar with an isentropic efficiency of 0.73. The power output of the turbine is 3 MW. The system is to be modified by fitting a new boiler, generating steam at 60 bar, 370°C, which supplies a new higher pressure turbine exhausting to the original turbine at 25 bar. The high pressure turbine has an isentropic efficiency of 0.76. Between the turbines the steam is reheated to 320°C at constant pressure. Using tables and chart as required calculate:

   (i)   the enthalpy change of the steam during reheating;

   (ii) the condition of the steam at condenser inlet;

   (iii) the percentage reduction in steam flow due to the plant modification when total power output is unchanged.

26. The power absorbed by a single-acting, single-stage reciprocating air compressor is 1358 kW when the mean piston speed is 2.8 m/s and rotational speed 3.5 rev/s. The air is compressed from 1.0 bar and delivered at 10 bar, the index of the law of compression being 1.32. Neglecting clearance, calculate (i) the stroke of the compressor piston, (ii) the cylinder diameter and (iii) the mean effective pressure.

27. In an engine working on the ideal diesel cycle, the temperature of the air at the beginning of compression is 37°C, compression takes place according to the law $pV^{1.4}$ = constant, and the volumetric compression ratio is 13:1. At the end of compression the air receives heat energy at constant pressure and is then expanded to the original volume. If 1 kg of fuel is burned per 35 kg of air compressed, the calorific value of the fuel being 42 MJ/kg, calculate (i) the temperature at the end of compression, (ii) the temperature at the end of heat reception and (iii) the volumetric expansion ratio. Take c_p = 1.02 kJ/kg K.

28. At a certain stage of a reaction turbine, the steam leaves the guide blades at a velocity of 135 m/s, the exit angle being 20°. The linear velocity of the moving blades is 87 m/s. Assuming the channel section of fixed and moving blades to be identical, and assuming ideal conditions, calculate (i) the entrance angle of the moving blades and (ii) the stage power per kg/s steam flow.

29. In an open cycle gas turbine plant, a heat exchanger is included to heat the air before entering the combustion chamber by the exhaust gases from the engine. The gases enter the heat exchanger at 300°C and 140 m/s and leave at 240°C and 10 m/s. The air enters the exchanger at 200°C and the air/fuel ratio is 84. Calculate the temperature of the air at the exchanger exit, taking c_p as 1.1 kJ/kg K for the gases and 1.005 kJ/kg K for air.

30. A CO_2 refrigerating machine produces 250 kg of ice per hour at −10°C from water at 15°C. The refrigerant enters the evaporator 0.2 dry and leaves 0.95 dry. The compressor is single-acting, runs at 4.15 rev/s, and the stroke/bore ratio is 2:1. Calculate (i) the mass flow of the refrigerant through the circuit and (ii) the diameter and stroke of the compressor piston. Take the following values:

Specific heat of ice $= 2.04$ kJ/kg K
Latent heat of fusion $= 335$ kJ/kg K
Specific heat of water $= 4.2$ kJ/kg K

CO_2 vapour at evaporator pressure:

$$h_{fg} = 290.2 \text{ kJ/kg}, \ v_g = 0.02168 \text{ m}^3\text{/kg}$$

31. Air is taken into a single-stage air compressor at 1 bar and delivered at 5 bar. The piston swept volume is 1440 cm³ and the clearance volume is 40 cm³. Taking the index of compression and expansion as 1.3, calculate (i) the fraction of the stroke when the delivery valves open, (ii) the fraction of the stroke when the suction valves open and (iii) the mean indicated pressure.

32. Steam is supplied to a turbine at 30 bar, 350°C and the condenser pressure is 0.045 bar. The power developed is 5 MW when the steam consumption is 22.5 Mg/h. Calculate (i) the ideal efficiency of the Rankine cycle, (ii) the actual efficiency of the engine and (iii) the efficiency ratio.

33. A sample of steam at 10 bar is tested by a combined separating and throttling calorimeter, the data obtained were:

Mass of water collected in separator $= 0.113$ kg
Mass of condensed water after throttling $= 3.03$ kg
Pressure of steam in throttling calorimeter $= 1.2$ bar
Temperature of steam in throttling calorimeter $= 109.8°C$

Take specific heat of superheated steam at calorimetric pressure as 2.02 kJ/kg K and calculate the dryness fraction of the sample.

34. The piston swept volume of an engine working on the ideal dual-combustion cycle is 0.1068 m³ and the clearance volume is 8900 cm³. At the beginning of compression the pressure is 1 bar and temperature 42°C. The maximum pressure in the cycle is 45 bar and maximum temperature 1500°C. Taking $\gamma = 14$ and $c_v = 0.715$ kJ/kg K for air and products of combustion, calculate the proportion of heat received at constant volume to that received at constant pressure.

35. A fuel has an analysis by mass of 85% C, 11% H_2 and 3% O_2. This fuel is burned in a combustion chamber with an air/fuel ratio by mass of 11:1. For every kg of fuel burned calculate:

(i) the mass of carbon burned to produce carbon monoxide;

(ii) the mass of carbon burned to produce carbon dioxide.

Atomic mass relationships are: $C = 12$, $O = 16$, $H = 1$. Air contains 23% O_2 by mass.

36. At one stage of a steam turbine the inlet velocity of the steam to the rotating blades is 590 m/s and the relative velocity at outlet is 620 m/s. The blade speed is 320 m/s and the inlet and outlet angles are 37° and 26°, respectively. Calculate the force on the blades and the power developed at this stage for a steam flow of 0.075 kg/s.

37. A single-acting air compressor takes in air at 1 bar and delivers it at 4 bar, the cylinder is 300 mm diameter, stroke 450 mm and it runs at 5 rev/s. Initially the index of compression was 1.15 and after running for some time the index of compression was found to be 1.35. Neglecting clearance, calculate the power absorbed in each case and the percentage increase in power.

38. A steam pipe 140 mm outside diameter and 23 m long is lagged with insulating material of thermal conductivity 0.13 W/m K. Steam passes along the pipe at the rate of 1200 kg/h, entering at 18 bar dry saturated and leaving at the same pressure 0.985 dry. The outside surface temperature of the lagging is 35°C and the inside surface may be taken as equal to the steam temperature. Calculate the thickness of the lagging taking the rate of heat transfer through the insulating material, in J/s per unit length of pipe, to be:

$$\frac{2\pi k\left(T_1 - T_2\right)}{\ln\left(r_2/r_1\right)}$$

39. Steam is supplied to a turbine at 20 bar 400°C and exhausts at 0.04 bar and 0.85 dry. At the stage in the turbine where the pressure is 1.4 bar, 13.4% of the steam is withdrawn and passed to the feed heater and this heats the feed water to the saturation temperature of the tapped off steam. Compare the thermal efficiencies with and without feed heating.

40. Air is compressed in a cylinder according to the law pV^n = constant. The initial condition of the air is 0.125 m³, 1.01 bar and 19°C, and the final condition is 36 bar and 508°C. Taking c_p = 1.005 kJ/kg K and c_v = 0.718 kJ/kg K, calculate (i) the index of compression, (ii) the mass of air compressed, (iii) the work done during compression, (iv) the change of internal energy and (v) the transfer of heat to or from the air.

41. The following data were taken during a test on a four-cylinder, four-stroke, compression ignition engine of cylinder diameter 320 mm and stroke 480 mm while running at 4 rev/s. Mean effective pressure 14.9 bar, brake load 12 kN on a radius of 960 mm, fuel consumption 99 kg/h, calorific value of fuel 44.5 MJ/kg, mass flow of engine cooling water 154 kg/min, water inlet and outlet temperatures 14°C and 47°C. Calculate the (i) indicated and (ii) brake thermal efficiencies and (iii) percentage of heat carried away in the cooling water, and (iv) draw up a heat balance. Specific heat of cooling water = 42 kJ/kg K.

42. In a simple gas turbine working on the ideal cycle, the pressure ratio of both the compressor and turbine is 4.3:1. The temperatures at inlet to the compressor and at inlet to the turbine are 16°C and 600°C, respectively. Calculate (i) the temperature at the outlet from the compressor, (ii) the temperature at the outlet from the turbine, (iii) the heat supplied per kilogram of working fluid and (iv) the thermal efficiency. Take $\gamma = 1.4$ and $c_p = 1.005$ kJ/kg K.

43. A simple Freon-12 refrigerator operates with evaporator at 1.509 bar, −20°C and condenser at 8.477 bar, 35°C. The compression between these two states is isentropic and the refrigerant leaving the compressor is dry saturated vapour. Calculate:

    (i) using thermodynamic tables:

      (a) the refrigerating effect per kg of refrigerant;

      (b) the coefficient of performance.

    (ii) the coefficient of performance of a reversed Carnot cycle operating between the same temperatures.

44. A two-pass oil cooler consists of 350 tubes and is required to cool 4 kg/s of oil from 50°C to 20°C. The overall heat transfer coefficient of the tubes is 70 W/m² K, cooling water temperature 15°C. Determine:

    (i) logarithmic mean temperature difference of the oil in the cooler;

    (ii) surface area of the tubes;

    (iii) length of the thin tubes if their diameter is 19 mm.

$$\text{Mean temperature difference } \theta_m = \frac{T_1 - T_2}{\ln\left(\dfrac{T_1 - T_c}{T_2 - T_c}\right)}$$

where T_1 = oil inlet temperature,

  T_2 = oil outlet temperature,

  T_c = cooling water temperature.

The specific heat capacity of oil is 1395.6 J/kg K.

45. Dry saturated steam at 7 bar is throttled to 0.5 bar and then passed through two oil heaters in series. The steam leaves the first heater 0.98 dry and is throttled to 0.16 bar before entering the second heater. If there is no pressure drop in the heaters determine using the enthalpy/entropy chart:

    (i) final steam condition if there has been no overall change in entropy;

    (ii) mass flow of steam if 0.72 kg/s of oil, specific heat capacity 21 kJ/kg K is raised in temperature through 72°C.

46. A marine boiler installation is fired with methane (CH_4). For stoichiometric combustion calculate:

    (i) the correct air-to-fuel mass ratio;

    (ii) the percentage composition of the dry flue gases by volume.

    Atomic mass relationships: hydrogen 1, oxygen 16, carbon 12, nitrogen 14. Air contains 23% oxygen and 77% nitrogen by mass.

47. Gas enters a rotary compressor at 15°C with a velocity of 75 m/s and a specific enthalpy of 80 kJ/kg. It is discharged at 200°C with a velocity of 175 m/s and a specific enthalpy of 300 kJ/kg. If the compressor loses 10 kJ/kg of gas flowing through the compressor, calculate:

    (i) the external work transfer per kilogram of gas;

    (ii) the datum temperature on which the specific enthalpies given are based.

48. Saturated steam at 50 bar is supplied through a pipe which has two layers of insulation. Outside diameter of the pipe is 200 mm, inner layer of insulation 100 mm thick with thermal conductivity 0.05 W/m K, outer layer 50 mm thick with thermal conductivity 0.15 W/m K, which has a surface heat transfer coefficient of 8 W/m² K. If the ambient temperature is 20°C, calculate the condensation rate per metre of pipe length.

49. Steam at 65 bar and 500°C is supplied to a three-stage turbine. It leaves the first stage at 15 bar, 330°C and then passes through a reheater, which it leaves at 500°C with specific enthalpy of 3475 kJ/kg. After leaving the second stage at 330°C and 3.5 bar, the steam passes through the third stage and exhausts at 0.05 bar, 0.94 dry.

    Determine using the enthalpy/entropy chart:

    (i) the pressure drop in the reheater;

    (ii) the isentropic efficiency of each stage;

    (iii) the ratio of powers developed in each stage.

50. The following successive processes are performed on 1 kg of air from a complete cycle:

    (a) isentropic compression from 120°C and 1 bar to 10 bar;

    (b) heating at constant volume to 800°C;

    (c) isentropic expansion;

    (d) heat rejection at constant pressure.

    Sketch the cycle on p–V and T–S axes and determine (i) cycle efficiency and (ii) mean effective pressure.

SOLUTIONS TO EXAMINATION QUESTIONS – HIGHER LEVEL

1. Refer to Figure 12.17:

$$\frac{T_3\pi}{T_4} = \left(\frac{p_3}{p_4}\right)^{\frac{r-1}{r}}$$

$$\frac{925}{T_4} = 5^{\frac{0.4}{1.4}}$$

$$T_4 = 637.8 \text{ K}$$

$$\frac{T_3 - T_4^1}{T_3 - T_4} = 0.86$$

$$925 - T_4^1 = (925 - 637.8) \times 0.86$$

$$T_4^1 = 678 \text{ K}$$

Temperature of exhaust gases leaving turbine = 403°C Ans. (i)

$$\frac{T_2}{T_1} = \left(\frac{p_2}{p_1}\right)^{\frac{r-1}{r}}$$

$$T_2 = 300 \times 5^{\frac{0.4}{1.4}}$$

$$= 475 \text{ K}$$

$$\frac{T_2 - T_1}{T_2' - T_1} = 0.82$$

$$475 - 300 = (T_2' - 300) \times 0.82$$

$$T_2' = 513.7 \text{ K}$$

Net work $= \dot{m}_g c_{pg}(925 - 678) - \dot{m}_a c_{pa}(513.7 - 300)$

$$5000 = \frac{51}{50} = \dot{m}_f(289.73 - 214.8), \text{ i.e., } \dot{m}_f = \frac{\dot{m}_a}{50}$$

$$\dot{m}_f = 1.334$$

Fuel mass $= 1.334$ kg/s Ans. (ii)

2. From steam tables, for a saturation temperature of 39°C, pressure is 0.07 bar, and spec. volume 20.53 m³/kg. Therefore mass of steam in volume of 15 m³

$$= \frac{15}{20.53} = 0.7307 \text{ kg Ans. (i)}$$

Partial pressure due to air

$$= \text{Total pressure} - \text{Steam pressure}$$

$$= 0.09 - 0.07 = 0.02 \text{ bar} = 2 \text{ kN/m}^2$$

$$pV = mRT$$

Mass of air

$$m = \frac{pV}{RT}$$

$$= \frac{2 \times 15}{0.287 \times 312} = 0.335 \text{ kg Ans. (ii)}$$

Ratio of air to steam:

Therefore to adjust the steam side to unity, divide both sides by 0.7307

$$= 0.335 \text{kg of air} : 0.7307 \text{ kg of steam}$$

$$= \frac{0.335}{0.7307} : \frac{0.7307}{0.7307}$$

$$= 0.4584 : 1 \text{ Ans. (iii)}$$

3. $h_1 + \frac{1}{2}c_1^2 + q = h_2 + \frac{1}{2}c_2^2 + w$

is the Steady Flow Energy Equation.

$$H_1 + \tfrac{1}{2}mc_1^2 + Q = H_2 + \tfrac{1}{2}mc_2^2 + w$$

$$
\begin{aligned}
Q &= H_2 - H_1 + \tfrac{1}{2}m\,(c_2^2 - c_1^2) + w \\
&= mc_p\,(T_2 - T_1) + \tfrac{1}{2}m\,(c_2^2 - c_1^2) + w \\
&= 0.4\left[1.005 \times 10^3\,(40 - 20) + 0.5\left(100^2 - 0^2\right)\right] - 13000 \\
&= 0.4\,(20100 + 5000) - 13000 \\
&= 10040 - 13000 \\
Q &= -2960 \text{ W}
\end{aligned}
$$

Q negative (heat out) and W negative (work on air)

$$Q = mc_p\left(T_o - T_i\right)$$
$$2960 = 0.1 \times 10^3 \times 4.18\,(T_o - 20)$$
$$T_o = 27.08°C$$

Cooling water outlet temperature $= 27.08°C$ Ans.

Note: The pressures given are not required.

4. Stoichiometric oxygen required per kg of fuel

$$
\begin{aligned}
&= 2\tfrac{2}{3}C + \left\{H_2 - \frac{O_2}{8}\right\} \\
&= 2\tfrac{2}{3}C + 8H_2 - O_2 \\
&= 2\tfrac{2}{3} \times 0.87 + 8 \times 0.11 - 0.02 \\
&= 3.18 \text{ kg}
\end{aligned}
$$

$$\text{Stoichiometric air} = \frac{100}{23} \times 3.18 = 13.82 \text{ kg}$$

Mass of nitrogen in 13.82 kg of air

$$= 0.77 \times 13.82 = 10.64 \text{ kg}$$
$$\left(\text{or, } 13.82 \text{ kg air} - 3.18 \text{ kg } O_2 = 10.64 \text{ kg } N_2\right)$$
$$CO_2 \text{ formed} = 3\tfrac{2}{3} \times 0.87 = 3.19 \text{ kg}$$
$$H_2O \text{ formed} = 9 \times 0.11 = 0.99 \text{ kg}$$

Total gases/kg fuel

$$= 10.64 + 3.19 + 0.99 = 14.82 \text{ kg}$$
$$\text{also, } 13.82 \text{ kg air} + 1 \text{ kg fuel} = 14.82 \text{ kg}$$

At 1.8 tonne of fuel per hour, fuel rate

$$= \frac{1.8 \times 10^3}{3600} = 0.5 \text{ kg/s}$$

Mass flow rate of each of the gases, Ans. (i):

$$\text{Nitrogen} = 0.5 \times 10.64 = 5.32 \text{ kg/s}$$
$$CO_2 = 0.5 \times 3.19 = 1.595 \text{ kg/s}$$
$$H_2O = 0.5 \times 0.99 = 0.495 \text{ kg/s}$$

As a percentage analysis, Ans. (ii):

$$\text{Nitrogen} = \frac{10.64}{14.82} \times 100 = 71.79\%$$
$$CO_2 = \frac{3.19}{14.82} \times 100 = 21.53\%$$
$$H_2O = \frac{0.99}{14.82} \times 100 = 6.68\%$$

5. From steam tables:

$$9 \text{ bar, } h_g = 2774$$
$$5 \text{ bar, } h_f = 640 \quad h_{fg} = 2109 \quad v_g = 0.3748$$
$$0.14 \text{ bar, } h_f = 220 \quad h_{fg} = 2376 \quad v_g = 10.69$$

Enthalpy drop from 9 bar to 5 bar
$$2774 - (640 + x \times 2109) = 107$$
$$2774 - 640 - 107 = x \times 2109$$
$$2027 = x \times 2109$$
Dryness at throat, $x = 0.9614$

Specific volume of steam at throat

$$= 0.9614 \times 0.3748 = 0.3603 \text{ m}^3/\text{kg}$$
$$\text{Velocity [m/s]} = \sqrt{2 \times \text{Spec. enthalpy drop [J/kg]}}$$

Velocity through throat

$$= \sqrt{2 \times 107 \times 10^3} = 462.6 \text{ m/s}$$

Area [m²] × Velocity [m/s] = Mass flow [kg/s] × Spec. vol. [m³/kg]

∴ Area at throat [mm²]

$$= \frac{23 \times 0.3603}{60 \times 462.6} \times 10^6 = 298.5 \text{ mm}^2 \quad \text{Ans. (i)}$$

Effective enthalpy drop from entrance to exit

$$= 0.92 \times 633 = 582.4 \text{ kJ/kg}$$

Enthalpy drop from 9 bar to 0.14 bar =

$$2774 - (220 + x \times 2376) = 582.4$$
$$2774 - 220 - 582.4 = x \times 2376$$
$$1971.6 = x \times 2376$$
$$\text{Dryness at exit, } x = 0.8299$$

Specific volume of steam at exit

$$= 0.8299 \times 10.69 = 8.869 \text{ m}^3/\text{kg}$$

Velocity at exit

$$= \sqrt{2 \times 582.4 \times 10^3} = 1079 \text{ m/s}$$

Area at exit [mm²]

$$= \frac{\text{Mass flow } [\text{kg/s}] \times \text{Spec. vol. } [\text{m}^3/\text{kg}]}{\text{Velocity } [\text{m/s}]} \times 10^6$$

$$= \frac{23 \times 8.869}{60 \times 1079} \times 10^6 = 3150 \text{ mm}^2 \quad \text{Ans. (ii)}$$

6. From R134a tables:

$$\text{At } 84.35 \text{ kPa}, \quad h_f = 160.89 \qquad h_g = 380.27$$
$$\therefore h_{fg} = 380.27 - 160.89 = 219.38$$

$$s_f = 0.849 \qquad s_g = 1.7512$$
$$\therefore s_{fg} = 1.7512 - 0.849$$

At 770 kPa, $h_f = 241.69$ and the sat. temp. = 30°C

∴ at 40°C the refrigerant 134 is superheated by 10°C
This means that $h = 425.21$ and $s = 1.7482$

For isentropic compression $s_1 = s_2$

$$0.849 + x_1 \times 0.9022 = 1.7482$$
$$x_1 \times 0.9022 = 0.8992$$
$$x_1 = 0.9967$$

$h_1 = h$ leaving evaporator $= h$ entering compressor
$$= 160.89 + 0.9967 \times 219.38$$
$$= 379.54$$

$h_4 = h$ entering evaporator $= h$ leaving condenser $\left(h_3\right)$
Refrigerating effect/kg $= h_1 - h_4$
$$= 379.54 - 241.69$$
$$= 137.85$$

Refrigerating effect for 15kg/min = 15 x 137.85 = 2067 kJ/min   Ans. (i)

Work transfer in the compressor/kg $= h_2 - h_1$
$$= 425.21 - 379.54$$
$$= 45.66 \text{ kJ/kg}$$

$$\text{Coeff. of performance} = \frac{\text{Refrigerating effect}}{\text{Work transfer}}$$
$$= \frac{137.85}{45.66}$$
$$= 3.018 \quad \text{Ans. (ii)}$$

7. Area = Work done $= \int pdV$

Constant temperature (isothermal) $pV = C$

$$p = \frac{C}{V}$$
$$\text{Area} = C\int_{V_1}^{V_2} \frac{dV}{V}$$
$$= C\ln\left(\frac{V_2}{V_1}\right)$$
$$= C\ln\left(\frac{p_1}{p_2}\right) \quad \text{as } p_1V_1 = p_2V_2$$
$$= pV\ln\left(\frac{p_1}{p_2}\right) \quad \text{as } pV = C$$

$$= RT \ln\left(\frac{p_1}{p_2}\right) \quad \text{as} \quad pV = RT$$

$$\therefore \text{Area under } p-V \text{ curve} = RT \ln\left(\frac{p_1}{p_2}\right) \quad \text{Ans. (i)}$$

Isothermal process, no change of internal energy

$$\text{Heat extracted} = \text{Work done on the gas}$$

$$\text{Work done} = \int p dV$$

$$= p_1 V_1 \ln\left(\frac{p_1}{p_2}\right)$$

$$= 1 \times 100 \times 0.25 \ln\left(\frac{1}{8}\right)$$

$$= -51.99 \text{ kJ}$$

Negative result confirming work done on the gas

$$\text{Heat energy transfer} = -51.99 \text{ kJ} \quad \text{Ans. (ii)}$$

8. Refer to Figures 12.15 and 12.17:

$$\frac{T_2}{T_1} = \left(\frac{p_2}{p_1}\right)^{\frac{r-1}{r}}$$

$$T_2 = 291 \times 4^{\frac{0.333}{1.333}}$$

$$= 432.4 \text{ K}$$

$$\frac{T_3}{T_4} = \left(\frac{p_3}{p_4}\right)^{\frac{r-1}{r}}$$

$$\frac{923}{T_4} = 4^{\frac{0.4}{1.4}}$$

$$T_4 = 652.8 \text{K}$$

$$\eta_c = \frac{T_2 - T_1}{T_2' - T_1} \times 100$$

$$= \frac{432.4 - 291}{473 - 291} \times 100$$

Compressor isentropic efficiency = 77.69%   Ans. (i)

$$= \frac{432.4 - 291}{473 - 291} \times 100$$

$$0.85 = \frac{923 - T_4'}{923 - 652.8}$$

$$T_4' = 693.3 \, K$$

$$\text{Turbine work} = c_p (T_3 - T_4')$$
$$= 1.15 (923 - 693.3)$$
$$= 264.2 \, kJ/kg$$

$$\text{Compressor work} = c_p (T_2' - T_1)$$
$$= 1.005 (473 - 291)$$
$$= 182.9 \, kJ/kg$$

Net turbine work = 81.3 kJ/kg Ans. (ii)

9.

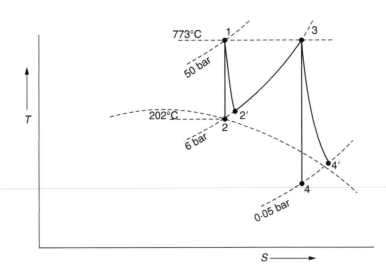

See the T–S diagram Ans. (i)

$$s_1 = 6.975 \quad h_1 = 3433 \, (\text{tables or } h - s \text{ chart})$$
$$s_2 = 6.975 \quad h_2 = 2854 \, (202°C, \text{ just superheated})$$

$$\Delta h = 3433 - 2854$$
$$= 579$$
$$0.9\,\Delta h = 521$$
$$h_2' = 2912, \text{ i.e. } 3433 - 521$$

$s_3 = 8.001 \quad h_3 = 3483 \quad \text{(tables or } h\!-\!s \text{ chart)}$
$s_4 = 8.001 \qquad\qquad\qquad \text{(0.05 bar, steam wet)}$

$$8.001 = 0.476 + 7.918\,x$$
$$x = 0.9504 \text{ (dryness fraction)}$$
$$h_4 = 138 + 0.9504 \times 2423$$
$$= 2441 \qquad \text{(or from } h\!-\!s \text{ chart directly)}$$

$$\Delta h = 3483 - 2441$$
$$= 1042$$
$$0.85\,\Delta h = 885.7$$
$$h_4' = 2598.7, \text{ i.e. } 3483 - 885.7$$

Heat supplied in boiler $= h_{500} - h_{f\,0.05}$
$$= 3433 - 138 = 3295$$

Heat supplied in reheat $= h_3 - h_2'$
$$= 3483 - 2912 = 571$$

Work output $= 521 + 885.7 = 1406.7$

Thermal efficiency $= \dfrac{1406.7}{3866} \times 100 = 36.39\%$ Ans. (i)

Note: The $h\!-\!s$ chart is more used; the $T\!-\!S$ sketch was specified here.

10. Temperature difference across thickness of wood

$$= \frac{QS_w}{k_w\,At}$$
$$= \frac{24 \times 0.03}{0.18 \times 1 \times 1} = 4\,\text{K} \text{ Ans. (i) (a)}$$

Temperature difference across thickness of cork

$$= \frac{QS_c}{k_c At}$$

$$= \frac{24 \times 0.07}{0.05 \times 1 \times 1} = 33.6K \quad \text{Ans. (i) (b)}$$

Temperature difference between inside and outside atmospheres and their respective exposed surfaces

$$= \frac{Q}{hAt}$$

$$= \frac{24}{10 \times 1 \times 1} = 2.4 \text{ K}$$

Total temperature difference between outside atmosphere and inside atmosphere

$$= 2.4 + 4 + 33.6 + 2.4 = 42.4 \text{ K or } 42.4°C \quad \text{Ans. (ii)}$$

Room temperature $= 20 - 42.4 = -22.4°C \quad \text{Ans. (iii)}$

11. Heat lost by steam = Heat gained by feed water

$$m_s \times x \times h_{fg} = m_w \times c_p \times \text{Temp. rise}$$

$$m_s = \frac{20000 \times 4.18 \,(105 - 55)}{3600 \times 0.95 \times 2202}$$

$$= 0.555 \text{ kg/s}$$

$$= 1998 \text{ kg/h} \quad \text{Ans. (i)}$$

Assumption: no undercooling of condensed steam

$$\theta_m = \frac{(120.2 - 55) - (120.2 - 105)}{\text{In} \dfrac{(120.2 - 55)}{120.2 - 105}}$$

$$= \frac{50}{1.456}$$

$$= 34.34°C$$

$$Q = m_w \times c_p \times \text{Temp. rise}$$

$$= \frac{20000 \times 4.18 \times 50}{3600}$$

$$= 1161.1 \text{ kW}$$

$$Q = UAt\theta_m$$
$$1161.1 = 4540 \times A \times 1 \times 34.34$$
$$A = 7.448 \text{ m}^2 \quad \text{Ans. (ii)}$$

$$\text{ip} = p_m LAn$$
$$= 8.9 \times 10^2 \times 0.7854 \times 0.36^2 \times 0.78 \times 2.3$$
$$= 162.5 \text{ kW} \quad \text{Ans. (i)}$$

$$\text{bp} = T\omega$$
$$= 8 \times 1.25 \times 2\pi \times 2.3$$
$$= 144.5 \text{ kW} \quad \text{Ans. (ii)}$$

Heat energy supplied per second

$$= \frac{0.251 \times 144.5 \times 41.5 \times 10^3}{3600} = 418.1 \text{ kJ/s}$$

Heat energy converted into work in cylinder per second

$$= 162.5 \text{ kJ/s}$$

Indicated thermal efficiency

$$= \frac{162.5}{418.1} = 0.3887 \text{ or } 38.87\% \quad \text{Ans. (iii)}$$

Total heat energy loss per second

$$= \text{Heat supplied} - \text{Heat converted into work}$$
$$= 418.1 - 162.5 = 255.6 \text{ kJ/s} \quad \text{Ans. (iv)}$$

13. Hydrogen $R_1 = \dfrac{8.3143}{2} = 4.1572$

Oxygen $R_2 = \dfrac{8.3143}{32} = 0.2598$

Ratio of partial pressures = Ratio of volumes

Hydrogen, partial pressure = $0.8 \times 3.5 = 2.8$ bar

Oxygen, partial pressure = $0.2 \times 3.5 = 0.7$ bar

$$pV = mRT$$

$$\text{Mass of hydrogen} = \frac{2.8 \times 100 \times 0.7}{4.1572 \times 311}$$

$$= 0.1516 \text{ kg Ans.}$$

$$\text{Mass of oxygen} = \frac{0.7 \times 100 \times 0.7}{0.2598 \times 311}$$

$$= 0.6064 \text{ kg Ans.}$$

14. From steam tables:

$$1.2 \text{ bar}, h_g = 2683$$
$$7 \text{ bar } 300°C, h = 3060$$

Enthalpy drop through nozzle

$$= 3060 - 2683 = 377 \text{ kJ/kg}$$

Velocity of steam at nozzle exit [m/s]

$$= \sqrt{2 \times \text{Spec. enthalpy drop [J/kg]}}$$
$$= \sqrt{2 \times 377 \times 10^3} = 868.4 \text{ m/s}$$

Refer Figure 12.8:

$$
\begin{aligned}
v_{w1} &= v_1 \cos \alpha_1 = 868.4 \times \cos 20° & = 816 &\quad \text{m/s} \\
8 &= 40\% \text{ of } v_1 = 0.4 \times 868.4 & = 347.4 &\quad \text{m/s} \\
x &= v_{w1} - u = 816 - 347.4 & = 468.6 &\quad \text{m/s} \\
v_{a1} &= v_1 \sin \alpha_1 = 868.4 \times \sin 20° & = 297 &\quad \text{m/s} \\
v_{r1} &= \sqrt{297^2 + 468.6^2} & = 554.8 &\quad \text{m/s} \\
v_{r2} &= 0.8 \times 554.8 = 443.8 \text{ m/s} \\
V_{a2} &= v_{r2} \times \sin\beta_2 = 443.8 \times \sin 35° & = 254.6 &\quad \text{m/s}
\end{aligned}
$$

Axial force on blades [N]

$$= \text{Mass flow [kg/s]} \times \text{Change of axial velocity [m/s]}$$
$$= 0.5 \times (297 - 254.6)$$
$$= 21.2 \text{ N Ans. (i)}$$

$$v_{w2} = v_{r2}\cos \beta_2 - u$$
$$= 443.8 \times \cos 35° - 347.4 = 16.2 \text{ m/s}$$

Effective change of velocity:

$$v_w = v_{w1} + v_{w2}$$
$$= 816 + 16.2 = 832.2 \text{ m/s}$$

Tangential force on blades

$$= \dot{m}V_w$$
$$= 0.5 \times 832.2 = 416.1 \text{ N}$$
$$\text{Power} [W] = \text{Force} [N] \times \text{Blade velocity} [m/s]$$
$$= 416.1 \times 347.4$$
$$= 1.445 \times 10^5 \text{ W} = 144.5 \text{ kW} \quad \text{Ans. (ii)}$$

15.

$$\frac{T_1}{T_2} = \left(\frac{p_1}{p_2}\right)^{\frac{r-1}{r}}$$

$$\frac{533}{T_2} = \left(\frac{6}{4}\right)^{\frac{0.4}{1.4}}$$
$$T_2 = 474.6 \text{ K}$$

$$h_1 - h_2 = c_P(T_1 - T_2) = \frac{1}{2}(c_2^2 - c_1^2)$$
$$1005 (533 - 474.6) = \frac{1}{2}(c_2^2 - 90^2)$$
$$c_2 = 354.2 \text{ m/s}$$
Air velocity at nozzle exit = 354.2 m/s Ans. (i)

$$pv_s = RT$$
$$4 \times 10^2 \times v_s = 0.2871 \times 474.6$$
$$v_s = 0.3406 \text{ m}^3/\text{kg}$$
$$\dot{m} = \frac{\text{Area} \times \text{Velocity}}{\text{Specific volume}}$$
$$= \frac{0.025 \times 354.2}{0.3406}$$
Mass flow rate of air = 26 kg/s Ans. (ii)

$$pv_s = RT$$
$$4 \times 10^2 \times v_s = 0.2871 \times 533$$
$$v_s = 0.255$$
$$26 = \frac{Area \times 90}{0.255}$$

Nozzle inlet area $= 0.0737 \ m^2$ Ans. (iii)

16. The ratio of partial pressures is the same as the ratio of partial volumes, therefore:

$$\left. \begin{array}{l} \text{Partial pressure of } CO_2 = 0.14 \times 1.01 = 0.1414 \text{ bar} \\ \text{Partial pressure of } N_2 = 0.86 \times 1.01 = 0.8686 \text{ bar} \end{array} \right\} \text{Ans. (i)}$$

For mass of CO_2, $pV = mRT$ where

$$p = 0.1414 \times 10^2 \ kN/m^2$$
$$V = 500 \times 10^{-6} \ m^3$$
$$m = \frac{pV}{RT} = \frac{0.1414 \times 10^2 \times 500 \times 10^{-6}}{0.189 \times 293}$$
$$= 1.276 \times 10^{-4} \ kg \ or \ 0.1276 \ g \quad Ans. \ (ii)$$

17.
$$\frac{T_2}{T_1} = \left(\frac{p_2}{p_1} \right)^{\frac{n-1}{n}}$$

$$\frac{693}{295} = \left(\frac{35}{1} \right)^{\frac{n-1}{n}}$$

$$n = 1.316 \quad Ans. \ (i)$$

$$\frac{p_1 V_1}{T_1} = \frac{p_2 V_2}{T_2}$$

$$\frac{1 \times 0.037}{295} = \frac{35 \times V_2}{693}$$

$$V_2 = 0.02483$$

$$\text{Work done} = \frac{p_1 V_1 - p_2 V_2}{n-1}$$

$$= \frac{1 \times 100 \times 0.037 - 35 \times 100 \times 0.02483}{0.316}$$

$$= -15.794 \ kJ \quad Ans. \ (ii)$$

Negative (compression) work done on the gas

$$pV = mRT$$
$$1 \times 100 \times 0.037 = m \times 282 \times 295$$
$$m = 0.0000444 \text{ kg}$$

Change of internal energy $= mc_v \left(T_2 - T_1\right)$
$$= 0.0000444 \times 718\left(693 - 295\right)$$
$$= 12.69 \text{ kJ} \quad \text{Ans. (iii)}$$

Heat transfer $= 12.69 - 15.794$
$$= -3.104 \text{ kJ} \quad \text{Ans. (iv)}$$

18. Refer to Figure 14.2:

i.e., from tables, page 14, at 25°C

$$h_3 = h_4 = 59.7 \text{ kJ/kg}$$

i.e., from tables, page 14, at −15°C

$$h_4 = h_f + x_4 \left(h_g - h_f\right)$$
$$59.7 = 22.33 + x_4 \left(180.97 - 22.33\right)$$

$$x_4 = 0.2356$$

Heat extracted $= \dot{m}\left(h_1 - h_4\right)$
$$= \dot{m}[h_g - h_f - x\left(h_g - h_f\right)]$$
$$73.3 = \dot{m}\left\{180.97 - 23.33 - 0.2356 \left(180.97 - 23.33\right)\right\}$$
$$= \dot{m} \left(180.97 - 59.71\right)$$
$$\dot{m} = 0.605 \text{ kg/s}$$

Refrigerant mass flow rate $= 0.605$ kg/s Ans. (i)
$$h_2 = 208.5 \text{ kJ/kg}$$

i.e., from tables, page 13, at 6.516 bar and 40°C:

Compressor power $= \dot{m}(h_2 - h_1)$
$$= 0.605 \left(208.5 - 180.97\right)$$
$$= 16.66 \text{ kW} \quad \text{Ans. (ii)}$$

$$\text{c.o.p.} = \frac{h_1 - h_4}{h_2 - h_1}$$

$$= \frac{180.97 - 59.71}{208.5 - 180.97}$$

$$= 4.405 \quad \text{Ans. (iii)}$$

19.

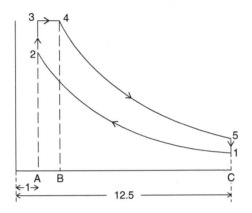

Representing volumes by ratio:

$$V_1 \text{ and } V_5 = 12.5 \qquad V_2 \text{ and } V_3 = 1$$

$$p_1 V_1^r = p_2 V_2^r$$

$$1.013 \times 12.5^{1.4} = p_2 \times 1^{1.4} \quad p_2 = 34.77 \text{ bar}$$

$$\frac{p_1 V_1}{T_1} = \frac{p_2 V_2}{T_2}$$

$$T_2 = \frac{308 \times 34.77 \times 1}{1.013 \times 12.5} = 845.9 \text{ K}$$

$$\frac{T_3}{T_2} = \frac{p_3}{p_2}$$

$$T_3 = \frac{845.9 \times 40}{34.77} = 973.4 \text{ K}$$

$$\frac{V_4}{V_3} = \frac{T_4}{T_3}$$

$$V_4 = \frac{1 \times 1698}{973.4} = 1.745$$

$$p_4 V_4^r = p_5 V_s^r$$

$$40 \times 1.745^{1.4} = p_5 \times 12.5^{1.4}$$

$$p_5 = 40 \times \left\{ \frac{1.745}{12.5} \right\}^{1.4} = 2.541 \text{ bar}$$

Areas representing positive work done:

Area during combustion = A 3 4 B

$$= 40 \times (1.745 - 1) = 29.8$$

Area during expansion = B 4 5 C

$$= \frac{p_4 V_4 - p_5 V_5}{\gamma - 1}$$

$$= \frac{40 \times 1.745 - 2.541 \times 12.5}{1.4 - 1} = 95.07$$

Gross area = A 3 4 5 C

$$= 29.8 + 95.07 = 124.87$$

Area representing negative work done:

Area during compression = C1 2 A

$$= \frac{p_1 V_1 - p_2 V_2}{\gamma - 1}$$

$$= \frac{1.013 \times 12.5 - 34.77 \times 1}{1.4 - 1} = -55.27$$

Net area representing useful work

$$= 124.87 - 55.27 = 69.6$$

Mean effective pressure = Mean height of net area

$$= \frac{\text{Area}}{\text{Length}} = \frac{69.6}{12.5 - 1}$$

$$= 6.052 \text{ bar} \quad \text{Ans.}$$

Note: M.e.p. and maximum pressure are low by modern standards.

20.

$$p = p_s + p_A$$

$$mR = m_s R_s + m_A R_A$$

$$21 \times R = 20 \times 0.462 + 1 \times 0.287$$

$$R = 0.4537 \text{ kJ/kg K}$$

$$p_s = 0.02 \times \frac{20 \times 0.462}{21 \times 0.4537}$$

$$p_s = 0.0194 \text{ bar}$$

$$p_A = 0.02 - 0.0194 = 0.0006 \text{ bar}$$

$$\left. \begin{array}{l} \text{Partial pressure, steam} = 0.0194 \text{ bar} \\ \text{Partial pressure, air} = 0.0006 \text{ bar} \end{array} \right\} \quad \text{Ans. (i)}$$

$$pv = RT$$

$$100 \times 0.02 \times v = 0.4537 \times 303$$

$$v = 68.735 \text{ m}^3/\text{kg}$$

$$\rho = 0.0145 \text{ kg/m}^3$$

Density of mixture $= 0.0145 \text{ kg/m}^3$ Ans. (ii)

21. Working on the basis of 1 kg fuel:

Stoichiometric air

$$= \frac{100}{23}\left\{2\tfrac{2}{3}C + 8\left(H_2 - \frac{O_2}{8}\right)\right\}$$

$$= \frac{100}{23}\left\{2\tfrac{2}{3}C + 8H_2 - O_2\right\}$$

$$= \frac{100}{23}\left\{2\tfrac{2}{3} \times 0.855 + 8 \times 0.135 - 0.1\right\}$$

$$= \frac{100}{23} \times 3.35 = 14.56 \text{ kg}$$

Excess air $= 0.25 \times 14.56 = 3.64 \text{ kg}$

Actual air $= 14.56 + 3.64 = 18.2 \text{ kg}$

Mass products of combustion per kg fuel:

$$CO_2 = 3\tfrac{2}{3} \times 0.855 = 3.135 \text{ kg}$$
$$H_2O = 9H_2 = 9 \times 0.135 = 1.215 \text{ kg}$$
$$O_2 = 23\% \text{ of excess air} = 0.23 \times 3.64 = 0.8372 \text{ kg}$$
$$N_2 = 77\% \text{ of all air} = 0.77 \times 18.2 = 14.014 \text{ kg}$$

Total products = 1 kg fuel + 18.2 kg air = 19.2 kg % mass analysis, Ans. (i)

$$\text{Dry flue gases} = \text{total gases} - H_2O$$
$$= 19.2 - 1.215 = 17.985 \text{ kg}$$

∴ % Volumetric analysis of dry flue gases, Ans. (ii)

DFG	m%	M	N	N%
CO_2	16.32	44	0.371	11.91
O_2	4.36	32	0.1362	4.38
N_2	72.99	28	2.606	83.71
Total			3.1132	

22. Heat passing across inner surface:

$$Q = h_i At\left(T_i - T_1\right) \therefore T_i - T_1 = \frac{Q}{h_i At}$$

$$-23 - T_1 = \frac{-30}{12 \times 1 \times 1}$$

Temperature, inner surface $(T_1) = -20.5°C$ Ans. (i)

i.e., heat flow *into* room is $-Q$

Heat passing across outer surface:

$$Q = h_o At\left(T_3 - T_o\right) \therefore T_3 - T_o = \frac{Q}{h_o At}$$

$$T_3 - 24 = \frac{-30}{12 \times 1 \times 1}$$

Temperature, outer surface $\left(T_3\right) = 21 - 5°C$ Ans. (i)

Heat conducted through inner material:

$$Q = \frac{kAt(T_1 T_2)}{S}$$

$$T_1 - T_2 = \frac{QS}{kAt}$$

$$-20.5 - T_2 = \frac{-30 \times 0.1}{0.115 \times 1 \times 1}$$

$$-20.5 - T_2 = -26.09$$

$$T_2 = 5.59°C$$

Temperature of the interface $(T_2) = 5.59°C$ Ans. (ii)

Heat conducted through cork:

$$T_2 - T_3 = \frac{QS}{kAt}$$

$$5.59 - 21.5 = \frac{-30 \times S}{0.06 \times 1 \times 1}$$

$$S = \frac{15.91 \times 0.06}{30}$$

$$S = 0.0318 \text{ m}$$

Thickness of the cork = 31.8 mm Ans. (iii)

23. Reference to Freon-12 tables, and Figure 14.2:

Compressor suction and evaporator exit:

1. 509 bar, sat. temp. = −20°C,

∴ at 5°C refrigerant is superheated by 15°

$h_1 = h$ at 1.509 bar, supht. 15° = 187.75

Compressor discharge:

4. 914 bar, sat. temp. = 15°C

∴ at 45°C refrigerant is superheated by 30°

$h_2 = h$ at 4.914 bar, supht. 30° = 214.35

Condenser outlet: $h_3 = h_f$ at 15°C = 50.1

Evaporator inlet: $h_4 = h_3 = 50.1$

$$\text{Coeff. of performance} = \frac{\text{Refrigerating effect in evaporat or [kJ/kg]}}{\text{Work transfer in compressor [kJ/kg]}}$$

$$= \frac{h_1 - h_4}{h_2 - h_1}$$

$$= \frac{187.75 - 50.1}{214.35 - 187.75} = 5.175 \ \text{Ans.}$$

24. Stoichiometric air $= \frac{100}{23}(2\tfrac{2}{3}C + 8H)$

$$= 4.348 \, (2.667 \times 0.82 + 8 \times 0.18)$$
$$= 15.77 \ \text{kg/kg fuel}$$

Air fuel ratio by mass $= 15.77$ Ans. (i)

Mass products of combustion per kg fuel:

$$CO_2 = 3\tfrac{2}{3} \times 0.8 = 3.007 \ \text{kg}$$
$$H_2O = 9 \times 0.18 = 1.62$$
$$N_2 = 0.77 \times 15.77 = \underline{12.143}$$
$$\text{Total} = 16.77 \ \text{kg}$$

% mass analysis of the wet flue gases:

$$CO_2 = \frac{3.007}{16.77} \times 100 = 17.93$$
$$H_2O = \frac{1.62}{16.77} \times 100 = 9.66 \ \text{Ans. (ii)}$$
$$N_2 = \frac{12.14}{16.77} \times 100 = 72.39$$

% volume analysis of the wet flue gases:

DFG	m%	M	N	N%
CO_2	17.93	44	0.4075	11.54
H_2O	9.66	18	0.5367	15.20
N_2	72.39	28	2.5854	73.26
			3.5296	

Ans. (ii)

25.

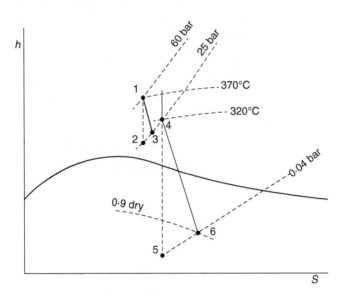

From the h–s chart; in kJ/kg:

$$h_4 = 3057$$
$$h_5 = 2030$$
$$h_4 - h_5 = 1027$$
$$h_4 - h_6 = 1027 \times 0.73$$
$$= 750$$

$$h_6 = 2307$$
$$P = \dot{m}_A (h_4 - h_6)$$
$$3 \times 10^3 = \dot{m}_A \times 750$$
$$\dot{m}_A = 4 \text{ kg/s}$$
$$h_1 = 3097$$
$$h_2 = 2900$$
$$h_1 - h_2 = 197$$
$$h_1 - h_3 = 197 \times 0.76$$
$$= 150$$

$$h_3 = 2947$$
$$h_4 - h_3 = 3057 - 2947$$
$$= 110$$

Enthalpy change during reheating is 110 kJ/kg   Ans. (i)

Condition of steam at condenser inlet (from chart) 0.9 dry   Ans. (ii)

$$P = \dot{m}_B \left\{ \left(h_1 - h_3 \right) + \left(h_4 - h_6 \right) \right\}$$
$$3 \times 10^3 = \dot{m}_B \left(150 + 750 \right)$$
$$\dot{m}_B = 3.33 \text{ kg/s}$$
$$\frac{100(4 - 3.33)}{4} = 16.8$$

Percentage reduction in steam flow is 16.8 Ans. (iii)

26.

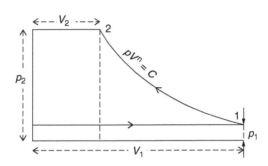

Mean piston speed $[\text{m/s}] = $ Distance $[\text{m}]$ moved by piston/second
$$2 \times \text{Stroke} \times \text{rev/s}$$

$$\therefore \text{Stroke} = \frac{2.8}{2 \times 3.5}$$
$$0.4 \text{ m} = 400 \text{ mm} \text{ Ans. (i)}$$

Work per cycle $[\text{kJ}] = \dfrac{\text{Work per second } [\text{kJ/s=kW}]}{\text{Cycles per second}}$

$$= \frac{13.38}{3.5} = 3.88 \text{ kJ}$$

Referring to sketch, neglecting clearance:

$$\text{Area under compression curve} = \frac{p_1V_1 - p_2V_2}{n-1}$$

This is the work done *by* the air.

$$\text{Work done } on \text{ the air} = \frac{p_2V_2 - p_1V_1}{n-1}$$

Work done on the air per cycle

$$= \text{Net area of diagram}$$
$$= \text{Compression area} + \text{Delivery area} - \text{Suction area}$$

$$\frac{p_2V_2 - p_1V_1}{n-1} + p_2V_2 - p_1V_1$$

$$\frac{n}{n-1} mR\,(T_2 - T_1)$$

$$= \frac{n}{n-1} mRT_1 \left[\frac{T_2}{T_1} - 1 \right]$$

$$= \frac{n}{n-1} p_1V_1 \left[\left\{ \frac{p_2}{p_1} \right\}^{\frac{n-1}{n}} - 1 \right]$$

$$\therefore 3.88 = \frac{1.32}{0.32} \times 10^2 \times V_1 \times \left[\left\{ \frac{10}{1} \right\}^{\frac{0.32}{1.32}} - 1 \right]$$

$$3.88 \times 0.32 = 1.32 \times 10^2 \times V_1 \times 0.748$$

$$V_1 = \frac{3.88 \times 0.32}{1.32 \times 10^2 \times 0.748} = 0.01257\,\text{m}^3$$

$$= \sqrt{\frac{\text{Volume}}{0.7854 \times \text{Stroke}}} = \sqrt{\frac{0.01257}{0.7854 \times 0.4}}$$

$$0.2\,\text{m} = 200\,\text{mm} \quad \text{Ans. (ii)}$$

$$\text{Mean eff. press.} = \frac{\text{Net area of diagram}}{\text{Length of diagram}}$$

$$= \frac{3.88}{0.01257} = 308.6\,\text{kN/m}^2$$

$$= 3.086\,\text{bar} \quad \text{Ans. (iii)}$$

27. Referring to Figure 8.3:

$$\frac{T_2}{T_1} = \left\{\frac{V_1}{V_2}\right\}^{n-1}$$

$$T_2 = 310 \times 13^{0.4} = 865 \text{ K}$$

Temp. at end of compression = 592°C Ans. (i)

Assume 35 kg of air is compressed and 1 kg of fuel is burned, mass of gases formed = 35 + 1 = 36 kg

$$\frac{\text{Heat energy given}}{\text{up by 1 kg fuel}} = \frac{\text{Heat energy received}}{\text{by 36 kg gases}}$$

$$\text{c.v.} = m \times c_p \times (T_3 - T_2)$$

$$T_3 - T_2 = \frac{42 \times 10^3}{36 \times 1.02} = 1144 \text{ K}$$

$$T_3 = 1144 + 865 = 2009 \text{ K}$$

$$\frac{V_3}{V_2} = \frac{T_3}{T_2}$$

$$V_3 = 1 \times \frac{2009}{865} = 2.322$$

$$\text{Ratio of expansion} = \frac{V_4}{V_3} = \frac{13}{2.322} = 5.598 \quad \text{Ans. (iii)}$$

28. Referring to Figure 12.10:

$$V_{a1} = v_1 \sin \alpha_1 = 135 \times 135 \times \sin 20° = 46.17 \text{ m/s}$$

$$V_{w1} = v_1 \cos \alpha_1 = 135 \times \cos 20° \qquad = 129.6 \text{ m/s}$$

$$x = v_{w1} - u = 126.9 - 87 \qquad = 39.9 \text{ m/s}$$

$$\tan \beta_1 = \frac{46.17}{39.9} = 1.157$$

∴ Entrance angle $\beta_1 = 49° \ 10'$ Ans. (i)

$$\beta_2 = \alpha_1 \quad v_{r2} = v_1 \quad v_{w2} = x$$

Effective change of velocity, $v_w = v_{w1} + v_{w2}$

$$= 126.9 + 39.9 = 166.8 \text{ m/s}$$

$$\text{Force on blades} = \text{Change of momentum per second}$$
$$\text{Force } [N] = \text{Mass flow } [kg/s] \times v_w \, [m/s]$$
$$\text{For 1 kg/s, force} = 1 \times 166.8 = 166.8 \, N$$
$$\text{Power } [W = J/s = Nm/s] = \text{Force}[N] \times \text{Blade velocity } [m/s]$$
$$= 166.8 \times 87$$
$$= 1.451 \times 10^4 \, W$$
$$= 14.51 \, kW \quad \text{Ans. (ii)}$$

29. Kinetic energy $= \frac{1}{2} \, mv^2$

Change in kinetic energy per kg of exhaust gases through exchanger due to change of velocity

$$= \tfrac{1}{2}(v_1^2 - v_2^2) = \tfrac{1}{2}(140^2 - 10^2) = 9750 \, J/kg$$
$$\text{which is converted into heat energy} = 9.75 \, kJ/kg$$

Mass of exhaust gases per kg of fuel

$$= 84 \, kg \, air + 1 \, kg \, fuel = 85 \, kg \, gases$$

Heat energy transferred $= \text{Mass} \times c_p \times \text{Temp. change}$

$\therefore$ Transfer of heat in exchanger, from 85 kg of gases, to 84 kg of air:

$$85 \times 1.1 \times (300 - 240) + 85 \times 9.75 = 84 \times 1.005 \times \text{Temp. change}$$
$$\text{Air temp. rise} = \frac{85(1.1 \times 60 + 9.75)}{84 \times 1.005} = 76.28°$$
$$\text{Air temp. outlet} = 200 + 76.28 = 276.28°C \quad \text{Ans.}$$

30. Heat to be taken from water to make ice [kJ/s]

$$= \frac{250}{3600}(4.2 \times 15 + 335 + 2.04 \times 10)$$
$$= 29.05 \, U/s$$
$$\text{Let } \dot{m} \, [kg/s] = \text{Mass flow of refrigerant}$$

Heat absorbed by refrigerant in evaporator [kJ/s]

$$= \dot{m} \times (0.95 - 0.2) \times 290.2$$
$$= \dot{m} \times 217.7 kJ/s$$

Assuming perfect heat transfer:

$$\dot{m} \times 217.7 = 2905$$
$$\dot{m} = 0.1335 \text{kg/s} \quad \text{Ans. (i)}$$

Volume flow of refrigerant leaving evaporator and entering compressor [m^3/s]
$= 0.1335 \times 0.02168 \times 0.95$

$$= 2.749 \times 10^{-3} \, m^3 /s$$

Let d = diameter, stroke = $2d$, assuming 100% volumetric efficiency, volume taken into compressor per second [m^3/s]

$$= 0.7854 \times d^2 \times 4.15 = 2.749 \times 10^{-3}$$
$$d = 3\sqrt{\frac{2.749 \times 10^{-3}}{0.7854 \times 2 \times 4.15}}$$
$$= 0.07499 \text{ m say 75 mm}$$
$$\text{Stroke} = 2 \times 75 = 150 \text{ mm} \quad \text{Ans. (ii)}$$

31. Referring to Figure 9.7:

$$V_1 = \text{Stroke volume} + \text{Clearance volume}$$
$$= 1440 + 40 = 1480 \text{ cm}^3$$
$$V_3 = \text{Clearance volume} = 40 \text{ cm}^3$$
$$p_1 V_1^n = p_2 V_2^n$$
$$1 \times 1480^{1.3} = 5 \times V_2^{1.3}$$
$$V_2 = \frac{1480}{\sqrt[1.3]{5}} = 429.1 \text{ cm}^3$$

Volume swept by piston from beginning of compression stroke to point where delivery valves open

$$= V_1 - V_2$$
$$= 1480 - 4291 = 1050.9 \text{ cm}^3$$

As a fraction of the stroke when delivery valves open

$$= \frac{1050.9}{1440} = 0.7298 \quad \text{Ans. (i)}$$
$$p_3 V_3^n = p_4 V_4^n$$
$$5 \times 40^{1.3} = 1 \times V_4^{1.3}$$
$$V_4 = 40 \times \sqrt[1.3]{5} = 138 \text{ cm}^3$$

Volume swept by piston from beginning of suction stroke to point when suction valves open

$$= V_4 - V_3$$
$$= 138 - 40 = 98 \text{ cm}^3$$

As a fraction of the stroke when suction valves open

$$= \frac{98}{1440} = 0.06805 \quad \text{Ans. (ii)}$$

$$\text{Area under compression curve} = \frac{p_3 V_3 - p_4 V_4}{n-1}$$

By using $p_2 V_2 - p_1 V_1$ instead of $p_1 V_1 - p_2 V_2$ the area represents work done *on* the air instead of work done *by* the air.

Note that the mean indicated pressure will be obtained from net area ÷ length, therefore, since the actual work is not required, it is not necessary to convert pressures into kN/m² and volumes into m³.

$$\text{Area under compression curve} = \frac{p_2 V_2 - p_1 V_1}{n-1}$$
$$= \frac{5 \times 429.1 - 1 \times 1480}{1.3 - 1} = 2218.3$$
$$\text{Area under delivery line} = p_2 \times (V_2 - V_3)$$
$$5 \times (429.1 - 40) = 1945.5$$
$$\text{Area under expansion curve} = \frac{p_3 V_3 - p_4 V_4}{n-1}$$
$$= \frac{5 \times 40 - 1 \times 138}{1.3 - 1} = 206.7$$
$$\text{Area under suction line} = p_1 \times (V_1 - V_4)$$
$$= 1 \times (1480 - 138) = 1342$$
$$\text{Net area of diagram} = 2218.3 + 1945.5 - 206.7 - 1342$$
$$4163.8 - 1548.7 = 2615.1$$
$$\text{Mean indicated press.} = \text{Mean height of diagram}$$
$$= \frac{\text{Net area}}{\text{Length}} = \frac{2615.1}{1440} = 1.816 \text{ bar} \quad \text{Ans. (iii)}$$

Alternatively, the net area of the diagram, representing work per cycle could be obtained from:

$$\frac{n}{n-1}P_1\left(V_1 - V_4\right)\left[\left\{\frac{P_2}{p1}\right\}^{\frac{n-1}{n}} - 1\right]$$

32. From steam tables:

$$30\,\text{bar}\ \ 350°\text{C},\ \ h = 3117 \quad s = 6.744$$
$$0.045\,\text{bar},\ \ h_f = 130 \quad s_f = 0.451$$
$$h_{fg} = 2428 \quad s_{fg} = 7.980$$

For the Rankine cycle, isentropic expansion from 30 bar 350°C to 0.045 bar:

Entropy after expansion = Entropy before
$$0.451 + x \times 7.98 = 6.744$$
$$x = 0.7886$$
$$h_2 = h \text{ at } 0.045\,\text{bar},\ 0.7886\ \text{dry}$$
$$= 130 + 0.7886 \times 2428 = 2044$$

$$\text{Rankine efficiency} = \frac{h_1 - h_2}{h_1 - h_{f2}} = \frac{3117 - 2044}{3117 - 130}$$

$$= \frac{1073}{2987} = 0.3592 \quad \text{or} \quad 35.92\% \ \ \text{Ans. (i)}$$

$$\text{Actual efficiency} = \frac{\text{Heat energy converted into work}}{\text{Heat energy supplied by steam}}$$

Heat energy into work [kJ/s = kW] = 5×10^3 kJ/s

Heat energy supplied [kJ/s] = Steam consumption [kg/s] × $(h_1 - hf_2)$ [kJ/kg]

$$= \frac{22.5 \times 10^3}{3600} \times 2987$$

$$\therefore \text{Engine effic.} = \frac{5 \times 10^3 \times 3600}{22.5 \times 10^3 \times 2987}$$

$$= 0.2679 \quad \text{or} \quad 26.79\% \ \ \text{Ans. (ii)}$$

$$\text{Efficiency ratio} = \frac{0.2679}{0.3592}$$

$$= 0.7457 \quad \text{or} \quad 74.57\% \ \ \text{Ans. (iii)}$$

33. From steam tables:

$$10 \text{ bar}, \quad h_f = 763 \qquad h_{fg} = 2015$$
$$1.2 \text{ bar, sat. temp.} = 104.8°C \qquad h_g = 2683$$
$$\therefore \text{ at } 1.2 \text{ bar } 109.8°C, \text{ steam is superheated } 5°$$

Dryness fraction by separating calorimeter:

$$x_1 = \frac{3.03}{3.03 + 0.113} = 0.964$$

Dryness fraction by throttling calorimeter:

$$\text{Enthalpy before throttling} = \text{Enthalpy after}$$
$$763 + x_2 \times 2015 = 2683 + 2.02 \times 5$$
$$x_2 \times 2015 = 1930.1$$
$$x_2 = 0.9579$$
$$\text{Dryness fraction of sample} = 0.964 \times 0.9579 = 0.9234 \quad \text{Ans.}$$

34. Referring to Figure 8.4:

$$T_1 = 315 \text{ K}$$
$$T_4 = 1773 \text{ K}$$
$$V_1 = V_5 = \text{Stroke volume} + \text{Clearance volume}$$
$$= 0.1068 + 0.0089 = 0.1157 \text{ m}^3$$
$$V_2 = V_3 = \text{Clearance volume} = 0.0089 \text{ m}^3$$
$$p_1 V_1^y = p_2 V_2^y$$
$$1 \times 0.1157^{1.4} = p_2 \times 0.0089^{1.4}$$

$$p_2 = \left\{ \frac{0.1157}{0.0089} \right\}^{1.4} = 36.26 \text{ bar}$$

$$\frac{p_1 V_1}{T_1} = \frac{p_2 V_2}{T_2}$$

$$T_2 = \frac{315 \times 36.26 \times 0.0089}{1 \times 0.1157} = 878.8 \text{ K}$$

$$\frac{T_3}{T_2} = \frac{p_3}{p_2}$$

$$T_3 = \frac{878.8 \times 45}{36.26} = 1090 \text{ K}$$

Heat received at constant volume $= m \times c_v \times (T_3 - T_2)$

$$\text{(per kg of air)} := 1 \times 0.715 \times (1090 - 878.8)$$

$$= 151 \text{ kJ/kg}$$

Spec. heat at constant press. $c_p = c_v \times \gamma$

$$= 0.715 \times 1.4 = 1.001$$

Heat received at constant pressure

$$= m \times c_p \times (T_4 - T_3)$$

$$= 1 \times 1.001 \times (1773 - 1090)$$

$$= 683.7 \text{ kJ/kg}$$

$$\text{Ratio} = \frac{\text{Heat received at const. vol}}{\text{Heat received at const. press}}$$

$$= \frac{151}{683.7} = 0.2209 : 1 \text{ Ans}$$

35. Per kg of fuel:

$$\text{Air supplied} = 11 \text{ kg}$$

$$\text{Gases} = 11 + 0.99 \text{ kg} \tag{1}$$

Let $x = $ C to CO_2

then $(0.85 - x) = $ C to CO

$$\text{Air suplied} = \frac{100}{23} \{ 2\tfrac{2}{3}x + 1\tfrac{1}{3}(0.85 - x) + 8 \times 0.11 - 0.03 \} \tag{2}$$

$$= 5.796x + 8.622 \text{ kg}$$

$$\text{Gases} = 5.796x + 8.622 + 0.99 \text{ kg}$$

From (1) and (2):

$$11 + 0.99 = 5.796x + 8.622 + 0.99$$

$$x = 0.41$$

$$(0.85 - x) = 0.44$$

Carbon to carbon monoxide $= 0.44$ kg Ans. (i)

Carbon to carbon dioxide $= 0.41$ kg Ans. (ii)

36.

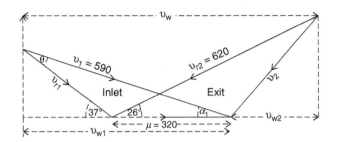

By sine rule referring to inlet triangle:

$$\frac{320}{\sin \theta} = \frac{590}{\sin(180° - 37°)}$$

$$\sin \theta = \frac{320 \times 0.6018}{590} = 0.3263$$

$$\theta = 19°3', \ \alpha_1 = 37° - 19°3' = 17° \ 57'$$
$$v_{w1} = 590 \times \cos 17°57' = 561.4 \text{ m/s}$$
$$v_{w2} = 620 \times \cos 26° - 320 = 237.3 \text{ m/s}$$

Effective change of velocity $v_w = vw_1 + vw_2$

$$= 561.4 + 237.3 = 798.7 \text{ m/s}$$

Force on blades $[\text{N}]$ = Mass flow $[\text{kg/s}] \times$ Change of velocity $[\text{m/s}]$
$$= 0.075 \times 798.7 = 59.91 \text{N} \quad \text{Ans. (i)}$$

Power $[\text{W} = \text{J/s} = \text{N m/s}]$ = Force $[\text{N}] \times$ Blade velocity $[\text{m/s}]$
$$= 59.91 \times 320$$
$$= 1.917 \times 10^4 \text{ W}$$
$$= 19.17 \text{ kW} \quad \text{Ans. (ii)}$$

37. $V_1 = 0.7854 \times 0.3^2 \times 0.45 = 0\,03181$ m^3

$$\text{Work per cycle} = \frac{n}{n-1} p_1 V_1 \left[\left\{ \frac{p_2}{p_1} \right\}^{\frac{n-1}{n}} - 1 \right]$$

When $n = 1.15$

$$\text{Work/cycle} = \frac{1.15}{0.15} \times 1 \times 10^2 \times 0.03181 (4^{\frac{0.15}{1.15}} - 1)$$

$$= 4.828 \ \text{kJ/cycle}$$

At 5 cycles per second:

$$\text{Power} \left[\text{kW} = \text{kJ/s} \right] = 4.828 \times 5 = 24.14 \ \text{kW} \quad \text{Ans. (i)}$$

When $n = 1.35$:

$$\text{Work/cycle} = \frac{1.35}{0.35} \times 1 \times 10^2 \times 0.03181 (4^{\frac{0.35}{1.35}} - 1)$$

$$= 5.299 \ \text{kJ/cycle}$$

$$\text{Power} = 5.299 \times 5 = 26.5 \ \text{kW} \quad \text{Ans. (ii)}$$

$$\% \text{ increase} = \frac{26.5 - 24.14}{24.14} \times 100 = 9.777\% \quad \text{Ans. (iii)}$$

An alternative solution to the above is to find the value of V_2 in each case, from $p_1 V_1 n = p_2 V_2 n$ then calculate the work per cycle from:

$$\frac{n}{n-1} (p_2 V_2 - p_1 V_1)$$

38. From steam tables,

$$18 \text{ bar}, \quad t_s = 207.1°\text{C} \quad h_{fg} = 1912$$

Enthalpy drop per kg of steam

$$= (1 - 0.985) \times 1912 = 28.68 \ \text{kJ/kg}$$

Rate of heat loss [J/s] per unit length

$$= \frac{1200 \times 28.68 \times 10^3}{3600 \times 25} = 382.4 \text{ J/s}$$

and this is to be equal to $\dfrac{2\pi k (T_1 - T_2)}{\ln (r_2/r_1)}$

$$\frac{2\pi \times 0.13 \times (207.1 - 35)}{\ln (r_2/r_1)} = 382.4$$

$$\ln \left(\frac{r_2}{r_1} \right) \frac{2\pi \times 0.13 \times 172.1)}{382.4} = 0.3676$$

$$\therefore \frac{r_2}{r_1} = 1.444 \quad r_2 = 1.444 \times 70 = 101.1 \text{mm}$$

Thickness $= r_2 - r_1 = 101.1 - 70 = 31.1 \text{ mm}$ Ans.

39. From steam tables:

$$20 \text{ bar } 400°C, \ h = 3248$$
$$1.4 \text{ bar}, \quad t_s = 109.3 \qquad h_f = 458 \quad h_{fg} = 2232$$
$$0.04 \text{ bar}, \quad h_f = 121 \qquad h_{fg} = 2433$$

Without feed heating:

$$h \text{ at } 0.04 \text{ bar } 0.85 \text{ dry}$$
$$= 121 + 0.85 \times 2433 = 2188$$

Enthalpy drop through turbine/kg

$$= 3248 - 2188 = 1060$$

Heat supplied to steam in boiler

$$= 3248 - 121 = 3127$$

$$\text{Thermal eff.} = \frac{\text{Heat energy converted into work}}{\text{Heat energy supplied}}$$

$$= \frac{1060}{2127} = 0.3389 \text{ or } 33.89\%$$

With feed heating:

0.134 kg of steam at 1.4 bar is tapped off per kg of supply steam, leaving (1 – 0.134) = 0.866 kg to complete its path through the engine.

Considering heater and hotwell as one system:

$$\text{Enthalpy at entry} = \text{Enthalpy at exit}$$
$$0.134\left(458 + x \times 2232\right) + 0.866 \times 121 = 1 \times 458$$
$$61.38 + 299x + 104.8 = 458$$
$$299x = 291.82$$
$$\text{Dryness fraction at 1.4 bar, } x = 0.976$$

h at 1.4 bar 0.976 dry

$$= 458 + 0.976 \times 2232 = 2636$$

Enthalpy drop through turbine/kg

$$= 1 \times (3248 - 2646) + 0.866\,(2636 - 2188)$$

$$= 612 + 388 = 1000$$

Heat supplied to steam in boiler

$$= 3248 - 458 = 2790$$
$$\text{Thermal eff.} = \frac{1000}{2790} = 0.3584 \quad \text{or} \quad 35.84\%$$
$$\left\{ \text{Comparison}: \begin{array}{l} \text{With feed heating, } \eta = 35.84\% \\ \text{Without feed heating, } \eta = 33.89\% \end{array} \right\} \text{Ans.}$$

40.
$$T_1 = 292 \text{ K}$$
$$T_2 = 781 \text{ K}$$
$$\frac{p_1 V_1}{T_1} = \frac{p_2 V_2}{T_2}$$
$$V_2 = \frac{1.01 \times 0.125 \times 781}{36 \times 292} = 0.00938 \text{ m}^3$$
$$p_1 V_1^n = p_2 V_2^n$$
$$1.01 \times 0.125^n = 36 \times 0.00938^n$$
$$n = 1.38 \quad \text{Ans. (i)}$$

Alternatively, n could be obtained from

$$\frac{T_1}{T_2} = \left\{\frac{p_1}{p_2}\right\}^{\frac{n-1}{n}}$$

$$R = c_p - c_v = 1.005 - 0.718 = 0.287 \text{ kJ/kg K}$$

$$p_1 V_1 = m R T_1$$

$$m = \frac{1.01 \times 10^2 \times 0.125}{0.287 \times 292}$$

$$= 0.1506 \text{ kg} \quad \text{Ans. (ii)}$$

$$\text{Work done by the air} = \frac{p_1 V_1 - p_2 V_2}{n-1}$$

$$= \frac{0.1506 \times 0.287 \times (292 - 781)}{1.38 - 1}$$

$$= -55.62 \text{ kJ} \quad \text{Ans. (iii)}$$

The minus sign indicates that the work is done *on* the air.

Increase in internal energy:

$$U_2 - U_1 = m \times c_v \times (T_2 - T_1)$$

$$= 0.1506 \times 0.718 \times (781 - 292)$$

$$= 52.86 \text{ kJ} \quad \text{Ans. (iv)}$$

Heat supplied	=	Increase in internal energy	+	External work done

$$= 52.86 + (-55.62) = -2.76 \text{ kJ} \quad \text{Ans. (v)}$$

The minus sign indicates that the transfer of heat is from the air to its surrounds.

41. I.p. of 4 cylinders $= p_m L A n \times 4$

$$\text{where } n = \text{rev/s} \div 2 \text{ for a four-stroke engine}$$

$$\text{i.p.} = 14.9 \times 10^2 \times 0.7854 \times 0.32^2 \times 0.48 \times 2 \times 4$$

$$= 460 \text{ kW}$$

$$\text{b.p.} [\text{kW}] = T [\text{kNm}] \times \omega [\text{rad/s}]$$

$$= 12 \times 0.96 \times 2\pi \times 4 = 289.6 \text{ kW}$$

$$\text{Ind. thermal effic.} = \frac{\text{Heat into work in cylinders [kJ/h]}}{\text{Heat energy supplied [kJ/h]}}$$

$$= \frac{460 \times 3600}{99 \times 44.5 \times 10^3}$$

$$= 0.3758 \quad \text{or} \quad 37.58\% \quad \text{Ans. (i)}$$

$$\text{Brake thermal effic.} = \frac{\text{Heat into work at brake [kJ/h]}}{\text{Heat energy supplied [kJ/h]}}$$

$$= \frac{289.6 \times 3600}{99 \times 44.5 \times 10^3}$$

$$= 0.2367 \quad \text{or} \quad 23.67\% \quad \text{Ans. (ii)}$$

Heat energy carried away by cooling water [kJ/h]

$$= \text{Mass} \times \text{Spec. ht.} \times \text{Temp. rise}$$
$$= 154 \times 60 \times 4.2 \times (47 - 14)$$

As a percentage of the heat supplied

$$= \frac{154 \times 60 \times 4.2 \times 33}{99 \times 44.5 \times 10^3} \times 100$$

$$= 29.07\% \quad \text{Ans. (iii)}$$

The remainder of the heat losses may be attributed to the heat carried away in the exhaust gases

$$= 100 - (37.58 + 29.07) = 33.35\%$$

Friction and pumping losses

$$= 37.58 - 23.67 = 13.91\%$$

42. Referring to Figure 12.5:

$$\frac{T_2}{T_1} = \left\{ \frac{p_2}{p_1} \right\}^{\frac{\gamma-1}{\gamma}} \quad \text{where} \quad \frac{\gamma-1}{\gamma} = \frac{0.4}{1.4} = \frac{2}{7}$$

$$T_2 = 4.3^{\frac{2}{7}} \times 289 = 438.4K$$

Temperature at compressor outlet = 165.4°C   Ans. (i)

$$\frac{T_4}{T_3} = \frac{T_1}{T_2} \text{ because pressure rations are equal}$$

$$T_4 = \frac{873 \times 289}{438.4} = 575.4K$$

Alternatively T_4 could be obtained from

$$\frac{T_4}{T_3} = \left\{ \frac{p_4}{p_3} \right\}^{\frac{\gamma-1}{\gamma}}$$

Temperature at turbine outlet

$$= 302.4°C \quad \text{Ans. (ii)}$$

Heat supplied per kg of working fluid

$$= \text{mass} \times \text{spec. heat} \times \text{temp. rise}$$
$$= 1 \times 1.005 \times (873 - 483.4)$$
$$= 436.7 \text{ kJ/kg} \quad \text{Ans. (iii)}$$

$$\text{Thermal effic.} = 1 - \frac{T_4 - T_1}{T_3 - T_2} \quad \text{or} \quad 1 - \frac{T_1}{T_2} \quad \text{or} \quad 1 - \frac{T_4}{T_3}$$

$$\text{or} \quad 1 - \frac{1}{r_p^{(\gamma-1)/\gamma}}$$

$$1 - \frac{T_1}{T_2} = 1 - \frac{289}{438.4} = 1 - 0.6592$$

$$= 0.3408 \quad \text{or} \quad 34.08\% \quad \text{Ans. (iv)}$$

43. Refer to Figure 12.9:

i.e., from tables, page 14, at 8.477 bar and 35°C:

$$h_3 = h_4 = 69.55 \text{ kJ/kg}$$

i.e. from tables, page 14, at 1.509 bar and −20°C:

$$h_4 = h_f + x_4 (h_g - h_f)$$
$$69.55 = 17.82 + x_4 (178.73 - 17.82)$$

$$x_4 = 0.3215$$
$$s_1 = s_2$$

i.e., isentropic compression

$$0.0731 + x_1(0.7087 - 0.0731) = 0.6839$$

$$x_1 = 0.961$$

$$h_1 = h_f + x_1(h_g - h_f)$$

$$= 17.82 + 0.961(178.73 - 17.82)$$

$$= 172.45 \text{ kJ/kg}$$

Refrigerating effect $= h_1 - h_4$

$$= 172.45 - 69.55$$

$$= 102.9 \text{ kJ/kg} \quad \text{Ans. (i)(a)}$$

$$\text{c.o.p.} = \frac{h_1 - h_4}{h_2 - h_1}$$

$$= \frac{102.9}{201.45 - 172.45}$$

$$= 3.548 \quad \text{Ans. (i)(b)}$$

$$\text{Reversed Carnot c.o.p.} = \frac{T_L}{T_H - T_L}$$

$$= \frac{253}{308 - 253}$$

$$= 4.6 \quad \text{Ans. (ii)}$$

44.

$$\theta_m = \frac{50 - 20}{\ln \dfrac{50 - 15}{20 - 15}}$$

$$= 15.42°C \text{ Ans. (i)}$$

$Q = UAt\theta_m$ also $Q = 4(50 - 20) \times 1395.6 / 10^3$

$$\therefore UA\theta_m = 167.5$$

$$A = \frac{167.5 \times 10^3}{70 \times 15.42}$$

$$= 155 \text{ m}^2 \quad \text{Ans. (ii)}$$

$$\pi dLn = A \quad n = 350 \text{ tubes}$$

$$L = \frac{A}{\pi dn}$$

$$= \frac{155 \times 10^3}{\pi \times 19 \times 350}$$

$$= 7.42 \text{ m} \quad \text{Ans. (iii)}$$

45. From h–s chart $\qquad\qquad h_1 = 2760 \text{ kJ/kg}$

$$h_2 = 2760 \text{ kJ/kg}$$
$$h_3 = 2595 \text{ kJ/kg}$$
$$h_4 = 2595 \text{ kJ/kg}$$
$$h_5 = 2180 \text{ kJ/kg}$$

Final condition of steam $= 0.823$ dry   Ans.

Enthalpy lost by steam $=$ Enthalpy gained by the oil

$$m\left[\left(h_2 - h_3\right) + \left(h_4 - h_5\right)\right] = 0.72 \times 2.1 \times 72$$

$$m = 0.186 \text{ kg/s} \quad \text{Ans.}$$

46. Mass of 1 mol of fuel $\qquad\qquad = 12 \times 1 + 1 \times 4$

$$= 16 \text{ kg}$$

$$H_2 \text{ by mass} = \frac{4}{16}$$
$$= 0.25 \text{ kg}$$
$$C \text{ by mass} = \frac{12}{16}$$
$$= 0.75 \text{ kg}$$

$$\text{Stoichiometric air required} = \frac{100}{23}\left(2\tfrac{2}{3}C + 8H\right)$$

$$= 4.348\left(2^{2/3} \times 0.75 + 8 \times 0.25\right)$$

$$= 17.39 \text{ kg/kg fuel}$$

Correct air-fuel mass ratio $= 17.39$ Ans. (i)

Mass of dry products of combustion per kg fuel burned:

$$CO_2 = 3^{2/3} \times 0.75 = 2.75 \text{ kg}$$
$$N_2 = 0.77 \times 17.39 = 13.39$$
$$\text{Total} = 16.14 \text{ kg}$$

% mass analysis of the dry flue gases:

$$CO_2 = \frac{275}{16.14} = 17.04$$

$$N_2 = \frac{1339}{16.14} = 82.96$$

DFG	m%	M	N	N%
CO_2	17.04	44	0.3873	11.56
	82.96	28	2.9629	88.44
			3.3502	

% volume analysis of the dry flue gases:

Ans. (ii)

47. Steady flow energy equation:

$$h_1 + \tfrac{1}{2}c_1^2 + q = h_2 + \tfrac{1}{2}c_2^2 + w$$

$$80 + \tfrac{1}{2} \times \frac{75^2}{10^3} - 10 = 300 + \tfrac{1}{2} \times \frac{175^2}{10^3} + w$$

from which $w = -242.5$ kJ/kg   Ans. (i)

If t_h = datum temperature on which the specific enthalpies are based

$$\begin{cases} \text{then } h_1 = c_p(t_1 - t_h) \\ \text{and } h_2 = c_p(t_2 - t_h) \end{cases} \text{divide}$$

$$\therefore \frac{h_1}{h_2} = \frac{c_p(t_1 - t_h)}{c_p(t_2 - t_h)}$$

i.e. $\dfrac{80}{300} = \dfrac{15 - t_h}{200 - t_h}$

i.e. $80(200 - t_h) = 300(15 - t_h)$

$16000 - 80t_h = 4500 - 300t_h$

$16000 - 4500 = 80t_h - 300t_h$

$\dfrac{11500}{220} = -t_h$

i.e. $t_h = -52.27°C$ or   220.7 K Ans. (ii)

48.

For insulation (1) $Q = \dfrac{2\pi k_1 (T_1 - T_2)}{\ln\left(\dfrac{D_2}{D_1}\right)}$ W/m

For insulation (2) $Q = \dfrac{2\pi k_2 (T_2 - T_3)}{\ln\left(\dfrac{D_3}{D_2}\right)}$ W/m

For surface film $Q = \dfrac{hA(T_3 - T_4)}{L}$ W/m

Total temperature drop $T_1 - T_2$

$$= (T_1 - T_2) + (T_2 - T_3) + (T_3 - T_4)$$
$$= \frac{Q \ln (D_2 / D_1)}{2\pi k_1} + \frac{Q \ln (D_3 / D_2)}{2\pi k_2} + \frac{Q}{h\pi D_3}$$

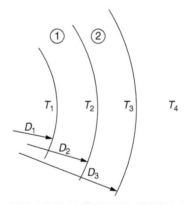

From steam tables: $T_1 = 263.9°C$

$$\therefore 263.9 - 20 = Q \left[\frac{\ln\left(\dfrac{400}{200}\right)}{2\pi \times 0.5} + \frac{\ln\left(\dfrac{500}{400}\right)}{2\pi \times 0.1} + \frac{10^3}{8\pi \times 500} \right]$$

$243.9 = Q[2.2\ 06 + 0.355 + 0.0796]$

$Q = 92.35$ W/m

Q = Mass of steam condensed/metre length of pipe/s × Latent heat of steam

$Q = m \times h_{fg}$

$m = \dfrac{92.35}{1639}$

$= 0.0536$ kg/s Ans.

49.

Using the h–s chart $h_1 = 3410$ kJ/kg

$h_2^1 = 3100$ kJ/kg

$h_2 = 2990$ kJ/kg

$h_3 = 3475$ kJ/kg

$h_4^1 = 3130$ kJ/kg

$h_4 = 3085$ kJ/kg

$h_5^1 = 2415$ kJ/kg

$h_5 = 2370$ kJ/kg

Pressure drop in reheater $= 15 - 13$

$= 2$ bar Ans. (i)

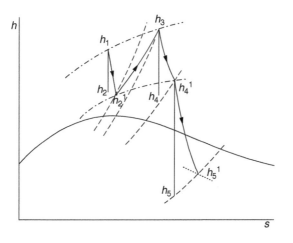

$$\text{Stage 1. Isentropic efficiency} = \frac{h_1 - h_2^1}{h_1 - h_2}$$
$$= 0.74 \quad \text{or} \quad 74\%$$

$$\text{Stage 2. Isentropic efficiency} = \frac{h_3 - h_4^1}{h_3 - h_4}$$
$$= 0.88 \quad \text{or} \quad 88\%$$

$$\text{Stage 3. Isentropic efficiency} = \frac{h_4 - h_5^1}{h_4 - h_5}$$
$$= 0.94 \quad \text{or} \quad 94\% \quad \text{Ans. (ii)}$$

Ratio of powers, $\quad h_1 - h_2^1 : h_3 - h_4^1 : h_4 - h_5^1$

i.e.,	310	:	345	:	715	
i.e.,	1	:	1.112	:	2.31	Ans. (iii)

50.

$$\frac{T_1}{T_2} = \left(\frac{p_1}{p_2}\right)^{\frac{\gamma-1}{\gamma}}$$

2. $\quad T_2 = 393 \times 10^{\frac{0.4}{1.4}} = 759 \text{ K} = 486°C$

3. $\quad T_3 = 800°C$

4. $\dfrac{T_3}{T_4} = \left(\dfrac{p_3}{p_4}\right)^{\frac{\gamma-1}{\gamma}}$

$$\frac{p_2}{T_2} = \frac{p_3}{T_3}$$

$$p_3 = 10 \times \frac{1073}{759} = 14.14 \text{ bar}$$

$$T_4 = 1073\left(\frac{1}{14.14}\right)^{\frac{0.4}{1.4}} = 503\,K = 230°C$$

Cycle efficiency $= 1 - \dfrac{\text{heat rejected}}{\text{heat rejected}}$

$$= 1 - \frac{c_p(T_4 - T_1)}{c_v(T_3 - T_2)}$$

$$= 1 - \gamma\frac{(T_4 - T_1)}{(T_3 - T_2)}$$

$$= 1 - 1.4\frac{(230 - 120)}{(800 - 486)} \times 100\%$$

$$= 50.96\% \quad \text{Ans. (i)}$$

$$\text{m.e.p.} = \frac{\text{Area}}{\text{Length}}$$

$$= \frac{\left(\dfrac{p_3V_3 - p_4V_4}{\gamma - 1}\right) - \left(\dfrac{p_2V_2 - p_1V_1}{\gamma - 1}\right) - (p_4V_4 - p_1V_1)}{V_4 - V_2}$$

as $pV = mRT$

$$\text{m.e.p.} = \frac{\dfrac{(T_3 - T_4)}{\gamma - 1} - \dfrac{(T_2 - T_1)}{\gamma - 1} - (T_4 - T_1)}{\dfrac{T_4}{p_4} - \dfrac{T_2}{p_2}}$$

$$= \frac{\left(\dfrac{800 - 230}{0.4}\right) - \left(\dfrac{486 - 120}{0.4}\right) - (230 - 120)}{\dfrac{503}{10^5} - \dfrac{759}{10^6}}$$

$$= 0.937 \text{ bar} \quad \text{Ans. (ii)}$$

INDEX